Lecture Notes in Artificial Intelligence 8891

Subseries of Lecture Notes in Computer Science

T0212760

Rajendra Prasath   Philip O'Reilly
T. Kathirvalavakumar (Eds.)

Mining Intelligence and Knowledge Exploration

Second International Conference, MIKE 2014
Cork, Ireland, December 10-12, 2014
Proceedings

 Springer

Volume Editors

Rajendra Prasath
National University of Ireland
Cork, Ireland
E-mail: drrprasath@gmail.com

Philip O'Reilly
National University of Ireland
Cork, Ireland
E-mail: philip.oreilly@ucc.ie

T. Kathirvalavakumar
V.H.N. Senthikumara Nadar College (Autonomous)
Tamil Nadu, India
E-mail: kathirvalavakumar@yahoo.com

ISSN 0302-9743   e-ISSN 1611-3349
ISBN 978-3-319-13816-9   e-ISBN 978-3-319-13817-6
DOI 10.1007/978-3-319-13817-6
Springer Cham Heidelberg New York Dordrecht London

Library of Congress Control Number: 2014955586

LNCS Sublibrary: SL 7 – Artificial Intelligence

Typesetting: Camera-ready by author, data conversion by Scientific Publishing Services, Chennai, India

Printed on acid-free paper

Springer is part of Springer Science+Business Media (www.springer.com)

Preface

This volume contains the papers presented at MIKE 2014: The Second International Conference on Mining Intelligence and Knowledge Exploration held during December 10–12, 2014, at University College Cork, National University of Ireland, Cork, Ireland (http://www.mike.org.in/2014/). There were 69 submissions from 35 countries and each qualified submission was reviewed by a minimum of two Program Committee members using the criteria of relevance, originality, technical quality, and presentation. The committee accepted 22 papers for oral presentation (acceptance rate: 31.88%) and 12 papers for short presentation at the conference. We also had eight presentations in the doctoral consortium.

The International Conference on Mining Intelligence and Knowledge Exploration (MIKE) is an initiative focusing on research and applications on various topics of human intelligence mining and knowledge discovery. The primary goal was to present state-of-art scientific results, to disseminate modern technologies, and to promote collaborative research in mining intelligence and knowledge exploration. At MIKE 2014, specific focus was placed on the "Business Intelligence" theme.

Human intelligence has evolved steadily over several generations, and today human expertise is excelling in multiple domains and in knowledge-acquiring artifacts. MIKE's primary objective is to focus on the frontiers of human intelligence mining toward building a body of knowledge in this key domain.

The accepted papers were chosen on the basis of their research excellence, which provides a body of literature for researchers involved in exploring, developing, and validating learning algorithms and knowledge-discovery techniques. Accepted papers were grouped into various subtopics including information retrieval, feature selection, classification, clustering, image procesing, network security, speech processing, machine learning, recommender systems, natural language processing, language, cognition and computation, and business intelligence. Researchers presented their work and had ample opportunity to interact with eminent professors and scholars in their area of research. All participants benefitted from discussions that facilitated the emergence of new ideas and approaches. The authors of short/poster papers presented their work during a special session and obtained feedback from thought leaders in the discipline.

A large number of eminent professors, well-known scholars, industry leaders, and young researchers participated in making MIKE 2014 a great success. We express our sincere thanks to University College Cork for allowing us to host MIKE 2014.

We were pleased to have Prof. Ramon Lopaz de Mantaras, Artificial Intelligence Research Institute, Spain, Prof. Mandar Mitra, Indian Statistical Institute (ISI), Kolkata, India, Prof. Pinar Ozturk, Norwegian University of Science and Technology, Norway, Prof. Anupam Basu, Indian Institute of Technology, Kharagpur, India, and Dr. Santiago Ontanon, Drexel University, USA, serving as advisory chairs for MIKE 2014.

Several eminent scholars, including Prof. Sudeshna Sarkar, Indian Institute of Technology, Kharagpur, India, Prof. Paolo Rosso, Universitat Politècnica de València, Spain, and Prof. Nirmalie Wiratunga, Robert Gordon University, Aberdeen, Scotland, delivered invited talks on learning and knowledge exploration tasks in various interdisciplinary areas of intelligence mining. Dr. Donagh Buckley, Head of EMC^2 Research Europe, delivered a keynote address on the leading-edge R&D being undertaken in his organization at present. Dr. Amitava Das, University of North Texas, USA, and Prof. Erik Cambria served as the co-organizers of a workshop focusing on the "Automatic Understanding of Creativity in Language (Creative-Lingo)," which was sponsored by Science Foundation Ireland. Leading practitioners from top-tier technology organizations participated in this very successful workshop.

We are very grateful to all our sponsors, especially EMC research Europe, Ireland, Scientific Foundation Ireland (SFI), SAS, the School of Business and Law, University College Cork, Ireland, Cork Convention Bureau for their generous support of MIKE 2014. We would especially like to thank the local organizing team for their efforts in making MIKE 2014 such a success.

We thank the Program Committee and all reviewers for their timely and thorough participation in the reviewing process. We also thank Mr. Muthuvijayaraja of VHNSN College for his great support. We appreciate the time and effort put in by the local organizers at Business Information Systems, especially the support from the members of the Financial Services Innovation Center (FSIC), who dedicated their time to MIKE 2014. Finally, we acknowledge the support of EasyChair in the submission, review, and proceedings creation processes. We are very pleased to express our sincere thanks to Springer, especially Alfred Hofmann, Anna Kramer, and Abier El-Saeidi for their faith and support in publishing the proceedings of MIKE 2014.

October 2014   Rajendra Prasath
                                                  Philip O'Reilly
                                          T. Kathirvalavakumar

Organization

Program Committee

Adeyanju Ibrahim	Robert Gordon University, Scotland, UK
Agnar Aamodt	Norwegian University of Science and Technology, Norway
Aidan Duane	Waterford Institute of Technology, Ireland
Alexander Gelbukh	Instituto Politécnico Nacional, Mexico
Amélie Cordier	Claude Bernard University of Lyon 1 (IUT A), France
Amitava Das	University of North Texas, USA
Anil Kumar Vuppala	International Institute of Information Technology, Hyderabad, India
Anupam Basu	Indian Institute of Technology, Kharagpur, India
Aradhna Malik	Indian Institute of Technology, Kharagpur, India
Arijit Sur	Indian Institute of Technology, Guwahati, India
Biswanath Barik	Tata Consultancy Services Limited (TCSL), Kolkata, India
Björn Gambäck	Norwegian University of Science and Technology, Norway
Chaman Sabharwal	Missouri University of Science and Technology, USA
Chhabi Rani Panigrahi	Indian Institute of Technology, Kharagpur, India
D.K. Lobiyal	Jawaharlal Nehru University, New Delhi, India
Debasis Ganguly	Dublin City University, Ireland
Diana Trandabat	A. I. Cuza University of IAŞI, Romania
Dipankar Das	Jadavpur University, Kolkata, India
Dipti Misra Sharma	International Institute of Information Technology, Hyderabad, India
Donagh Buckley	EMC^2 Research Europe, Cork, Ireland
Erik Cambria	Nanyang Technological University, Singapore
Geetha T.V.	Anna University, Chennai, India
Gethsiyal Augasta	Sarah Tucker College, Tirunelveli-7, India
Gladis Christopher	Presidency College, Chennai, India
Gloria Inés Alvarez	Pontificia Universidad Javeriana Cali, Colombia
Guru D.S.	University of Mysore, India

Hojjat Adeli	The Ohio State University, Columbus, USA
Huayu Wu	Institute for Infocomm Research (I2R), Singapore
Inah Omoronyia	University of Glasgow, Scotland, UK
Isis Bonet Cruz	Antioquia School of Engineering, Medellin, Colombia
Jaiprakash Lallchandani	International Institute of Information Technology Bangalore, India
Jian-Yun Nie	University of Montreal, Canada
Jiaul Paik	University of Maryland, College Park, USA
Joe Feller	University College Cork, Ireland
John McAvoy	University College Cork, Ireland
Joydeep Chandra	Indian Institute of Technology, Patna, India
Juan Recio-Garcia	Universidad Complutense de Madrid, Spain
Kamal Kumar Choudhary	Indian Institute of Technology, Ropar, India
Kathirvalavakumar Thangairulappan	VHNSN College (Autonomous), Virudhunagar, India
Kumaran. T.	Government Arts College, Krishnagiri, India
M. Gethsiyal Augasta	Sarah Tucker College, Tirunelveli, India
Maciej Ogrodniczuk	Institute of Computer Science, Polish Academy of Sciences, Poland
Mandar Mitra	Indian Statistical Institute, Kolkata, India
Manoj Chinnakotla	Microsoft, Hyderabad, India
Marco Palomino	University of Exeter, UK
Maunendra Sankar Desarkar	Samsung R & D Institute India, Bangaluru, India
Mohamed K. Watfa	University of Wollongong in Dubai, UAE
Murugan A.	Dr. Ambedkar Government Arts College, Chennai, India
Muthu Rama Krishnan Mookiah	Ngee Ann Polytechnic, Singapore
Niamh O'Riordan	University College Cork, Ireland
Nirmalie Wiratunga	Robert Gordon University, Aberdeen, UK
Niloy Ganguly	Indian Institute of Technology, Kharagpur, India
Paolo Rosso	Universitat Polytechnic de Valencia, Spain
Parnab Kumar Chanda	University College Cork, Ireland
Pattabhi R.K. Rao	AU-KBC Research Centre, Anna University, Chennai, India
Philip O'Reilly	University College Cork, Ireland
Pinaki Bhaskar	IIT - CNR, Italy
Pinar Ozturk	Norwegian University of Science and Technology, Norway

Plaban Kumar Bhowmick	Indian Institute of Technology, Kharagpur, India
Prasenjit Majumdar	DAIICT, Gandhinagar, India
Rajarshi Pal	Institute for Development and Research in Banking Technology, Hyderabad, India
Rajendra Prasath	University College Cork, Ireland
Rajib Ranjan Maiti	IIT - CNR, Italy
Rajkumar P.V.	University of Texas at San Antonio, USA
Rakesh Balabantaray	International Institute of Information Technology, Bhubaneswar, India
Ramakrishnan K.	Pondicherry Engineering College, Pondicherry, India
Ramon Lopaz de Mantaras	IIIA - CSIC, Barcelona, Spain
Ranjani Parthasarathi	Anna University, India
Rob Gleasure	University College Cork, Ireland
Sangwoo Kim	University of California, San Diego, USA
Sanjay Chatterji	Samsung R & D Institute India, Bangaluru, India
Santiago Ontanon	Drexel University, USA
Saptarshi Ghosh	Max Planck Institute for Software Systems, Saarbrücken-Kaiserslautern, Germany
Saurav Sahay	Intel Research Labs, USA
Shashidhar G. Koolagudi	National Institute of Technology Karnataka, Surathkal, India
Simon Woodworth	University College Cork, National University of Ireland, Ireland
Sivaji Bandyopadhyay	Jadavpur University, Kolkata, India
Sobha Lalitha Devi	AU-KBC Research Centre, Anna University, Chennai, India
Srinivasan T.	Rajalakshmi Institute of Technology, Chennai, India
Subba Reddy	Gyeongsang National University, Jinju, Korea
Sudeshna Sarkar	Indian Institute of Technology, Kharagpur, India
Sudip Roy	Indian Institute of Technology Roorkee, India
Sudipta Saha	National University of Singapore, Singapore
Sujan Kumar Saha	Birla Institute of Technology, Mesra, India
Sukomal Pal	Indian School of Mines, Dhanbad, India
Sumit Goswami	Defence Research and Development Organisation, New Delhi, India
Sylvester Olubolu Orimaye	Monash University, Malaysia
Thangaraj V.	Vel-Tech University, Chennai, India
Udayabaskaran S.	VelTech University, Chennai, India
Uttama Lahiri	Indian Institute of Technology, Gandhinagar, India

V. Pallavi   Philips Innovation Campus, Bangalore, India
Vahid Jalali   Samsung Research America, USA
Vamsi Krishna Velidi   ISRO Satellite Centre (ISAC), ISRO,
                                 Bangalore, India
Vasudha Bhatnagar   University of Delhi, India
Vijay Kumar T.V.   Jawaharlal Nehru University, New Delhi, India
Vijay Sundar Ram   AU-KBC Research Centre, Anna University,
                                 India
Wei Lee Woon   Masdar Institute, Abu Dhabi, UAE
Xiaolong Wu   California State University, USA
Yutaka Maeda   Kansai University, Japan
Zeyar Aung   Masdar Institute of Science and Technology,
                                 UAE

Additional Reviewers

Alvarez, Gloria Inés   M., Anbuchelvi
Barik, Biswanath   Mcavoy, John
Baskar, Pinaki   Oconnor, Yvonne
Bhattacharyya, Malay   Ozturk, Pinar
Christopher, Gladis   Panigrahi, Chhabi Rani
Das, Amitava   Patel, Dhaval
Duane, Aidan   Patra, Braja Gopal
Feller, Joseph   Poddar, Sudip
Gleasure, Rob   Roy, Sudip
Heavin, Ciara   T., Kathirvalavakumar
Kumar, Naveen   T., Kumaran
Lohar, Pintu   Yadav, Jainath
Lopez De Mantaras, Ramon

Table of Contents

An Effective Term-Ranking Function
for Query Expansion Based on Information
Foraging Assessment

Ilyes Khennak, Habiba Drias, and Hadia Mosteghanemi

Laboratory for Research in Artificial Intelligence,
USTHB, Algiers, Algeria
{ikhennak,hdrias,hmosteghanemi}@usthb.dz

Abstract. With the exponential growth of information on the Internet and the significant increase in the number of pages published each day have led to the emergence of new words in the Internet. Owning to the difficulty of achieving the meaning of these new terms, it becomes important to give more weight to subjects and sites where these new words appear, or rather, to give value to the words that occur frequently with them. For this reason, in this work, we propose an effective term-ranking function for query expansion based on the co-occurrence and proximity of words for retrieval effectiveness enhancement. A novel efficiency/effectiveness measure based on the principle of optimal information forager is also proposed in order to evaluate the quality of the obtained results. Our experiments were conducted using the OHSUMED test collection and show significant performance improvement over the state-of-the-art.

Keywords: Information retrieval, information foraging theory, query expansion, term proximity, term co-occurrence.

1 Introduction

The performance and quality of a retrieval system is measured on the basis of effectiveness and efficiency [1]. From the theoretical standpoint, the effectiveness is indicated by returning only what user needs and efficiency is indicated by returning the results to the user as quickly as possible [7]. From the practical standpoint the effectiveness, or relevance, is determined by measuring the precision, recall, etc., and efficiency is determined by measuring the search time [5]. Reliance on these two measures varies from one community to another. The information retrieval community, for example, is focusing too much on the quality of the top ranked results while the artificial intelligence community, which started paying attention to information retrieval, ontologies and the Semantic Web [9], is focusing on the retrieval process cost. Accordingly, and in order to establish consensus between communities and adopt both effectiveness and efficiency measures, the Information Foraging Theory has been proposed to do so.

R. Prasath et al. (Eds.): MIKE 2014, LNAI 8891, pp. 1–10, 2014.

We introduce in the first part of this study the basic principles and concepts of information foraging theory which is employed later in the experimental section to evaluate the retrieval systems. The second part is devoted to presenting and discussing our suggested approach for improving retrieval performance. The main goal of this suggested method is to propose a novel term-ranking function for query expansion based on the co-occurrence and proximity of words. This approach assigns importance, during the retrieval process, to words that frequently occur in the same context. Relying on this principle came as a result of the studies carried out recently concerning the evolution and growth of the Web. All of these studies have shown an exponential growth of the Web and rapid increase in the number of new pages created. This revolution, which the Web is witnessing, has led to the appearance of two points:

- The first point is the entry of new words into the Web which is mainly due to: neologisms, first occurrences of rare personal names and place names, abbreviations, acronyms, emoticons, URLs and typographical errors [6].
- The second point is that the users employ these words during the search [3].

Out of these two points which the web is witnessing and due to the difficulty, or better, the impossibility to use the meanings of these words, we proposed a method based on finding the locations and topics where these words appear, and then trying to use the terms which neighbor and occur with the latter in the search process. We will use the best-known instantiation of the Probabilistic Relevance Framework system: Okapi BM25, and the Blind Relevance Feedback: Robertson/Sparck Jones' term-ranking function as the baseline for comparison, and evaluate our approach using OHSUMED test collection.

In the next section, we will introduce the information foraging theory. The BM25 model and the Blind Feedback approach are presented in Section 3. In Section 4 we will explain our proposed approach and finally we will describe our experiments and results.

2   Information Foraging Theory

The information foraging is a theory proposed by [8]. It is becoming a popular theory for characterizing and understanding web browsing behavior [4]. The theory is based on the behavior of an animal deciding what to eat, where it can be found, the best way to obtain it and how much energy the meal will provide. The basis of foraging theory is a *cost* and *benefit* assessment of achieving a goal where cost is the amount of resources consumed when performing a chosen activity and the benefit is what is gained from engaging in that activity. Conceptually, the *optimal forager* finds the best solution to the problem of maximizing the rate of benefit returned by effort expended given the energetic profitabilities of different habitats and prey, and the costs of finding and pursuing them. By analogy, the *optimal information forager* finds the best solution to the problem of maximizing the rate of valuable information gained per unit cost. Pirolli and

Card [8] expressed the rate of valuable information gained per unit cost, by the following formula:

$$R = \frac{G}{T}$$

(1)

Where:

R, is the rate of gain of valuable information per unit cost,
G, is the ratio of the total amount of valuable information gained,
T, is the total amount of time spent.

In order to adopt both effectiveness and efficiency measures during the experimentation phase, we propose to evaluate and compare the quality of the results obtained by our suggested approach relying on the principle of optimal information forager and using the basis of the rate R, where G and T represent respectively the total number of relevant documents returned by the retrieval system, and the total amount of search time.

3   Probabilistic Relevance Framework

The probabilistic Relevance framework is a formal framework for document retrieval which led to the development of one of the most successful text-retrieval algorithms, Okapi BM25. The classic version of Okapi BM25 term-weighting function, in which the weight w_i^{BM25} is attributed to a given term t_i in a document d, is obtained using the following formula:

$$w_i^{BM25} = \frac{tf}{k_1((1-b) + b\frac{dl}{avdl}) + tf} w_i^{RSJ}$$

(2)

Where:
tf, is the frequency of the term t_i in a document d;
k_1, is a constant;
b, is a constant;
dl, is the document length;
$avdl$, is the average of document length;
w_i^{RSJ}, is the well-know Robertson/Sparck Jones weight [10]:

$$w_i^{RSJ} = \log \frac{(r_i + 0.5)(N - R - n_i + r_i + 0.5)}{(n_i - r_i + 0.5)(R - r_i + 0.5)}$$

(3)

Where:
N, is the number of documents in the whole collection;
n_i, is the number of documents in the collection containing t_i;
R, is the number of documents judged relevant;
r_i, is the number of judged relevant documents containing t_i.

The RSJ weight can be used with or without relevance information. In the absence of relevance information (the more usual scenario), the weight is reduced to a form of classical *idf*:

$$w_i^{IDF} = \log \frac{N - n_i + 0.5}{n_i + 0.5} \tag{4}$$

The final BM25 term-weighting function is therefore given by:

$$w_i^{BM25} = \frac{tf}{k_1((1-b) + b\dfrac{dl}{avdl}) + tf} \log \frac{N - n_i + 0.5}{n_i + 0.5} \tag{5}$$

The similarity score between the document d and a query q is then computed as follows:

$$Score_{BM25}(d, q) = \sum_{t_i \in q} w_i^{BM25} \tag{6}$$

During the interrogation process, the relevant documents are selected and ranked using this similarity score.

3.1   Blind Relevance Feedback for Query Expansion

One of the most successful techniques to improve the retrieval effectiveness of document ranking is to expand the original query with additional terms. Many approaches have been proposed to generate and extract these additional terms. The Blind Relevance Feedback (or the Pseudo-Relevance Feedback) is one of the suggested approaches. It uses the pseudo-relevant documents to select the most important terms to be used as expansion features. In its simplest version, the approach starts by performing an initial search on the original query using the BM25 term-weighting and the previous document-scoring function (formula 6), suppose the best ranked documents to be relevant, assign a score to each term in the top retrieved documents using a term-scoring function, and then sort them on the basis of their scores. One of the best-known functions for term-scoring is the *Robertson/Sparck Jones* term-ranking function, defined by formula 3. The original query is then expanded by adding the top ranked terms, and re-interrogated by using the BM25 similarity score (formula 6), in order to get more relevant results.

4   The Closeness and Co-occurrence of Terms for Effectiveness Improvement

The main goal of this work is to return only the relevant documents. For that purpose, we have introduced the concept of co-occurrence and closeness, during the search process. This concept is based on finding for each query term the documents where it appears, and then assess the relevance of the terms contained in these documents to the query term on the basis of:

1. The co-occurrence, which gives value to words that appear in the largest possible number of those documents.

2. The proximity and closeness, which gives value to words in which the distance separating them and the query term within a document is small.

We started our work by reducing the search space through giving importance to documents which contain at least two words of the initial query. The following formula allows us to select the documents that contain at least two words of the query, i.e. to pick out any document d whose $Score_{Bi-BM25}$ to a query q is greater than zero:

$$Score_{Bi-BM25}(q,d) = \sum_{\substack{(t_i,t_j)\in q}}^{i\neq j} (w_i^{BM25} + w_j^{BM25}) \tag{7}$$

As mentioned previously, we will find the terms which often appear together with the query terms. Finding these words is done via the measurement of the external distance of each term t_i of the R_c' vocabulary to each term $t_{j_{(q)}}$ of the query q (R_c, is the set of documents returned by using the formula (7)). This distance computes the rate of appearance of t_i with $t_{j_{(q)}}$ in the collection of documents R_c. In the case where t_i appears in all the documents in which $t_{j_{(q)}}$ occurs, the value of the external distance will be 1.0; and in the case where t_i does not appear in any of the documents in which $t_{j_{(q)}}$ occurs, the value of the external distance will be 0.0. Based on this interpretation, the external distance $ExtDist$ of t_i to $t_{j_{(q)}}$ is calculated as follows:

$$ExtDist(t_i, t_{j_{(q)}}) = \frac{1}{\sum_{d_k \in R_c} x_{(j,k)}} \left[\sum_{d_k \in R_c} x_{(i,k)} * x_{(j,k)} \right] \tag{8}$$

Where:
$$x_{(i,k)} = \begin{cases} 1 \text{ if } t_i \in d_k \ , \\ 0 \text{ else.} \end{cases}$$

d_k, is a document that belongs to R_c.

The total external distance between a given term t_i and the query q is estimated as follows:

$$ExtDist(t_i, q) = \sum_{t_{j_{(q)}} \in q} ExtDist(t_i, t_{j_{(q)}}) \tag{9}$$

In the second step, we will find the terms which are often neighbors to the query terms via the measurement of the internal similarity between each term t_i of V_R (V_R is the vocabulary of R_c) and each term $t_{j_{(q)}}$ of the query q. This similarity, which takes into consideration the content of documents, computes the similarity between t_i and $t_{j_{(q)}}$ within a given document d in terms of the number of words separating them. The more t_i is close to $t_{j_{(q)}}$, the greater is its internal similarity. For this purpose, we used the well-known Gaussian kernel function to measure the internal similarity $IntDist$ between t_i and $t_{j_{(q)}}$ within a given document d:

$$IntDist_{(d)}(t_i, t_{j_{(q)}}) = exp\left[\frac{-(i-j)^2}{2\sigma^2}\right] \tag{10}$$

Where:
i (resp. j) , is the position of the term t_i (resp. $t_{j_{(q)}}$) in d;
σ, is a parameter to be tuned.

The terms t_i and $t_{j_{(q)}}$ may appear more than once in a document d. Therefore, the internal distance between the term pair $(t_i, t_{j_{(q)}})$ is estimated by summing all possible $IntDist_{(d)}$ between t_i and $t_{j_{(q)}}$. Thus, the preceding formula becomes:

$$IntDist_{(d)}(t_i, t_{j_{(q)}}) = \sum_{occ(t_i, t_{j_{(q)}})} exp\left[\frac{-(i-j)^2}{2\sigma^2}\right] \tag{11}$$

Where:
$occ(t_i, t_{j_{(q)}})$, is the number of occurrences of t_i in d multiply by the number of occurrences of $t_{j_{(q)}}$.

The average internal similarity between t_i and $t_{j_{(q)}}$ in the whole R is then determined as follows:

$$IntDist(t_i, t_{j_{(q)}}) = \frac{\displaystyle\sum_{d_k \in R} IntDist_{(d_k)}(t_i, t_j)}{\displaystyle\sum_{d_k \in R_c} x_{(j,k)}} \tag{12}$$

The following formula calculates the total internal similarity between a given term t_i and the query q :

$$IntDist(t_i, q) = \sum_{t_{j_{(q)}} \in q} IntDist(t_i, t_{j_{(q)}}) \tag{13}$$

Finally, in order to compute the total similarity $(Dist)$, the values of $ExtDist$ and $IntDist$ were normalized between 0 and 1. The overall similarity between t_i and q is obtained using the following formula:

$$Dist(t_i, q) = \lambda ExtDist(t_i, q) + (1 - \lambda)IntDist(t_i, q) \tag{14}$$

Where:
λ, is a parameter to adjust the balance between the external and internal similarities ($\lambda \in [0, 1]$).

Using formula (14), we evaluate the relevance of each term $t \in V_R$ with respect to the query q. Then we rank the terms on the basis of their relevance and add the top ranked ones to the original query q. Based on the BM25 similarity score, we retrieve the new relevant documents.

5 Experiments

In order to evaluate the effectiveness of the proposed approach, we carried out a set of experiments. First, we describe the dataset, the software, and the effectiveness measures used. Then, we present the experimental results.

5.1 Dataset

Extensive experiments were performed on OHSUMED test collection. The collection consists of 348 566 references from MEDLINE, the on-line medical information database. The OHSUMED collection contains a set of queries, and relevance judgments. In order that the results be more accurate and credible, we divided the OHSUMED collection into 6 sub-collections. Each sub-collection has been defined by a set of documents, queries, and a list of relevance documents. Regarding the queries, the OHSUMED collection includes 106 queries. Each query is accompanied by a set of relevance judgments chosen from the whole collection of documents. Partitioning the collection of documents into sub-collections leads inevitably to a decrease in the number of relevant documents for each query. Table 1 shows the number of queries (*Nb Queries*) for each sub-collection, the average query length in terms of number of words (*Avr Query Len*), the average number of relevant documents (*Avr Rel Doc*).

Table 1. Some statistics on the OHSUMED sub-collections queries

#documents	50000	100000	150000	200000	250000	300000
Nb Queries	82	91	95	97	99	101
Avr Rel Doc	4.23	7	10.94	13.78	15.5	19.24
Avr Query Len	6.79	6.12	5.68	5.74	5.62	5.51

5.2 Effectiveness Measures

The precision and the Mean Average Precision (MAP) have been used as measures to evaluate the effectiveness of the systems and to compare the different approaches. As indicated is Section 2, the principle of optimal information forager has been employed to assess the performance of the search methods.

5.3 Results

Before proceeding to compare the quality of the suggested approach with the BM25 and the Pseudo-Relevance Feedback method, we fixed the parameter σ of the internal similarity (formula (10)). For this aim, we considered the internal similarity as the total similarity ($Dist$), i.e. $\lambda = 0$, and systematically tested a set of fixed σ values from 1 to 30 in increments of 5. Table 2 presents the precision values after retrieving 10 documents ($P@10$) and MAP reached by the proposed approach, while using the sub-collection of 50000 documents. The number of pseudo-relevant documents (denoted by PSD) was tuned at 10, 20 and 50.

Table 2. The best performance of the proposed approach for different σ

PSD	10		20		50	
σ	P@10	MAP	P@10	MAP	P@10	MAP
1	0.1060	0.2110	0.1048	0.2208	0.1073	0.2193
5	0.1109	**0.2265**	0.1121	0.2252	0.1146	**0.2241**
10	0.1109	0.2253	0.1121	**0.2255**	0.1146	0.2231
15	0.1109	0.2231	0.1121	0.2245	0.1146	0.2228
20	0.1109	0.2230	0.1121	0.2245	0.1146	0.2235
25	0.1109	0.2230	0.1121	0.2245	0.1146	0.2233
30	0.1109	0.2230	0.1121	0.2245	0.1146	0.2233

From Table 2, we can conclude that the appropriate values of σ, which bring the best performance, are 5 and 10. For all the following experiments, the parameters σ and λ were set to 5 and 0.5, respectively. Moreover, the number of expansion terms added to the initial query for the proposed system and the Pseudo-Relevance Feedback approaches was set to 10, which is a typical choice [2]. In the first stage of testing, we evaluated and compared the results of the suggested approach (EXT/INT), which use both the external and internal similarities, with those of BM25 and RSJ (Robertson/Sparck Jones algorithm for Relevance Feedback); where we computed the precision values after retrieving 10 documents ($P@10$). Figure 1 shows the precision values for the EXT/INT, the BM25 and the RSJ techniques. In the second stage of testing, we computed the Mean Average Precision (MAP) score to evaluate the retrieval performance of the EXT/INT, the BM25, and the Relevance Feedback method.

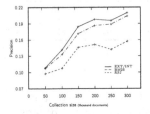 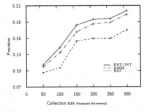 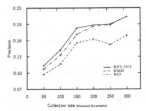

(a) Precision after retrieving 10 documents (P@10), (PSD=10).

(b) Precision after retrieving 10 documents (P@10), (PSD=20).

(c) Precision after retrieving 10 documents (P@10), (PSD=50).

Fig. 1. Effectiveness comparison of the EXT/INT to the state-of-the-art

From Figures 1a and 1b, we note an obvious superiority of the suggested approach EXT/INT compared with the BM25, and this superiority was more significant in comparison to the RSJ technique. Despite the superiority shown in Figure 1c, the result was not similar to that observed in 1a and 1b, however, the precision values of the proposed approach were the best in all the sub-collections.

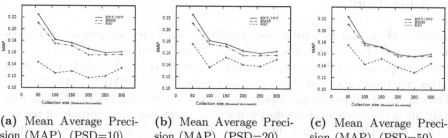

(a) Mean Average Precision (MAP), (PSD=10).

(b) Mean Average Precision (MAP), (PSD=20).

(c) Mean Average Precision (MAP), (PSD=50).

Fig. 2. Mean Average Precision (MAP) results of the EXT/INT approach, the BM25 and the RSJ methods

Figure 2 shows a clear advantage of the EXT/INT approach compared to the RSJ. It also shows a slight superiority over the BM25. As previously explained in Section 2, we propose to use the principle of optimal information forager in order to evaluate the quality of the obtained results. For this purpose, we calculate for each query the rate R, illustrated in formula 1. The different rates, each of which is linked to a query, are then summed and divided by the total number of queries. As a result, we obtain an average rate $R(Q)$ defined as follows:

$$R(Q) = \frac{1}{|Q|} \sum_{i=1}^{|Q|} \frac{G_{q_i}}{T_{q_i}} \tag{15}$$

Where:

$|Q|$, is the total number of queries,
G_{q_i}, is the number of relevant documents retrieved for query q_i,
T_{q_i}, is the total time spent in processing q_i.

Table 3. R(Q)-score achieved by EXT/INT, BM25 and RSJ

#documents	50000	100000	150000	200000	250000	300000
EXT/INT	14.9049	10.7110	10.1994	9.9643	9.0079	9.2276
BM25	16.0251	11.7612	11.2681	10.9097	10.0364	10.2747
RSJ	11.1365	7.3431	7.0448	6.6212	5.7615	5.9524

It can be seen from Table 3 that the BM25 overcame the EXT/INT approach in terms of $R(Q)$. This superiority is mainly due to the short time taken by BM25 during the search as it used only the original query words. However, it is clear that the proposed method produces the best results over the RSJ in all cases.

6   Conclusion

In this work, we put forward a novel term-term similarity score based on the co-occurrence and closeness of words for retrieval performance improvement. We have incorporated in this work, the concept of the External/Internal similarity of terms. We thoroughly tested our approach using the OHSUMED test collection. The experimental results show that the proposed approach EXT/INT achieved a significant improvement in effectiveness. Although the main purpose of relying on the principle of optimal information forager, and in particular the $R(Q)$ Score, in assessing the quality of retrieval systems was not to get better results compared to BM25 and RSJ methods, but rather to introduce a new measure in order to compare the performance of retrieval systems, taking into account both effectiveness and efficiency measures.

References

1. Cambazoglu, B.B., Aykanat, C.: Performance of query processing implementations in ranking-based text retrieval systems using inverted indices. Information Processing & Management 42(4), 875–898 (2006)
2. Carpineto, C., Romano, G.: A survey of automatic query expansion in information retrieval. ACM Computing Surveys 44(1), 1–50 (2012)
3. Chen, Q., Li, M., Zhou, M.: Improving query spelling correction using web search results. In: EMNLP-CoNLL 2007: Proceedings of the 2007 Joint Conference on Empirical Methods in Natural Language Processing and Computational Natural Language Learning, pp. 181–189. ACL, Stroudsburg (2007)
4. Dix, A., Howes, A., Payne, S.: Post-web cognition: evolving knowledge strategies for global information environments. International journal of Web engineering and technology 1(1), 112–126 (2003)
5. Dominich, S.: The modern algebra of information retrieval. Springer, Heidelberg (2008)
6. Eisenstein, J., O'Connor, B., Smith, N.A., Xing, E.P.: Mapping the geographical diffusion of new words. In: NIPS 2012: Workshop on Social Network and Social Media Analysis: Methods, Models and Applications (2012)
7. Frøkjær, E., Hertzum, M., Hornbæk, K.: Measuring usability: Are effectiveness, efficiency, and satisfaction really correlated? In: CHI 2000: Proceedings of the SIGCHI Conference on Human Factors in Computing Systems, pp. 345–352. ACM, New York (2000)
8. Pirolli, P., Card, S.: Information foraging. Psychological Review 106(4), 643–675 (1999)
9. Ramos, C., Augusto, J.C., Shapiro, D.: Ambient intelligence the next step for artificial intelligence. IEEE Intelligent Systems 23(2), 15–18 (2008)
10. Robertson, S.E., Jones, K.S.: Relevance weighting of search terms. Journal of the American Society for Information science 27(3), 129–146 (1976)
11. Robertson, S., Zaragoza, H.: The probabilistic relevance framework: BM25 and beyond. Foundations and Trends in Information Retrieval 3(4), 333–389 (2009)

Personalized Search over Medical Forums

Arohi Kumar, Amit Kumar Meher, and Sudeshna Sarkar

Department of Computer Science and Engineering, Indian Institute of Technology
Kharagpur, Kharagpur, India
{arokumar90,amitcet.05,shudeshna}@gmail.com

Abstract. The Internet is used extensively to obtain health and medical information. Many forums, blogs, news articles, specialized websites, research journals and other sources contain such information. However, though current general purpose and special purpose search engines cater to medical information retrieval, there are many gaps in this domain because the information retrieved is not tailored to the specific needs of the user. We argue that personalized requirement of the user needs to take into account the medical history and the demographics of the user in addition to the medical content that the user desires. For example, one user may desire factual information whereas another may desire a vicarious account of another person with respect to the same medical condition. Medical forums contain medical information on several medical topics and are rich in the demographics and the type of information that is available due to the large number of people participating throughout the world. However, present search engines on medical forums do not incorporate the type of information that the user is looking for and their demographics. In this paper we propose a novel approach for facilitating search on medical forums that provides medical information specially tailored to the user's needs. Our experiments show that such an approach considerably outperforms the search presently available on medical forums. . .

Keywords: Medical Forum, Semantic Search, Natural Language Processing, Personalized retrieval.

1 Introduction

Medical forums allow people to obtain highly personalized information about their medical conditions. Questions posted in medical forums generally describe the demographics, medical history and concerns of the user along with her medical query. In order to effectively facilitate search on medical forums, understanding the information requirements of the users is a must. These requirements can be gathered by the study of existing content available on medical forums. Here is a typical question posted on Medhelp Forums (www.medhelp.org/forums/list): *I found out that I an RH Negative. And im really worried about everything. I read an article that it makes it a lot harder to keep a pregnancy your second time around. My doctor said ill just need a shot. But my mom was negative also, and she never had any issues. Was that because I ended up being*

R. Prasath et al. (Eds.): MIKE 2014, LNAI 8891, pp. 11–20, 2014.

the same as her. My husband is Rh positive, and I read if he is positive then it will take after the father. Sorry for all the questions im just so confused. And I cant really call my doctor, were in the military, and its difficult to get answers (source : http://www.medhelp.org/posts/Pregnancy-18-24/Worriedand-confused-helpplease/show/1956129). We can infer from the above question that the gender of the author is female and her age is 18-24 (this can be inferred from the post url). Her blood type is Rh-negative which is the same as her mother. Her medical query is related to pregnancy and the complications associated with childbirth. Her concerns tell us that she is looking for advice or the experience of someone with a similar situation. A suitable answer to this query will address the complications of childbirth and provide insights to the user with respect to her specific demographic information and medical history. It is important to note that though the reliability of content on medical forums may be questionable at times, it sometimes can still be useful in emergency situations. PatientsLikeMe[1] is a website where users share their personal health information and a study has revealed a range of benefits for users. According to [2], users who shared information about Epilepsy on PatientsLikeMe benefitted through improved understanding of their seizures, finding other patient like them, and learning more about symptoms or treatments. Facilitating search on medical forums will allow us to tap into an enormous repository of medical knowledge and provide valuable information and insights to users.

2   Previous Work

The rapid growth in medical information has led to the development of numerous health web search engines. All of these search engines operate on input of specific health related keywords. One of these search engines is Healthline[3] which provides symptom search, treatment search, drug search and also discusses top trending medical topics. SearchMedica[4] comprises of a series of free medical search engines built by a group of doctors and is particularly intended for medical professionals. Apart from these, there exists other intelligent web based medical search engines which try to effectively capture the users information need and improve the search results by using iterative and feedback-based search. The MedSearch[5] transforms users' lengthy free text queries into shorter queries by extracting important and representative keywords thereby increasing the query processing speed and improving the quality of search results. It also provides diverse search results and suggests related medical phrases to help the user quickly digest search results and refine the query. [6] describes the challenges faced during iterative medical search and proposes the design of an intelligent medical Web search engine called iMed[7] that extensively uses medical knowledge and questionnaire to facilitate ordinary internet users to search for medical information. It uses a questionnaire based query interface to guide searchers to provide the most important information about their situations. Secondly, it uses medical knowledge to automatically form multiple queries from a searcher's answers to the questions. Another medical search engine called MedicoPort[8] incorporates the domain knowledge obtained from Unified Medical Language System (UMLS)[9] to increase the effectiveness of the searches. The power of the system is based on the ability to understand the semantics of web pages and the user queries. This is mainly intended for people having no medical expertise. However, searching over medical forums is unique in that the corpus of information is anecdotal and therefore understanding the semantics of the information would require us to go beyond incorporating domain knowledge.

Similarly, on the stylometric analysis of age and gender from free text sources such as blogs, there has been extensive work in the past. The study of the variation of linguistic characteristics in texts according to the authors' age or gender has already become feasible on literary corpora [15-19]. For authors' gender prediction, [20] uses both function words distribution and parts-of-speech n-gram on a large corpus of formally written text (fiction and non fiction) to get an accuracy of approximately 80%. With regard to Computer-mediated Communication (CMC), there have been several studies on gender prediction of blogs [22-24]. In [25], they gathered a corpus of over 71,000 blogs and extracted style-based features (non-dictionary words, parts-of-speech, function words and hyperlinks) and content-based features (content-based single words with the greatest information gain) for both age and gender prediction of the blogs' authors. Their research showed that, regardless of gender, language usage in blogs are highly correlated with the author's age, i.e, people tend to use less pronouns and assent/ negation when they grow older whereas they prefer frequent usage of prepositions and determiners. Blog words were found to be a clear hallmark of youth, while the use of hyperlinks was typical for older bloggers. [22,26] use non-dictionary words and the average sentence length as features. They found that teenage bloggers tend to use more non-dictionary words than adult bloggers do. Furthermore, the stylistic difference in usage of non-dictionary words combined with content words showed to predict the age group (10s, 20s, 30s or higher) with an accuracy of 80.32% and gender with an accuracy of 89.18%.

3 System Architecture

In order to provide a personalized search experience to the user, the system must enhance the content that is available from posts. Apart from the medical content, a post gives information about the type of content that the user is looking for and demographic information (like age, sex and gender) of the user. Furthermore a portion of the content of a post may refer to the user's past medical history/family history in addition to the posed query. Identifying this set of information in both the user's query and the corpus of documents from which results are found will improve the quality of the results that are returned to users. Figure 1 depicts the overall system architecture.

4 Data Annotation

In order to personalize search results we need to understand the genre and the demographics that the post caters to.

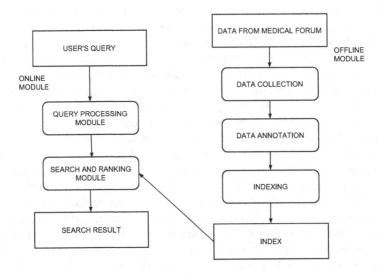

Fig. 1. System Architecture

4.1   Genre Classification

We broadly categorize the content of posts into three genres, providing information, offering advice or sharing experience. Posts that provide information primarily contain medical facts, whereas those offering advice present one or more alternative choice of actions and those sharing experience talk about a particular decision and its repercussions and/or advantages. It is important to note that the content of a post can belong to more than one category within information, advice and experience. As an example consider the following post from the MedHelp Diabetes Forum which provides information, offers advice and shares experience. Please note that the following example has been annotated with (I)...(/I), (A)...(/A) and (E)...(/E) blocks to highlight information, advice and experience respectively.

(I)What you are feeling is the beginning of diabetic neuropathy, a fairly common effect of diabetes. It's mostly felt in the feet and can be very mild or extremely painful(/I). (A)Fortunately you can do some things to help prevent or slow it down. Main thing is to control your blood sugars. You have to change the way you eat. Trust me, I have learned that lesson the hard way. Next, go to the vitamin store right now and get an anti-oxidant called alpha lipoic acid and also get the methyl version of vitamin b12. The people in the store will know what you mean when you say methyl b12. Ala is prescribed in germany to treat neuropathy(/A). (E)I was starting to have stabbing pains, cramps, etc. That has stopped and I am now trying to achieve regeneration of the damaged nerves, but that is a long process(/E). (A)Take my advice, jump on this now and you will be much better off(/A).

For genre classification we assign three scores P_{inf}, P_{adv}, P_{exp} to each post based on the probability of the post content providing information, offering advice and sharing experience respectively. The data annotation module selects top K frequently occurring bigrams (based on a manually selected threshold frequency of 5) in the respective classes as features and trains a Decision Tree Classifier on 360 posts, 120 of each type. We have trained the model on a balanced training set since we are not sure about the actual distribution of each type of post in the forum. The Weka Machine Learning toolkit is used for classification and Decision Tree Classifier, with an accuracy of 89.64% using 10-fold cross validation performs superior to Naive Bayes and Support Vector Machine (SVM) classifiers.

4.2   Age and Gender Prediction

We are considering the age and gender as the demographic information of the post author. In some posts, age and gender are explicitly mentioned and thus we can mine the content of these posts to extract the age and gender. In other posts, the age and gender are not explicitly mentioned, that is, mining these posts does not provide any information. In this case, we can obtain the age and gender of the author by either crawling the author profile for this information or by stylistic analysis of the post content (if the information is not found by crawling/mining). We have used a set of standard features namely Pronoun count, Adjective count, Adverb count, Average Sentence Length, Average Word Length, WH Count, Conjuction Count, Modal Verbs Count, Smiley Count, No. of Words starting with Uppercase, No. of Unique Words, Parts-Of-Speech (POS) Bigram Count, Non Dictionary Words Count, Past-tensed Words Count which are commonly used in the stylometric analysis [15-22] to train the Naive Bayes, SVM and Decision Tree classifiers on 12659 annotated posts. The training set for age consisted of age groups '20-40' (6214 posts) & '40+' (6445 posts) and for gender, female (9181 posts) & male (3478 posts). On 10-fold cross validation, we obtained a maximum accuracy of 72.04% (for age) and 73.51% (for gender) using the Decision Tree classifier.

5   Ranking

5.1   Ranking Algorithm

Suppose that the user desires to find the k most relevant posts given a query Q. Our ranking algorithm works in two steps :

1. Finding a set of k' medically relevant posts ($k' > k$) with respect to query Q.
2. Using user demographics and genre information along with medical relevance of the documents to give a relevance score to each of the k' medically relevant posts retrieved in step 1. However, unlike the medical information need, which was the essential part of a query, the demographic preferences and post genre may or may not be available. This leads to two different ranking algorithms:

(a) *Implicit Ranking* is the ranking algorithm that is followed when the user does not provide any demographic preferences and post genre that she is looking for.
(b) *Explicit Ranking* is the ranking algorithm that is followed when the user specifies one or more of her demographic/genre preferences.

For the rest of the Section, let k and k' refer to the number of posts required by the user issuing the query and the number of documents retrieved after the first step of ranking respectively.

Implicit Ranking. In the implicit ranking algorithm, the user does not specify any demographic preferences. We use the following two characteristics of a medical forum post to rank the k' posts :

1. **The Trustworthiness of the Post Author.** Information from a trustworthy author is likely to be reliable. The problem lies in measuring trustworthiness because it is a subjective metric. We can say that the author of a post is trustworthy if:
   – He/she is a doctor.
   – He/she is rated as an expert by the Medical Forum Community. For example, authors that have more than 10 Best Answers on Medhelp Forums[14] can be considered to be expert users.
2. **The Length of the Post.** Posts of longer lengths are expected to be more informative.

We will now describe the ranking algorithm that will be used to rank the k' posts obtained after the first step of ranking. As an input to this step, we are given k' posts that have the highest medical relevance scores. Let S_{max} be the maximum medical relevance score of a document out of these k' documents. For a document D_i ($1 \leq i \leq k'$), the advanced score, S_i is calculated as follows :

$$S_i = S_m + S_{au} + S_l \tag{1}$$

where S_m is the medical relevance of the document obtained from the first step of the ranking algorithm, S_{au} is the trustworthiness associated with the author of the post and S_l is the score associated with the length of the post.

S_l indicates the importance that is attributed to the document based on the length of the post content. Let l_{max} be the maximum length of post content in all the k' documents. Let l_i be the length of the post in document D_i . Since no post can be more relevant due to length than the medically most relevant post,

$$S_l = \frac{l_i}{l_{max}} * S_{max} \tag{2}$$

for document D_i.

S_{au} indicates the relevance attributed to a document based on the author (*auth*) of the document. Posts written by doctors are expected to be a more reliable source of information than posts authored by novice users. Let S_{doc} be the reliability associated with a post that is authored by a doctor. Let Th be a threshold recognition criterion above which an author who is not a doctor is recognized as an expert. Let r be the user's recognition on the medical forum. In this case,

$$S_{au} = \delta(auth) * S_{doc} + (1 - \delta(auth)) * S_{user} * S_{doc} \qquad (3)$$

where,

$$\delta(auth) = \begin{cases} 1 : & \textit{if auth is doctor} \\ 0 : & \textit{otherwise} \end{cases} \qquad (4)$$

and,

$$S_{user} = \begin{cases} 1 : & \textit{if r is greater than Th} \\ \frac{r}{Th} : & \textit{otherwise} \end{cases} \qquad (5)$$

The value of the variable S_{doc}, that is, the relevance boost of a document authored by a doctor, is subjective and can be approximated to S_{max} (which is the maximum medical relevance of any of the k' documents). Furthermore, the approximation and quantification of Th and r are subjective and need to be defined based on the forum being indexed. We have taken Th as '10 Best Answers'.

Explicit Ranking. The user can specify the type of the post that she wants (information, advice and experience) along with her age and gender. We calculate a relevance ratio, R, for each of the k' posts with respect to the user's demographic and genre preferences. The user can choose from the five preferences, inf, adv, exp, age, gen corresponding to whether the user is looking for posts talking about information, advice, experience, or for posts related to a particular age and gender respectively. This relevance ratio is a value between 0 and 1 depending on how well a post satisfies a user's demographic preference. The user can specify a subset of these five preferences and the formula for calculating the relevance ratio is,

$$R = \frac{\displaystyle\sum_{i \in (inf,adv,exp,age,gen)} \delta(i) * P(i)}{\displaystyle\sum_{i \in (inf,adv,exp,age,gen)} \delta(i)} \qquad (6)$$

$\delta(i)$ represents whether the user has chosen to specify the particular demographic preference or not. As an example if the user wants posts giving advice $\delta(adv)$ is 1, else 0. $P(i)$ represents a score (between 0 and 1) of the post with respect to the particular demographic preference.

After the second step of the ranking algorithm, each of the k' documents has been assigned an advanced relevance score based on the medical relevance of the post content, demographic information of the post and post genre. The top k of these documents are the results and are most likely to satisfy the user's information need.

6 Results

In this section, we will compare the performance of our search engine with current search available on forums at www.medhelp.org.

6.1 Dataset

The dataset consists of posts crawled from medhelp.org. For each post the medical information consists of the topic, content and keywords associated with the post. The author information consists of the age, gender, profession (doctor or not) and the number of best answers given by this author (if not a doctor). The dataset is comprised of 6,94,745 posts from 24 different medical topics, authored by 1,18,246 authors.

6.2 Experiments

The query interface at Medhelp.org allows users to provide free text queries and in response to those queries, we are given posts, articles, user journals, blogs, health pages etc. In this way, we query both Medhelp.org and our search system with 20 queries and consider the top 10 results for each query. We have considered the values of k and k' to be 10 and 100 respectively. Ten human judges (1,2,3,4,5,6,7,8,9,10) evaluate each result as being Extremely Relevant, Relevant or Non Relevant. For each user, we then calculate the average number of posts (out of top 10) that the user finds Extremely Relevant, Relevant and Non Relevant across the 20 queries. The responses of the ten users are shown in Table 1 and Table 2.

Table 1. Evaluation of Medhelp.org Search Results

User	Extremely Relevant	Relevant	Not Relevant
1	3.00	3.55	3.45
2	2.15	3.60	4.25
3	3.15	3.00	3.85
4	2.10	3.50	4.40
5	2.50	3.40	4.10
6	2.80	3.50	3.70
7	3.00	3.35	3.55
8	2.85	3.65	3.45
9	3.25	3.35	3.40
10	2.15	3.50	4.35
Average	2.69	3.44	3.87

We see that our search engine gives approximately 5.1 extremely relevant results per query whereas Medhelp.org Search gives approximately 2.7 extremely relevant results per query. If we consider, (extremely relevant + relevant) results to be the criteria for relevance, our system gives 8.94 relevant results per query and Medhelp.org Search gives approximately 6.13 relevant results per query.

Table 2. Evaluation of Our Search Results

User	Extremely Relevant	Relevant	Not Relevant
1	5.70	3.25	1.05
2	4.85	4.05	1.10
3	5.60	3.85	0.55
4	5.00	3.50	1.50
5	4.80	3.50	1.70
6	4.50	4.25	1.20
7	5.20	4.20	0.60
8	5.00	4.10	0.85
9	4.75	4.25	1.00
10	5.50	3.50	1.00
Average	5.09	3.86	1.06

7 Conclusion

In this paper we have presented an approach for searching medical forums which goes beyond keyword search, incorporating the demographic information of the questioner and the type of information that the user is looking for. We identified the requirements that such a system needs to fulfill and presented the system design of such a system. Our two step ranking algorithm uses Apache Solr/Lucene to rank the documents based on medical relevance and heuristics to incorporate the demographic preferences and medical content along with medical relevance to give an overall relevance score to the documents. Using this algorithm, a user will be shown posts that are not only medically relevant but written by people who are trustworthy, provide the type of content the user is looking for and have similar demographics as the user. We have seen that considering the post genre and user demographics significantly improves the performance of search in terms of user satisfaction with the information. One may argue that the quality of results a user gets will be better if she posts the question on medical forums instead of searching. While that may be true, search is many orders of magnitude faster than posting a question on medical forum and waiting for a reply. This will be critical in cases where medical information is desired urgently.

References

1. PatientsLikeMe, http://www.patientslikeme.com
2. Wicks, P., Keininger, D.L., et al.: Perceived benefits of sharing health data between people with epilepsy on an online platform. In: Epilepsy & Behavior (2014)
3. Healthline, http://www.healthline.com
4. SearchMedica, http://www.searchmedica.com
5. Luo, G., Tang, C., Yang, H., et al.: MedSearch: a specialized search engine for medical information retrieval. In: CIKM (2008)
6. Luo, G., Tang, C.: Challenging issues in iterative intelligent medical search. In: ICPR (2008)

7. Luo, G., Tang, C.: On iterative intelligent medical search. In: SIGIR, pp. 3–10 (2008)
8. Can, B., Baykal, N.: MedicoPort: A medical search engine for all. J. Computer Methods and Programs in Biomedicine (86), 73–86 (2007)
9. Lindberg, D.A., Humphreys, B.L., McCray, A.T.: Unified Medical Language System. J. Am Med. Inform. Assoc. 5(1), 111 (1998)
10. Apache Solr, http://solr.apache.org
11. Apache Lucene, http://lucene.apache.org
12. Aronson, A.R.: Effective Mapping of Biomedical Text to the UMLS Metathesaurus: The MetaMap Program. In: Annual AMIA Symposium, pp. 17–21 (2001)
13. Savova, G.K., Masanz, J.J., Orgen, P.V., et al.: Mayo clinical Text Analysis and Knowledge Extraction System(cTAKES): architecture, component evaluation and applications. J Am Med. Inform. Assoc. 17(5), 507–513 (2010)
14. Medhelp, http://www.medhelp.org
15. Pennebaker, J.W., Graybeal, A.: Patterns of natural language use: disclosure, personality, and social integration. Current Directions in Psychological Science 10(3), 90–93 (2001)
16. Pennebaker, J.W., Stone, L.D.: Words of wisdom: Language use over the lifespan. J. Personality and Social Psychology 85(2) (2003)
17. Holmes, J., Meyerhoff, M.: The Handbook of Language and Gender. Blackwell, Oxford, UK (2004)
18. Kelih, E., Antic, G., Grzybek, P., Stadlober, E.: Classification of author and/or genre? The impact of word length, pp. 498–505. Springer, Heidelberg (2005)
19. Burger, J.D., Henderson, J.C.: An exploration of observable features related to blogger age. In: AAAI Spring Symposia on Computational Approaches to Analyzing Weblogs, California, USA (2006)
20. Holmes, J., Meyerhoff, M.: The Handbook of Language and Gender. Blackwell, Oxford (2004)
21. Argamon, S., Koppel, M., Fine, J., Shimoni, A.: Automatically Categorizing Written Texts by Author Gender. Literary and Linguistic Computing, pp. 401–412 (2002)
22. Argamon, S., Koppel, M., Pennebaker, W., Schler, J.: Mining the Blogosphere: Age, gender and the varieties of self expression (2007)
23. Yan, X., Yan, L.: Gender classification of weblog authors. In: AAAI Spring Symposia on Computational Approaches to Analyzing Weblogs (2006)
24. Nowson, S., Oberlander, J.: Identifying more bloggers. Towards large scale personality classication of personal weblogs. In: 1st International Conference on Weblogs and Social Media, International AAAI Conference on Weblogs and Social Media (2007)
25. Schler, J., Koppel, M., Argamon, S., Pennebaker, J.: Effects of age and gender on blogging. In: AAAI Spring Symposia on Computational Approaches to Analyzing Weblogs (2006)
26. Goswami, S., Sarkar, S., Rustagi, M.: Stylometric analysis of bloggers' age and gender. In: Third International ICWSM Conference. International AAAI Conference on Weblogs and Social Media (2009)

Using Multi-armed Bandit to Solve Cold-Start Problems in Recommender Systems at Telco

Hai Thanh Nguyen[1] and Anders Kofod-Petersen[1,2]

[1] Telenor Research
7052 Trondheim, Norway
{HaiThanh.Nguyen,Anders.Kofod-Petersen}@telenor.com
[2] Department of Computer and Information Science (IDI),
Norwegian University of Science and Technology (NTNU),
7491 Trondheim, Norway
anderpe@idi.ntnu.no

Abstract. Recommending best-fit rate-plans for new users is a challenge for the Telco industry. Rate-plans differ from most traditional products in the way that a user normally only have one product at any given time. This, combined with no background knowledge on new users hinders traditional recommender systems. Many Telcos today use either trivial approaches, such as picking a random plan or the most common plan in use. The work presented here shows that these methods perform poorly. We propose a new approach based on the multi-armed bandit algorithms to automatically recommend rate-plans for new users. An experiment is conducted on two different real-world datasets from two brands of a major international Telco operator showing promising results.

Keywords: multi-armed bandit, cold-start, recommender systems, telecom, and rate-plan.

1 Introduction

The Telco industry do not at first glance appear to be of particular interest from a recommender system perspective. Telcos do not commonly supply a lot of services; most general they supply subscriptions, or rate-plans; either pre-paid or post-paid. However, recommending the optimal rate-plans for users in general, and new users in particular can be challenging.

Suggesting a rate-plan for a new user is a typical *cold-start user problem* (following the separation suggested by Park et al., [1]). This problem has also been identified under slightly different names, such as: the *new user problem* [2], the *cold start problem* [3] or *new-user ramp-up problem* [4]. However, the fact that a customer traditionally only has one rate-plan at any given time increases the difficulty of this problem. Comparing this to a more traditional recommender problem where a user-item matrix might be sparse; in this example the matrix will be completely sparse.

To solve this cold-start problem, given the fact that no prior information on the new user exists, one might think of a random recommendation of rate-plans.

R. Prasath et al. (Eds.): MIKE 2014, LNAI 8891, pp. 21–30, 2014.

However, the chance that the recommended plan be accepted by the new user is small. In fact, given n available rate-plans the probability that a random pick-up plan is accepted is only $1/n$. We say this approach has too much randomness in its recommendations.

Another possibility for solving this problem is to use the distribution of selected plans from existing users. Concretely, it is sensible to recommend the most popular rate-plan to the new user. By doing this, we assume that there is a fixed distribution behind the choice of rate-plans by the new users. However, in reality and also in the experiment below, we can observe that this is not the case. We say this method exploits too much the most popular rate-plan.

The idea now is to have a better solution to control the randomness in the exploration of different rate-plans while keeping the exploitation of the most popular rate-plan at a time. This is the usual dilemma between *Exploitation* (of already available knowledge) versus *Exploration* (of uncertainty), encountered in sequential decision making under uncertainty problems. This has been studied for decades in the multi-armed bandit framework. The work presented here, attempts to tackle the cold-start user problem by recommending a plan that will appeal to the user in question, rather than *the best* plan. We approach this by applying the multi-armed bandit algorithms.

The multi-armed bandit (MAB) is a classical problem in decision theory [5,6,7]. It models a machine with K arms, each of which has a different and unknown distribution of rewards. The goal of the player is to repeatedly pull the arms to maximise the expected total reward. However, since the player does not know the distribution of rewards, he needs to explore different arms and at the same time exploit the current optimal arm (i.e. the arm with the current highest cumulative reward).

To evaluate our MAB approach in solving the cold-start user problem at Telco, we conduct experiments on real-world datasets and compare it with trivial approaches, which include the random and most popular method. Experimental results show that our proposed approach improves upon the trivial ones.

The paper is organised as follows: Section 2 gives an overview of related works; Section 3 provides a formal definition of the cold-start problem in the rate-plan recommender system at Telco. We describe our proposed approaches in Section 4. Section 5 presents some experimental results and discussions. The paper ends with a summary of our findings and a discussion on future work.

2   Related Work

Unfortunately, there are very few examples of research regarding rate-plan recommender systems for Telco, in particular with respect to the cold-start problem. Examples include, Thomas et al., who describe how to recommend best-fit recharges for pre-paid users [8]. Soonsiripanichkul et al., employes a naïve Bayes classifier to infer which rate-plan to suggest to existing users [9]. Both use existing data on customers' usage patterns and do not address the cold-start problem.

In general, one common strategy for mitigating the cold-start user problem is to gather demographic data. It is assumed that users who share a common background also share a common taste in products. Examples include Lekakos and Giaglis [10], where lifestyle information is employed. This includes age, marital status and education, as well as preferences on eight television genres. The authors report that this approach is the most effective way of dealing with the cold-start user problem in sparse environments.

A similar thought underlies the work by Lam et al., [11] where an *aspect model* (see e.g. [12]) including age, gender and job is used. This information is used to calculate a probability model that classifies users into user groups and the probability how well liked an item is by this user group.

Other examples of applying demographic information for mitigating the cold-start user problem exists, e.g. [13,14,15]. All the solutions above use similar demographic information; most commonly age, occupation and gender. Most of the solutions ask for less than five pieces of information. Even though five is a comparatively small number, the user must still answer these questions. Users do generally not like to answer a lot of questions, yet expect reasonable performance from the first interaction with the system [16].

Zigoris and Zhang [16], suggests to use a two part Bayesian model, where the prior probability is based on the existing user population and *data likelihood*, which is based on the data supplied by the user. Thus, when a new user enters the system, little is know about that user and the prior distribution is the main contributor. As the user interacts with the system the data data likelihood becomes more and more important. This approach performs well for cold-start users. Other similar approaches can by found in [17], suggesting a Markov mixture model, and [18] who suggests a statical user language model that integrates an individual model, a group model and a global model.

Our study differs from previous research on the cold-start problem, as no demographic information is taken into account. Only the information on selected plans of previous users is available to the recommender engine. This assumption makes the cold-start problems even harder to solve. However, we leave the issue of collecting more information from users and how to use it for cold-start recommender systems for future works.

3 Problem Definition

Recommending a rate-plan for a new mobile telephony user differs from traditional recommender systems. Traditionally, recommender systems are in a context where users can purchase and own several products, such as books; Rate-plans are different in the sense that one user can have any number of rate-plans, but typically only one plan at any given time. Further, the user will typically have the same product for an extended time period. Finally, no explicit rating for the rate-plans exist. We call this problem the *Cold Start Alternative Recommendation* (CSAR) problem and below is its formal definition.

Let $U = \{u_1, \ldots, u_T\}$ be the set of T new users. Assume that we have a set P of n rate-plans to recommend to a new user: $P = \{p_1, \ldots, p_n\}$ where each plan $p_i (i = 1 \ldots n)$ is described by m features $\{f_1, \ldots, f_m\}$, such as price, number of included SMS, number of voice minutes included and so on. Among k $(k \geq 1)$ suggested rate-plans, the new user can only select one plan at any given time.

Assume that at a given time t a new user u_t comes and the system recommends a rate-plan p_t without any knowledge on the new user. Depending on the user's needs, she will accept the offer or select another rate-plan. We want to design an algorithm that can find a best-fit rate-plan for the new user. Let $need_t$ be a vector described the user's demand: $need_t = (need_{t1}, \ldots, need_{tm})$, where each feature $need_{tj}$ corresponds to each feature f_j of the rate-plans. If we denote the similarity value between the recommended plan p_t and the actual demand of the new user u_t by a $similarity(need_t, p_t)$, then the objective when solving the CSAR problem is to select the rate-plans p_t that maximizes the following so called "cumulative reward" $(Reward)$ over all T new users:

$$Reward_T = \sum_{t=1}^{T}(similarity(need_t, p_t))$$

The CSAR problem would be easy to solve if we knew about the user's needs $need_t$. The task then becomes straightforward by selecting the rate-plan that provides the maximal value of the $similarity(need_t, p_t)$ over all available plans.

As mentioned, it is not possible to calculate $similarity(need_t, p_t)$ since $need_t$ is not available. We suggest to study an approximated problem to the CSAR problem where we consider the similarity value between the recommended plan p_t and the actual selection of the new user p_t^*. By doing this, we wish to achieve a recommendations as close as possible to the actual choice made by the user. The actual choice is also considered as her temporary best-fit plan. Formally, we want to maximize the following so called "reward":

$$Reward_T = \sum_{t=1}^{T}(similarity(p_t, p_t^*))$$

There are many ways to define the similarity value between two vectors p_t and p_t^*. Below, we suggest to take into account the two most popular measurements which are $i)$ the indicator function and $ii)$ the correlation value.

Indicator Function. If we use the indicator function as the similarity measurement, then the problem becomes to design an algorithm that predict the rate-plan p_t^* chosen by the new user. The cummulative reward now is the following: $Reward_T^{(1)} = \sum_{t=1}^{T}(\mathbb{I}(p_t^* \neq p_t))$, where $\mathbb{I}(p_t^* \neq p_t)$ is an indicator function which is equal to 0 if $p_t^* \neq p_t$ and to 1, otherwise. To evaluate any algorithm solving this problem, we can use the classical precision measurement: $Precision_T = \frac{1}{T}Reward_T^{(1)}$

Correlation Value. In the second case, we study how similar is the recommended rate-plan p_t to the actual selection p_t^* of the new user in terms of the features. Generally, when a new user purchases a rate-plan, she looks at the features describing the different rate-plans including the recommended rate-plan p_t. Finally she picks up a plan p_t^* that we can assume is perceived as the temporary best-fit for her. Therefore, it is sensible to choose the correlation coefficient as a similarity measurement between plans and the task is to maximizes the following so called "cumulative reward": $Reward_T^{(2)} = \sum_{t=1}^{T}(Corr(p_t^*, p_t))$, where $Corr(p_t^*, p_t)$ is the correlation value between two vectors p_t^* and p_t.

Possible correlation values can be Pearson correlation or Kendall correlation. Since the actual demand of new users is not available at the time when they enter, it is fair to treat all the features equally in the correlation calculation. While solving this problem, we try to recommend a rate-plan that is sufficiently good in terms the features and that the user will accept. Thus, classic precision measurements are not applicable. We, therefore, define the Average-Feature Prediction (AFP) as a new evaluation measurement of how much of the features of the rate-plan chosen by T new users are predictable on average:

$$AFP_T = \frac{1}{T}Reward_T^{(2)}$$

4 Bandit Algorithms for the CSAR Problem

Based on the idea of the multi-armed bandit [5,6,7], in the following we translate the new CSAR problem into a bandit problem.

Let us consider a set P of n available rate-plans to recommend to T completely new users. Each plan is associated with an unknown distribution of being selected by users. The game of the recommender system is to repeatedly pick up one of the rate-plans and suggest to a new user whenever she enters the system. The ultimate goal is to maximize the cumulative reward. As defined in previous section, the reward for our recommender system is the similarity value $similarity(p_t, p_t^*)$. Note that the setting in present context is slightly different from traditional MABs. In a traditional MAB only the reward of the selected arm is revealed. In our case all the non-selected arms also get rewards after the recommendation is made. In fact, in the case of using the *indicator function*, then the non-selected rate-plans by users will get a zero reward. In the case of using the *correlation value*, the rewards of the non-selected rate-plans will be the correlation value between the two vectors p and p^*. However, since the distributions of the rate-plans being selected are still unknown, the idea of using the MAB algorithms for the CSAR problem is still valid. The following three MAB algorithms are being used:

ε-greedy [7] aims at picking up the rate-plan that is currently considered the best (i.e. the rate-plan that has the maximal average reward) with probability ϵ (exploit current knowledge), and pick it up uniformly at random with probability $1 - \epsilon$ (explore to improve knowledge). Typically, ϵ is varied along time so that the plans get greedier and greedier as knowledge is gathered.

UCB. [7] consists of selecting the rate-plan that maximises the following function: $UCB_{tj} = \hat{\mu}_j + \sqrt{\frac{2\ln t}{t_j}}$ where t is the current time-step, μ_j is the average reward obtained when selecting plan j, t_j is the number of times the plan j has been selected so far. In this equation, $\hat{\mu}_j$ favours a greedy selection (exploitation) while the second term $\sqrt{\frac{2\ln t}{t_j}}$ favours exploration driven by uncertainty; it is a confidence interval on the true value of the expectation of reward for plan j.

EXP3. [19] selects a rate-plan according to a distribution, which is a mixture of the uniform distribution and a distribution that assigns each plan a probability mass exponential in the estimated cumulative rewards for that plan.

5   Experiments and Results

This section details the datasets used in the experiments; the experimental settings, in which the detail implementation of the proposed methods and of the competing algorithms are provided; and contains an analysis and discussion of the experimental results.

5.1   Dataset

We use two different real-world client datasets from two brands of a major international Telco operator. These two datasets were collected during the first quarter of 2013. The first brand's dataset contains the descriptive features of 16 rate-plans, as well as information about the plans used by 3066 users. The second dataset contains the descriptive features of 13 rate-plans, as well as information about the plans used by 1894 users. In this work we have assumed that users have not picked their rate-plan at random – that is, they have each chosen a rate-plan that fits their need. Figure 1 shows the distributions of rate-plans.

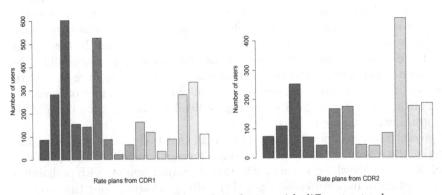

Fig. 1. Distributions of number of users with different rate-plans

Table 1. Features describing the rate-plans

Rate-plan Features	Description
Price per month	The fixed price of rate-plan per month (including 0)
Voice (minutes)	Number of voice minutes included per month (including 0)
Text (SMS)	Number of (SMS) messages included per month (including 0)
MMS	Number of MMS messages included per month (including 0)
MB	Number of MB included per month (including 0)
Voice (post cap, start)	Starting price per voice call after included
Voice (post cap, min)	Price of voice per minutes after included
SMS (post cap)	Price per SMS after included
MMS (post cap)	Price per MMS after incuded
Data (post cap, MB)	Price per MB after included
Speed (Mbit/s)	Speed of Internet allowed

Each of the used rate-plans are described by the 11 most important features, which are shown in Table 1. It is worth noticing that rate-plans broadly fall into three categories: *i*) pre-paid, where the customer pays a certain amount and receives a certain number of services that must be consumed within a certain time frame (e.g. 100 minutes, 100 SMS, 100 MB valid for 30 days); *ii*) traditional post-paid, where the customer pays a certain amount per month and pays for consumption; and *iii*) post-paid flat-rate, where all voice, SMS and MMS is included and the customer pays a certain amount depending on how much data is available (e.g. 100 voice minutes, 100 SMS/MMS and 1 GB of data per month).

5.2   Experimental Settings

Trivial Approaches. The first and the most naïve approach for the cold-start recommendation systems at Telcos is to choose randomly a rate-plan to recommend to a new user. This algorithm is very efficient, especially, when we do not have any description on users and the algorithm seems to be reasonable.

The second trivial approach is to recommend the most popular rate-plan (Most common) to the new user. This is a sensible approach in terms of the efficiency and many operators apply this.

The third trivial approach is to pick up the best-average-reward rate-plan (Best average) at a time (i.e. the current rate-plan that has the maximal average reward value) to recommend to a new user. In this case, we choose the Pearson correlation as a similarity measurement for the reward $similarity(p_t, p_t^*)$ value.

Multi-armed Bandit Algorithms. ϵ-greedy estimates the average reward of each rate-plan. It then selects a random plan with probability ϵ_t, and choose plan of the highest average value of rewards with probability $1 - \epsilon_t$. The parameter ϵ_t is decreased over time t. In fact, the ϵ_t is calculated as follows: $\epsilon_t = min(1, (cn)/(d^2(t - n - 1)))$, where n is the number of rate-plans; c and d are chosen constants. In our experiment, we selected $c = 0.001$ and $d = 0.01$, which provided the best results.

The UCB algorithm estimates the value UCB_{tj} for each plan. It then choose the plan with the highest UCB_{tj} value to recommend to the new user.

Finally, EXP3 selects a plan according to a give distribution, as described in [19]. We select $\gamma = 0.01$ before drawing the probability to select the best plan to recommend to the new user.

Each of the six algorithms are run five times with different choices of parameters. The best results recorded are shown in Table 2.

5.3   Results and Analysis

Table 2 shows the performances of the six different approaches for the cold-start problem on the two different real-world client datasets DS1 and DS2. We present the precision result *Precision* that indicates the accuracy of our recommendation and the prediction value AFP, which is the closeness of our recommended plan to the actual selected one in terms of the features.

Table 2. Precision and AFP for the two datasets

Method	Precision$_{DS1}$	AFP$_{DS1}$	Precision$_{DS2}$	AFP$_{DS2}$
Random	6.80	44.50	7.98	47.60
Most common	20.29	52.70	25.68	70.20
Best average	20.29	54.00	25.68	70.70
ϵ-greedy	20.28	54.00	25.60	70.07
EXP3	10.48	46.80	12.28	61.50
UCB	**43.04**	**69.20**	**45.08**	**75.30**

It can be seen from the table that the random approach provided very poor results in both datasets. In fact, it has only 6.80 percent precision and 44.50 percent prediction in the case of the DS1 dataset. This can be explained by the fact that the probability when a randomly recommended rate-plan being accepts is only $\frac{1}{16} = 6.25\%$. The is also true for the second dataset, where a randomly recommended rate-plan only has a $\frac{1}{13} = 7.67\%$ probability of being correct.

The most common (Most common) and the best average (Best average) performed better than the random one. Yet, the results are still not good. To explain this, we look at Figure 1. Clearly, both the DS1 and DS2 datasets have the most common rate-plans which has the maximal number of being selected by users. Beside, users also chose a variety of other different rate-plans. So following the most common, or the best average rate-plan would not be a good strategy.

The ϵ-greedy approach provides almost the same results as the most common and the best average approaches. The reason is that it was too greedy when setting up the ϵ_t to a too small value. This forces the ϵ-greedy algorithm to follow the best rate-plan (i.e. the rate-plan has the maximal average reward value) all the time.

The case of EXP3 shows even worse performance than the ϵ-greedy. This is probably because of a wrong assumption on the distribution of the selected rate-plans, which is a mixture of the uniform distribution and a distribution that assigns to each plan a probability mass exponential. This exemplifies the fact of being careful when selecting appropriate strategies for the MAB.

The UCB gave us a surprisingly good precision and prediction results. In fact, it increased the precision of the random approach to 39 percent and could predict more than 75 percent of the features of actual selected rate-plan by new users. The reason is that the UCB approach has a good strategy in balancing the exploitation of the best rate-plan at a time and the exploration of other different rate-plans which are also interest for the new users. To have a better explanation, by looking at the UCB algorithm as described in previous section, we see that the recommendation of a rate-plan is a result of solving the trade-off between the average reward and the number of times the plan has been selected so far by users. Therefore, beside the current best rate-plan, other good ones have a chance to be recommended, as well as the other rate-plans that already have been selected a few times. This UCB strategy resulted the distribution of recommended rate-plans closer to the real distribution, as shown in Figure 1, compared to most common and random approaches.

6   Conclusions and Future Research

This work approaches recommending rate-plans to completely new users at Telco, without any prior information on them. An experiment was conducted on two different real-world client datasets from two brands of a major international Telco operator. From the experimental results, we observed that the UCB algorithm clearly outperforms traditional naïve approaches, as well as other classical multi-arm bandit algorithms. This is still work in progress, and as such many issues still needs to be tackled. Improving the precision and AFP would still be preferable. Demographical information is likely to be required to improve this.

Acknowlededgements. The authors would like to extend our gratitude to Professor Helge Langseth at the Department of Computer and Information Science, at the Norwegian University of Science and Technology (NTNU), and Dr. Humberto N. Castejón Martínez and Dr. Kenth Engø-Monsen at Telenor Research; without whom this work would not have been possible.

References

1. Park, S.T., Pennock, D., Madani, O., Good, N., DeCoste, D.: Naïve filterbots for robust cold-start recommendations. In: Proceedings of the 12th ACM SIGKDD International Conference on Knowledge Discovery and Data Mining, pp. 699–705 (2006)
2. Adomavicius, G., Tuzhilin, A.: Towards the next generation of recommender systems: A survey of the state-of-the-art and possible extensions. IEEE Transactions on Knowledge and Data Engineering 17, 734–749 (2005)

3. Massa, P., Bhattacharjee, B.: Using trust in recommender systems: An experimental analysis. In: Jensen, C., Poslad, S., Dimitrakos, T. (eds.) iTrust 2004. LNCS, vol. 2995, pp. 221–235. Springer, Heidelberg (2004)
4. Burke, R.: Hybrid recommender systems: Survey and experiments. user modeling and user-adapted interaction. User Modeling and User-Adapted Interaction 12, 331–370 (2002)
5. Lai, T.L., Robbins, H.: Asymptotically efficient adaptive allocation rules. Advances in Applied Mathematics 6, 4–22 (1985)
6. Katehakis, M., Veinott, J.A.: The multi-armed bandit problem: decomposition and computation. Mathematics of Operations Research 12, 262–268 (1987)
7. Auer, P., Cesa-Bainchi, M., Fischer, P.: Finite-time analysis of the multiarmed bandit problem. Machine Learning 47, 235–256 (2002)
8. Thomas, S., Wilson, J., Chaudhury, S.: Best-fit mobile recharge pack recommendation. In: National Conference on Communications (NCC), pp. 1–5 (2013)
9. Soonsiripanichkul, B., Tongtep, N., Theeramunkong, T.: Mobile package recommendation using classification with feature discretization and threshold-based ensemble technique. In: Proceedings of the International Conference on Information and Communication Technology for Embedded Systems, ICICTES 2014 (2014)
10. Lekakos, G., Giaglis, G.M.: A hybrid approach for improving predictive accuracy of collaborative filtering algorithms. User Modeling and User-Adapted Interaction 17, 5–40 (2007)
11. Lam, X.N., Vu, T., Le, T.D., Duong, A.D.: Addressing cold-start problem in recommendation systems. In: Proceedings of the 2nd International Conference on Ubiquitous Information Management and Communication, pp. 208–211. ACM (2008)
12. Marlin, B.: Collaborative filtering: A machine learning perspective. Technical report, University of Toronto (2004)
13. Gao, F., Xing, C., Du, X., Wang, S.: Personalized service system based on hybrid filtering for digital library. Tsinghua Science & Technology 12, 1–8 (2007)
14. Agarwal, D., Chen, B.C.: Regression-based latent factor models. In: KDD 2009: Proceedings of the 15th ACM SIGKDD International Conference on Knowledge Discovery and Data Mining, ACM SIGKDD, pp. 19–28. ACM (2009)
15. Park, S.T., Chu, W.: Pairwise preference regression for cold-start recommendation. In: RecSys 2009: Proceedings of ACM conference on Recommender systems, pp. 21–28 (2009)
16. Zigoris, P., Zhang, Y.: Bayesian adaptive user profiling with explicit & implicit feedback. In: CIKM 2006: Proceedings of the 15th ACM International Conference on Information and Knowledge Management, pp. 397–404. ACM (2006)
17. Manavoglu, E., Pavlov, D., Giles, C.L.: Probabilistic user behavior models. In: ICDM 2003: Proceedings of Third IEEE International Conference on Data Mining (2003)
18. Xue, G.R., Han, J., Yu, Y., Yang, Q.: User language model for collaborative personalized search. ACM Transactions on Information Systems 27, 1–28 (2009)
19. Auer, P., Cesa-Bianchi, N., Freund, Y., Schapire, R.E.: The nonstochastic multiarmed bandit problem. SIAM Journal on Computing 32, 48–77 (2002)

An Improved Collaborative Filtering Model Based on Rough Set

Xiaoyun Wang and Lu Qian

Management Department, Hangzhou, China
15158113182@163.com

Abstract. Collaborative filtering has been proved to be one of the most success-ful techniques in recommender system. However, a rapid expansion of Internet and e-commerce system has resulted in many challenges. In order to alleviate sparsity problem and recommend more accurately, a collaborative filtering model based on rough set is proposed. The model uses rough set theory to fill vacant ratings firstly, then adopts rough user clustering algorithm to classify each user to lower or upper approximation based on similarity, and searches the target user's nearest neighborhoods and make top-N recommendations at last. Well-designed experiments show that the proposed model has smaller MAE than traditional collaborative filtering and collaborative filtering based on user clustering, which indicates that the proposed model performs better, and can improve recommendation accuracy effectively.

Keywords: Collaborative Filtering, Rough Set, Lower or Upper Approximation.

1 Introduction

Collaborative filtering (CF) has been proved to be one of the most successful tech-niques among all recommender systems [1, 2]. However, with rapid increase of items and users online, the traditional CF method suffers from a rage of serious problems, such as data sparsity [3], scalability [4], and cold start [5], etc, which leads to low quality of recommendations. In order to solve these issues, researchers have proposed extensive improved methods, among which clustering technique [6, 7, 8] is widely used and proved to be effective. Collaborative filtering based on user clustering (UC-CF) [7] classifies similar users to a same cluster, which narrows neighborhoods' search space and helps improve recommender speed and scalability [9, 10]. However, we find that some defects still remain in this algorithm: (1) low quality of recommen-dation for users on the edge of cluster; (2) ignorance of users' multi-interests, which cannot reflect the fact that most customers belong to multiple consumer groups; (3) unable to effectively alleviate sparsity problem.

In order to solve these critical problems, this paper introduces rough set theory, which is widely used to solve problems of incompleteness and uncertainty, into UC-CF. For the problem of incompleteness, Rough Set Theory based Incomplete Data Analysis Approach (ROUSTIDA) [11] is widely used in traditional CF to alleviate sparsity problem. Therefore, the paper first adopts ROUSTIDA to fill vacant ratings.

R. Prasath et al. (Eds.): MIKE 2014, LNAI 8891, pp. 31–41, 2014.

On the other hand, the paper puts forward a rough user clustering algorithm for the problem of uncertainty. Instead of classifying a user to only one cluster, this algorithm classifies a user to lower or upper approximation of a cluster by whether he or she positively or possibly belongs to the cluster [12]. It allows one user to be a candidate for more than one cluster, which reflects customers' multi-interests, and avoids users being on the edge of cluster at the same time. Establishing an improved collaborative filtering model based on rough set is expected to solve the problems above and to improve recommender accuracy.

2   Related Work

2.1   Collaborative Filtering

The term 'collaborative filtering' was coined by Goldberg [13], who was the first to use collaborative filtering techniques in filtering information. Generally, it can be divided to user-based CF and item-based CF. The former approach predicts a target user's interest on one item based on rating information from similar users. It can be implemented by the following steps [14].

Create User-Item Rating Matrix. User-Item Rating Matrix is shown in Table 1. R_{ij} represents the rating for user i on item j. R_{ij} is a integer, and rating scale is varied depending on different situation. The ratings reflect users' liking tendencies towards items. Usually the higher R_{ij} is, the more user i likes or is satisfied with item j.

Table 1. User-Item Rating Matrix

Item / User	1	...	j	...	n
1	R_{11}	...	R_{1j}	...	R_{1n}
...	...	...	...	...	...
i	R_{i1}	...	R_{ij}	...	R_{in}
...	...	...	...	...	...
m	R_{m1}	...	R_{mj}	...	R_{mn}

Calculate Similarity between Users. How to properly quantify the similarity between users is one of the most important problems in CF. Standard approaches are using the correlation between the ratings that two users give to a set of objects, such as Cosine index and Pearson correlation coefficient [15]. However, Cosine index doesn't consider the rating scales among users, and Pearson correlation coefficient doesn't perform well in sparse data. Therefore, a modified cosine index is adopted.

$$sim(i, j) = \frac{\sum_{c \in I_{ij}} (R_{i,c} - \overline{R_i})(R_{j,c} - \overline{R_j})}{\sqrt{\sum_{c \in I_i} (R_{i,c} - \overline{R_i})^2 \sum_{c \in I_j} (R_{j,c} - \overline{R_j})^2}} \tag{1}$$

Where $sim(i, j) \in [0,1]$. $R_{i,c}$ is the rating for user i on item c. $\overline{R_i}$ and $\overline{R_j}$ represent average ratings of user i and user j respectively. I_{ij} is set of items that user i and user j both rate. I_i and I_j represent the sets of items that user i and user j rate respectively.

Select Neighborhoods and Make Recommendation. It is useful, both for accuracy and performance, to select a subset of users (neighborhoods) with higher similarity to use in computing a prediction for a target user, instead of using the entire database. Once the neighborhood has been selected, ratings from those neighbors are combined to calculate a prediction. At last, N items with the highest predicted ratings were recommended to the target user. User i's predicted rating on item c can be calculated as:

$$P_{i,c} = \overline{R_i} + \frac{\sum |sim(i, j)| \times (R_{j,c} - \overline{R_j})}{\sum |sim(i, j)|} \qquad (2)$$

2.2 Rough Set Theory

Rough set theory is proposed by Professor Z.Pawlak in 1982 [16], and is widely used in the field of data mining, machine learning, and expert system, etc. Here give some related definitions.

Information System. Let $S = (U, A, V, f)$ be an information system (attribute-value system), where U is a non-empty set of finite objects (the universes) and A is a non-empty, finite set of attributes such that $a: U \rightarrow V_a$ for every $a \in A$. V_a is the set of values that attribute a may take. The information table assigns a value $a(x)$ from V_a to each attribute a and object x in the universe U. $f: U \times A$ is an information function and attribute value of each object x.

Indiscernibility Relation. With any $B \subseteq A$, there is an associated equivalence relation $IND(B)$: $IND(B) = \{(x, y) | (x, y) \in U^2, \forall b \in B(b(x) = b(y))\}$. $IND(B)$ is called an indiscernibility relation. If $(x, y) \in IND(B)$, then x and y are indiscernible (or indistinguishable) by attributes from B.

Discernibility Matrix. For information system $S = (U, A, V, f)$, the discernibility matrix is a $n \times n$ square matrix. The definition of M is as follows:

$$M(i, j) = \left\{ a_k \mid a_k \in A \wedge a_k(x_i) \neq a_k(x_j) \wedge a_k(x_i) \neq * \wedge a_k(x_j) \neq * \right\}$$

Where * express the lack. The element in extended discernibility matrix is a set of attribute subscripts. And the discernibility concept between attribute values is also extended to adapt for incomplete information system.

Missing Attribute and No-Different Object. For Information system $S = (U, A, V, f)$, if $x_i \in U$, then the missing attribute set MAS_i of object x_i, no-difference object set NS_i for object x_i and missing object set MOS_i are defined as:

$$MAS_i = \left\{ a_k \mid a_k(x_i) = *, k = 1, \cdots, m \right\}$$

$$NS_i = \left\{ j \mid M(i,j) = \Phi, i \neq j, j = 1, \cdots, n \right\}$$

$$MOS_i = \left\{ i \mid MAS_i \neq \Phi, i = 1, \cdots, n \right\}$$

Lower and Upper Approximations. Let $X \subseteq U$ be a target set that we wish to represent using attribute subset B. In general, X cannot be expressed exactly, because the set may include and exclude objects which are indistinguishable on the basis of attributes B. However, the set X can be approximated using only the information contained within B by constructing the lower and upper approximations of X:

$$\underline{B}(X) = \bigcup \left\{ Y_i \mid (Y_i \in U \mid IND(B) \wedge Y_i \subseteq X) \right\}$$

$$\overline{B}(X) = \bigcup \left\{ Y_i \mid (Y_i \in U \mid IND(B) \wedge Y_i \wedge X \neq \Phi) \right\}$$

3   The Proposed Model

In order to alleviate sparsity problem and improve recommender accuracy, rough set theory has been introduced in this paper, and an improved collaborative filtering model based on rough set (RS-CF) is established. The model can be divided into two phases: When in the off-line phase, it uses ROUSTIDA algorithm, which is based on rough set theory, to fill vacant ratings in User-Item rating matrix. Then the model adopts rough K-means user clustering algorithm to classify all users to each cluster's lower or upper approximation by whether the user positively or possibly belongs to the cluster based on user similarity, and thus generates each user's initial neighborhoods. When in the on-line phase, the model starts searching the target user's nearest neighborhoods from his initial ones, predicts his ratings and make top-N recommendations. The framework of proposed model is shown in Fig. 1.

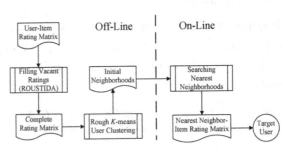

Fig. 1. Framework of RS-CF

3.1   Filling Vacant Ratings

ROUSTIDA is a widely used algorithm to analyze and replace incomplete data. Because of multiple vacant ratings and their different distribution, supplementing the vacant data cannot be achieved after a single computation of initiative extended discernibility matrix and a completeness analysis. It calls for multiple times computation of initiative extended discernibility matrix and completeness analysis. Let r be the number of times of computation and based on extended discernibility matrix above, ROUSTIDA is given as follows [11].

Input: Incomplete User-Item Rating Matrix $S^0 = (U^0, I, R, r^0)$

Output: Complete User-Item Rating Matrix $S^r = (U^r, I, R, r^r)$

Steps:

1. Compute the initiative discernibility matrix M^0, MAS^0, MOS^0, let $r=0$

2. For all $i \in MOS^r$, calculate NS_i^r

    (a) Generate S^{r+1}. For all $i \notin MOS^r$, let $r_{i,k}^{r+1} = r_{i,k}^r, k = 1, 2, \cdots n$;

    (b) For all $i \in MOS^r$, make loops for all $i_k \in MAS_i^r$

        (i) If $|NS_i^r| = 1$, let $j \in NS_i^r$, if $r_{i,k}^{r+1} = *$, then $r_{i,k}^{r+1} = *$; otherwise $r_{i,k}^{r+1} = r_{j,k}^r$

        (ii) Otherwise

            (1) If exist j_0 and $j_1 \in NS_i^r$, with the condition $r_{j_0,k}^r \neq *$ and $r_{j_1,k}^r \neq *$ and $r_{j_0,k}^r \neq r_{j_1,k}^r$, then $r_{i,k}^{r+1} = *$.

            (2) Otherwise, if exist $j_0 \in NS_i^r$, with the condition $r_{j_0,k}^r \neq *$, then $r_{i,k}^{r+1} = r_{j_0,k}^r$. Otherwise, $r_{i,k}^{r+1} = *$.

    (c) If $S^{r+1} = S^r$, finish the recycle and turn to Step 3, Otherwise compute M^{r+1}, MAS^{r+1} and MOS^{r+1}; $r=r+1$; turn to Step 2.

3. If there are still vacant ratings in User-Item Rating Matrix, combination completeness approach is adopted for further process.

3.2   Rough K-means User Clustering

Bringing in the thought of rough set in user clustering, rough K-means user clustering algorithm determines a user's membership based on his similarity with the centroid of each cluster. A user who positively belongs to one user cluster is classified to its lower approximation. Reversely, a user who possibly belongs to one user cluster is classified to its upper approximation. The following three properties given by Lingras [12] may help understand this algorithm.

- **Property 1.** An object can be part of at most one lower approximation.
- **Property 2.** If an object is part of one lower approximation, then it also belongs to the upper approximation.
- **Property 3.** If an object is not part of any lower approximation, then it belongs to two or more upper approximation.

Here give concrete steps of rough K-means user clustering algorithm.

Input: Complete User-Item Rating Matrix, the Number of User Clusters (K), Approximation threshold (*threshold*), adjusting parameters of centroids of user clusters (w_l, w_u).

Output: K User Clusters with Lower and Upper Approximations, Each User's Cluster Label.

Steps:

1. Select K users randomly and use their n-dimensional rating vectors $c_1, c_2, \cdots, c_K$ as initial centroids of clusters.

2. Calculate similarities between each user and K centroids by Equation (1). $sim(u, c_i)$ represents the highest similarity between user u and centroid c_i.

3. **Determine Each User's Membership.** Modified cosine index (Equation 1) subtracts user's average rating, which means that it eliminates the problem of users' different rating scales. Therefore, the discrepancy of similarities between user and centroids can be expressed directly by their absolute difference. Let *threshold* be an approximation threshold and U be one user cluster. The following two rules can classify all users to lower or upper approximations of K clusters. For the set:

$$T = \{ j : d = sim(u, c_i) - sim(u, c_j) \le threshold, 1 \le i, j \le k, i \ne j \}$$

Rule 1. If $T \ne \Phi$, then $x \in \overline{B}(U_i)$ and $x \in \overline{B}(U_j)$. In other words, user u shares high similarities with more than one centroid. Thus the membership of user u is ambiguous and u cannot be classified to any lower approximation. According to Property 3, user u belongs to upper approximation of these clusters.

Rule 2. If $T = \Phi$, then $x \in \underline{B}(U_i)$. In other words, user u only shares high similarities with one centroid, but shares low similarities with the others. Therefore, the membership of user u is clear and u can be classified to lower approximation.

4. **Adjust Centroids of Clusters.** Adjustment of centroid depends on users both in lower and upper approximations of the cluster. Centroid can be recalculated by the weighted arithmetic mean of vectors of users who are positively belongs to this cluster and who are possibly belong to it. The Equation is:

$$c_i = w_l \bullet \sum_{x_n \in \underline{B}(U_i)} \frac{x_n}{|\underline{B}(U_i)|} + w_u \bullet \sum_{x_n \in \overline{B}(U_i)} \frac{x_n}{|\overline{B}(U_i)|} \quad w_l + w_u = 1 \tag{3}$$

Parameters w_l and w_u show the importance of lower and upper approximation to user clustering. $|\underline{B}(U_i)|$ is the number of users in lower approximation of U_i, and $|\overline{B}(U_i)|$ is that in upper approximation of U_i.

5. Repeat step 2, 3, 4 until criterion function $J = 1/\sum_{i=1}^{K} \sum_{u \in U_i} sim(u, c_i)$ converges.

In a word, rough K-means user clustering algorithm can roughly classify all users to lower and upper approximations of K user clusters offline, which can be regarded as each user's initial neighborhoods.

3.3   Searching Nearest Neighborhoods

When a special user is online, the following search algorithm can help find the target user's nearest neighborhoods from his initial neighborhoods.

Input: Target User, Initial Neighborhoods, Threshold of the Number of nearest neighbors (N_u).

Output: Target User's Nearest Neighborhoods.

Steps (Two Conditions):

Condition 1. The target user belongs to lower approximation of U_i.

1. Let N be the number of nearest neighbors. Put all users in lower approximation of U_i into the nearest neighborhoods. If $N \geq N_u$, then finish. Otherwise, turn to next step.
4. Put all users in upper approximation of U_i into the nearest neighborhoods. If $N \geq N_u$, then finish. Otherwise, turn to next step.
5. Count the cluster labels of users in the upper approximation of U_i and find out the highest frequency of cluster label U_j. Add users in U_j to the nearest neighborhoods. If $N \geq N_u$, then finish. Otherwise, turn to next step.
6. Find out the second highest frequency of cluster label U_k. Add users in U_k to the nearest neighborhoods. And so on, until $N \geq N_u$ is valid.

Condition 2. The target user belongs to at least two clusters' upper approximation.

1. Let N be the number of nearest neighbors. Combine all the users in the upper approximation that the target user belongs to, and let the combination be the nearest neighborhoods. If $N \geq N_u$, then finish. Otherwise, turn to next step.
7. Extend the number of users in the nearest neighborhoods by following Step 3 and Step 4 in Condition 1, until $N \geq N_u$ is valid.

In a word, the search algorithm above can output the target user's nearest neighborhoods. At last, this model uses Equation (2) to predict the user's ratings based on the Nearest Neighbor-Item rating matrix, and recommends N items with highest predicted ratings to the online user.

4   Experimental Evaluation

4.1   Data Set

The testing data set that the following experiments use is provided by Movielens [18], which is a research site run by GroupLens Research at the University of Minnesota and is widely used in the study of personalized recommendation. The data set downloaded from Movielens web site consists of 100,000 ratings (1-5) from 943 users on 1682 movies. In this paper, we choose 150 users randomly for simple to implement in the experiments. Moreover, data set u1.base and u1.test are 80%/20% splits of the whole data into training and test data. As shown in Table 2, spares level of the experimental data reaches up to 92.31%, which means the data is extremely sparse.

Table 2. Experimental Data Statistics

	Users	Items	Ratings	Sparse Level
Data	150	1296	14951	92.31%

4.2   Evaluation Metric

A plenty of metrics exist to evaluate recommender system's performance. This paper adopts Mean Absolute Error (often referred to as *MAE*), which is one of the most effective and extensively used metrics [19, 20], to evaluate the proposed model. *MAE* measures the average absolute deviation between a predicted rating and the user's true rating. The smaller the *MAE*, the higher recommendation accuracy the model will be. Let $\{p_1, p_2, \cdots p_N\}$ be the set of predicted ratings, and let $\{q_1, q_2, \cdots q_N\}$ be the set of true ratings. *MAE* can be calculated as:

$$MAE = \frac{\sum_{i=1}^{N}|p_i - q_i|}{N} \qquad (4)$$

4.3   Experimental Scenario and Results

Experiment 1. The values of parameters w_l and w_u will have great influence on recommender accuracy because they show the importance of lower and upper approximation to clustering when adjusting centroids. Observe the changes of *MAE* by increasing the value of w_l from 0 to 1.0 (0.1 each time), namely decreasing the value of w_u from 1.0 to 0 (0.1 each time). The aim of experiment 1 is to test the influence of w_l and w_u to *MAE*, thus the following parameters are controlled and fixed. Let K be 20, N_u be 15, and *threshold* be 0.05 (empirical value).

The result is shown in Fig. 2. It is obvious that the proposed model works best when $w_l = 0.8, w_u = 0.2$. In other words, users in lower approximation are more important than that in upper one to the adjusting centroids.

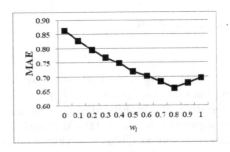

Fig. 2. The Influence of w_l and w_u to *MAE*

Experiment 2. In order to verify the proposed model (RS-CF) can make better recommendation, this experiment compares it with traditional collaborative filtering (CF) and collaborative filtering based on user clustering (UC-CF). Because the Number of User Clusters (K) and Threshold of the Number of nearest neighbors (N_u) will influence the recommendation, values of them are assigned as in Table 3.

Table 3. Parameter Setting in Experiment 2

Parameter	Value	In Algorithm
N_u	10, 15, 20, 25	CF, UC-CF, RS-CF
K	15, 20	UC-CF, RS-CF
Threshold0	0.5	UC-CF
Threshold	0.05	RS-CF
w_l, w_u	$w_l = 0.8$, $w_u = 0.2$	RS-CF

The experimental result is shown in Fig. 3. It shows that the RS-CF has smaller *MAE* than CF and UC-CF, which indicates that the proposed model performs better, and can improve recommender accuracy effectively.

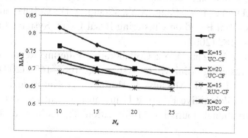

Fig. 3. CF, UC-CF and RUC-CF Comparison

5   Conclusions and Future Work

The proposed RS-CF is an improved approach in recommender system. The innovation of this paper is to introduce rough set into CF to solve problems in traditional CF. The proposed model firstly adopts ROUSTIDA to fill vacant ratings, which greatly alleviate sparsity problem and then puts forward a rough user clustering algorithm. It allows one user to be a candidate for more than one cluster, which reflects customers' multi-interests, and avoids users being on the edge of cluster at the same time. At last, the effectiveness of the model is verified by two well-designed experiments, which indicates that the RS-CF performs better than CF and UC-CF, and is of high feasibility and practical significance.

With an increase of users' demands, the study of recommender system is continuously developing. The next research focus should be put on how to bring in techniques in other research or other application fields to solve problems of CF.

References

[1] Lu, L.Y., Medo, M., Yeung, C.H., Zhang, Y.C., Zhang, Z.K., Zhou, T.: Recommender system. Physics Reports Review Section of Physics Letters 519, 1–49 (2012)

[2] Park, D.H., Kim, H.K., Choi, I.Y., Kim, J.K.: A Literature Review and Classification of Re-commender System Research. Expert Systems with Applications 39, 10059–10072 (2012)

[3] Anand, D., Bharadwaj, K.K.: Utilizing Various Sparsity Measures for Enhancing Accuracy of Collaborative Recommender Systems Based on Local and Global Similarities. Expert Systems with Applications 38, 5101–5109 (2011)

[4] Takacs, G., Pilaszy, I., Nemeth, B., Tikk, D.: Scalable Collaborative Filtering Approaches for Large Recommender System. Journal of Machine Learning Research 10, 623–656 (2009)

[5] Lika, B., Kolomvatsos, K., Hadjiefthymiades, S.: Facing the Cold Start Problem in Recom-mender Systems. Expert Systems with Application 41, 2065–2073 (2014)

[6] Ungar, L.H., Foster, D.P.: Clustering Methods for Collaborative Filtering. In: Proceedings of 1998 Workshop on Recommender Systems, pp. 114–129. AAAI (1998)

[7] Li, T., Wang, J.D., Ye, F.Y., Feng, X.Y., Zhang, Y.D.: Collaborative Filtering Recommen-dation Algorithm Based on Clustering Basal Users. Systems Engineering and Electronics 29, 1178–1182 (2007)

[8] Deng, A.I., Zuo, Z.Y., Zhu, Y.Y.: Collaborative Filtering Recommendation Algorithm Based on Item Clustering. Mini-Micro Systems 25, 1665–1670 (2004)

[9] Abdelwahab, A., Sekiya, H., Matsuba, I., Horiuchi, Y., Kuroiwa, S.: Alleviating the Sparsity Problem of Collaborative Filtering Using an Efficient Iterative Clustered Prediction Technique. International Journal of Information Technology & Decision Making 11, 33–53 (2012)

[10] Georgiou, O., Tsapatsoulis, N.: Improving the Scalability of Recommender Systems by Clustering Using Genetic Algorithms. In: Diamantaras, K., Duch, W., Iliadis, L.S. (eds.) ICANN 2010, Part I. LNCS, vol. 6352, pp. 442–449. Springer, Heidelberg (2010)

[11] Zhu, W.H., Zhang, W., Fu, Y.Q.: An Incomplete Data Analysis Approach Using Rough Set Theory. In: 2004 International Conference on Intelligent Mechatronics and Automation, pp. 332–338. IEEE, Piscataway (2004)

[12] Lingras, P., West, J.: Interval Set Clustering of Web Users with Rough K-means. Journal of Intelligent Information Systems 23, 5–16 (2004)

[13] Goldberg, D., Nichols, D., Oki, B.M., Terry, D.: Using Collaborative Filtering to Weave an Information Tapestry. Communications of the ACM 35, 61–70 (1992)

[14] Herlocker, J.L., Konstan, J.A., Borchers, A., Riedl, J.: An Algorithmic Framework for Performing Collaborative Filtering. In: Hearst, M., Tong, R. (eds.) SIGIR 1999, pp. 230–237. ACM, New York (1999)

[15] Zhang, Q.M., Shang, M.S., Zeng, W., Chen, Y., Lu, L.Y.: Empirical Comparison of Local Structural Similarity Indices for Collaborative-Filtering-Based Recommender Systems. In: Wang, B.H., Zhang, Y.C., Zhou, T., Castellano, C. (eds.) China-Europe 2010. Physics Procedia, vol. 3, pp. 1887–1896. Elsevier Science, Netherlands (2010)

[16] Pawlak, Z.: Rough sets. International Journal of Computer and Information Sciences 11, 341–356 (1982)
[17] Pawlak, Z.: Rough Sets: Theoretical Aspects of Reasoning about Data. Kluwer Academic Publishers, Poland (1991)
[18] Movielens Movie Rating Data Set, http://movielens.umn.edu/login
[19] Herlocker, J.L., Konstan, J.A., Terveen, K., Riedl, J.T.: Evaluating Collaborative Filtering Recommender Systems. ACM Transactions on Information Systems 22, 5–53 (2004)
[20] Sanchez, J.L., Serradilla, F., Martinez, E., Bobadilla, J.: Choice of Metrics Used in Collaborative Filtering and Their Impact on Recommender Systems. In: Second IEEE International Conference on Digital Ecosystems and Technologies, pp. 432–436. IEEE Press, Piscataway (2008)

Exploring Folksonomy Structure
for Personalizing the Result Merging Process
in Distributed Information Retrieval

Zakaria Saoud, Samir Kechid, and Radia Amrouni

LRIA, Computer Science Department, USTHB Algiers, Algeria
zakaria.saoud@live.fr, skechid@usthb.dz, radia_amrouni@yahoo.fr

Abstract. In this paper we present a new personalized approach that integrates a social profile in distributed search system. The proposed approach exploits the social profile and the different relations between social entities to : (i) make a query expansion, (ii) personalize and improve the result merging process in distributed information retrieval.

1 Introduction

With the increase of the web size, the amount of information covered by a centralized search engine decreases [1]. Thus, the centralized Information retrieval is no longer sufficient to satisfy the needs of users. To solve the problems of centralized Information Retrieval, distributed Information Retrieval appeared, which consist to use meta-search engines to increase the research coverage and by combining the results from several centralized search engines. The distributed information retrieval gave birth to two major problems: (1) the source selection and (2) the result merging. In this paper, we are interested in the result merging problem. The large amount of results returned by meta-search engines is a great disadvantage, and from there the custom meta-search engines have emerged. Personalized meta-search engines use profiles of users to filter the results and return the relevant documents that better meet to user needs. Many kinds of personal data can be used for the construction of the user profile such as user manually selected interests , search engine history , etc . Internet growth and the advent of web2.0 gave birth to different types of social networks on a large scale, which are now recognized as an important means for information dissemination[2]. Many of social networks are considered as social tagging systems, these systems allow the users to provide annotations (tags) to resources, to give their opinions about resources. Several social bookmarking services as Flickr[1] and Delicious[2] are considered an online folksonomy services, and their social tagging data, also known as folksonomies [3]. The set of tags can be used as a source of personal data to build the user profile . In this work we propose a new approach that exploit the social and the semantic relations among items and tags through the use of user

---

[1] Flickr - Photo sharing, http://www.flickr.com/

[2] Delicious - Social bookmarking, http://delicious.com/

R. Prasath et al. (Eds.): MIKE 2014, LNAI 8891, pp. 42–50, 2014.

profile defined within a social tagging system to make a query expansion and to improve the result merging process in order to improve the quality of search of a meta-search engine. The rest of the paper is organized as follows: In section 2 we briefly review the related work. In Section 3 we introduce our personalization approach. Finally, we conclude our work and list some future work in Section 4.

2   Related Work

2.1   Social Information Retrieval

Social tagging systems are web-based systems that allow internet users to add, edit, and share bookmarks of web documents. In social bookmarking services, such as Delicious and Flickr, the users can annotate their bookmarks with arbitrary keywords called tags. The collection of a user's tags constitutes their personomy, and the collection of all users' personomy constitutes the folksonomy.

A folksonomy is a tuple $F := (U, T, D, Y)$ where $U = \{u_1, \ldots, u_M\}$ is the set of users, $T = \{t_1, \ldots, t_L\}$ is the set of tags, and $D = \{d_1, \ldots, d_N\}$ is the set of resources or web documents, and Y is a ternary relation between U and T and D, i. e., $Y \subseteq U \times T \times D$, whose elements are called tag assignments [3] [4]. In our case, the elements of D represents the different web resources and are identified by a URL. Users are identified by a user ID. In social information retrieval, many studies have proposed in the context of search personalization. Most of these studies are based on the folksonomy structure.

Schenkel et al. [5] developed a framework for harnessing such social relations for search and recommendation. They created a scoring model that exploits social relations and semantic/statistical relations among items and tags; this scoring model gives a great importance to users who have a high score of strength friendship with the query initiator. The score of strength friendship is a linear combination of the spiritual strength friendship, the social strength friendship, and the global strength friendship. Rather than item recommendation, our personalized retrieval models, applicable to meta-search engine, is composed of several web search engines to re-rank the lists of search results according to the user profile.

Bender et al. [6] exploits the different entities of social networks (users, documents, tags) and social relations between these entities, to make a query expansion by adding the similar tags to the query keywords, and to make a social expansion to give an advantage to documents tagged by the user's close friends.

Hochul et al. [7] developed an approach that use the links and similarities between the user profiles in the filtering algorithm results. This approach is called collaborative filtering. The principal advantage of this approach is the enlargement of the coverage of research using similar profiles. For example, in the case where P does not obtain satisfactory results for a query, we can then use the most similar profiles to P to enlarge the search and retrieve more relevant results.

Vallet et al. [4] present a personalization model that exploits folksonomy structure. For this they developed two measures to calculate the relevance between a user profile and a document to re-rank the list of results returned by a search engine.

2.2 Result Merging Approach

The most referenced work was performed by Fox and Shaw [8]. The authors proposed several combination methods based on the min (CombMIN), max (CombMAX), medium (CombMED), sum (CombSUM) or average (ComANZ). The principle rules of these methods are the same, which provides to merge the documents based on the scores given by the server (the source). Merging process is an important step in a distributed information system, it allows the system to select and order the documents coming from different selected information sources into a single result list to be returned to the user. Several works addressed the results merging problem. In each work they tried to find a measure to rank the documents. For that, different strategies have been proposed . Yumono and al [9] assign a score for each server in the server selection phase. During the merging phase, the rank of each document is combined with the score of the server to calculate the score of this document. Documents are thus merged and sorted according to this score. Si and Callan [10] [11] propose another merging approach called SSL (Semi-Supervised Learning). This approach associates for each document returned by a source two scores. The first score obtained by the source, and the second score calculated by applying an effective weighting technique on the list of documents returned by the source. The documents are then merged according to the combined scores. Kechid and Drias [12] [13] integrates a user profile into the ranking process. For that, they associate for each document returned by a source three scores:

1. A score that represents the similarity between the document and the user query.
2. A score that represents the similarity between the document and the user profile.
3. A score that represents the accuracy degree of the document according to the user preferences.

The documents are then merged according to the combined scores.

3 Our Approach

In our approach, we define a new user social profile and exploiting it for personalizing and improving the result merging process in distributed information retrieval. Similar to the studies of Hochul et al [7] and Bender et al. [6], we exploit the folksonomy structure, and we use both the friendship measure and the similarity between two tags to make a query expansion rather than the use of the friendship and the similarity measure directly in the scoring model. Similar to the studies of Vallet et al [4], we follow the same personalization model but we exploit other relations between the folksonomy entities (users, documents, tags), such as the relation between user and user and the relation between tag and tag to expand the research coverage and to improve the search quality of a distributed information system, particularly, to improve the result merging process. In our approach, the documents are ordered according a score $ScoreDoc_{d,s}(u_m)$. This score is calculated by the combination of: 1- the score between the user and the document, 2- the score of the source document. The computation of score $ScoreDoc_{d,s}(u_m)$ is detailed in next sections.

3.1 User Profile Definition

The user profile is defined by the user's set of tags, we suppose that frequently used tags are more relevant to a user than rarely used tags, hence, we use just the tags who have a frequency greater than or equal to the average of user's tags. We note:

$$profile(u_m) = \sum_{l,tf_{u_m}(t_l) \geq avgtags_{u_m}} (t_l, tf_{u_m}(t_l))$$

where:

$tf_{u_m}(t_l)$: is the User-based tag frequency, which mean how many time the user use u_m the tag t_l .

and : $avgtags_{u_m}$ represent the average of all tags used by the user u_m, we calculate this average as follow:

$$avgtags_{u_m} = \frac{\sum_{l} tf_{u_m}(t_l)}{tags\_number_{u_m}}$$

where $tags\_number_{u_m}$: represent the number of tags used by the user u_m. The user profile in our approach is used to:

1. Work a query expansion by adding similar tags to the keywords that appear in the query.
2. Merge the results of sources.

3.2 Query Expansion Process

According to Barry Smyth [14], 66% of our new research is similar to those made by our colleagues. These figures show the importance of taking into account the social factor in information retrieval. Based on this idea, for our query expansion process, we propose the use of the profiles of the user's close friends, to find the list of tags that are similar to the query tags. To find this list of tags $\bar{T}$, we use the friendship measure and the tags expansion measure inspired from the work of Bender et al. [6]. The user query is reformulated by adding the tags of the list $\bar{T}$. The new query is weighted by the following formula inspired by Rocchio [15].

$$q^{\text{new}} = a.q^{\text{old}} + \frac{b}{|\bar{T}|} \sum_{t \in \bar{T}} \bar{t}$$

With q^{new} is the expanded query, q^{old} is the old query by the user, a and b are constants, $a, b \in [0, 1]$. The list of tags $\bar{T}$ is established as follows:

$$T = \sum_{t_i \in q^{\text{old}}} \bar{t}, \bar{t} = \{\bar{t} \in \text{sim}(t_i) \,|\, \text{TagSim}(\bar{t}, t_i) = \max_{\text{sim}(t_i)}\}$$

$\text{sim}(t_i)$: refers to the list of tags that are similar to the query tag t_i (i.e., all tags where $\text{TagSim}(\bar{t}, t_i) > 0$ and $\bar{t} \in friends_{\text{close}}(u_m)$.)

$\text{TagSim}(\bar{t}, t_i)$: represent the similarity between two tags $\bar{t}, t_i$. We compute the Dice coefficient defined as:

$$\text{TagSim}(\bar{t}, t_i) = \frac{2 \times df_{\bar{t}, t_i}}{df_{\bar{t}} + df_{t_i}}$$

where:

$df_{\bar{t}, t_i}$: is the number of documents which belongs to the list $friends(u_m)$, and that contain both term $\bar{t}$ and t_i.

$df_{\bar{t}}, df_{t_i}$: are the number of documents which belongs to the list $friends(u_m)$, and that have been tagged with $\bar{t}$ and t_i, respectively.

$\max_{\text{sim}(t_i)}$: represent the tag that have the high similarity in the list $\text{sim}(t_i)$.

$friends_{\text{close}}(u_m)$: refers the list of close friends of the query initiator u_m, the close friend of user u_m must have a friendship score greater than or equal to the average score of friendship. This list is appointed as follows:

$$friends_{\text{close}}(u_m) = \sum_{y \in friends(u_m)} y, friendship(u_m, y) \geq avg\_friendship(u_m)$$

where:

$avg\_friendship(u_m)$: the average friendship score of the user u_m.

$$avg\_friendship(u_m) = \frac{\sum\limits_{y \in friends(u_m)} friendship(u_m, y)}{|friends(u_m)|}$$

$friendship(u_m, y)$: represent the friendship score. This score is the number of common friends between the user u_m and y.

3.3 Results Merging Process

This process consists of ordering and merge all documents returned by the selected sources into a single ranked list. This list will be presented to the user. In our approach the documents are ordered according to a score $ScoreDoc_{d,s}(u)$, this score is calculated using our following formula:

$$ScoreDoc_{d,s}(u_m) = \alpha.sim(u_m, d) + (1 - \alpha).ScoreSource_s(u_m).\frac{1}{rank_s(d)}$$

$\alpha \in [0.1]$

where:

$rank_s(d)$: is the rank of the document d in the returned list of the source s.

$ScoreSource_s(u_m)$: (number of web pages tagged and returned by the source) / (total number of pages returned by the source).

$sim(u_m, d)$: Is the similarity between the user profile and the document profile[3]. To calculate this similarity we prefer the use of the approach developed by Vallet et al [4], because the experiment's results show that the approach has better performance than the other one's approaches. This personalization approach is defined as followed:

$$sim(u_m, d) = \sum_l (tf_{u_m}(t_l).iuf(t_l).tf_d(t_l).idf(t_l))$$

where:

$tf_{u_m}(t_l)$: is the number of times the user u_m has used the tag t_l.

$tf_d(t_l)$: is the number of times the document d has been annotated with tag t_l.

$iuf(t_l)$: measures the popularity of a tag t_l is across all users U.

$$iuf(t_l) = \log\frac{|U|}{nu(t_l)}, nu(t_l) = |\{u_x \in U, tf_{u_m}(t_l) > 0|\}|$$

$idf(t_l)$: measures the popularity of a tag t_l is across all documents D.

$$idf(t_l) = \log\frac{|D|}{nd(t_l)}, nd(t_l) = |\{d_n \in D, tf_d(t_l) > 0|\}|$$

4   Experiments

4.1   Experimental Setup

For our experiments, we used a dataset from the del.ico.us social bookmarking system, this dataset is released in the framework of the 2nd International Workshop on Information Heterogeneity and Fusion in Recommender Systems (HetRec 2011)[4]. It contains 69226 URLs (resources), 1867 users, and 53388 distinct tags.

To evaluate our approach in a distributed environment, we considered the 5 following search engines: Google, Yahoo, Bing, Ask, and AOL. Each search engine represents a source of information, and for each source we have downloaded the top 20 retrieved documents, hence, in our experiments the final result list

---

[3] The document profile is the set of tags used to annotate the document.

[4] http://grouplens.org/datasets/hetrec-2011/

contains 20 documents. In our calculations, to define the user profile we use just the top 15 tags, and, to make the query expansion, our list of close friends contains just the top 5 close friends. To examine the benefit of our approach for individual users, we allowed 5 participants to evaluate the results list, and each user ran 4 queries. In total, we have tested 20 different queries. These queries are made using the most popular tags in the dataset. To solve the problem of subjectively assessing (relevance judgment) [6], we follow the method of Bender et al [6], and that by selecting a fictitious profile for each query initiator. In our test, the fictitious profile is extracted from the set of profiles of the social bookmarking dataset del.ico.us, and this profile must contain the greatest sum of tags frequency of the query.

4.2   Experimental Results

Results of Merging Process Personalization Approach. In this section, we analyze the performance of our personalization approach when only the personalization scores are used to merge and reorder the documents returned by the distributed information system. In this section the query expansion is not taken into account, and for each query we considered the result in various cases, by varying the parameters α and in the interval $[0, 1]$.

We vary the parameters α evaluate the influence of the use of the social profile, and evaluate the importance between the two measures of the merging process score. Each participant sends a query and then judges the relevance of the 5, 10, 15, and 20 top ranked documents. A precision value is computed for each retrieval session according to the usual formula:

$$precision = \frac{number\ of\ selected\ relevant\ documents}{number\ of\ selected\ documents}$$

The various cases obtained by varying the parameters α in the interval $[0, 1]$ are described as followed:

- $\alpha = 0$: This case means that the social profile is not used, the relevance of the result is related just to the source score.
- $\alpha = 0.5$: In this case here, the relevance of the result is related to the user profile and the source score, in an equitable way.
- $\alpha = 0.7$: In this case here, the relevance of the result is related to the user profile and the source score, but the social profile is more significant than the source score.
- $\alpha = 1$: This case means that the source score is not used, the relevance of the result is related just to the social profile.

Table 1 shows the precision (at 5, 10,15, and 20) values of the personalization approaches.

These results shows that the combination of the social profile with the score source gives the best results of the merging process.

Table 1. Precision values of the merging process personalization approach

Metric	$\alpha = 0$	$\alpha = 0.5$	$\alpha = 0.7$	$\alpha = 1$
P5	0.77	0.82	0.87	0.81
P10	0.66	0.76	0.825	0.74
P15	0.69	0.72	0.81	0.73
P20	0.695	0.87	0.825	0.705
P average	0.703	0.792	0.832	0.746

Results of Query Expansion. In this section, we study the performance of the personalization approaches when we apply our query expansion method for each query. To realize that we preferred the use of queries that contains just a single tag, and that to avoid the changing of the meaning of the queries, after the collection of the new queries we apply our personalization approach to merge and order the documents such as the previous section, and choose the value of the parameter $\alpha = 0.7$. For each query, we have computed the precision values for a top 5 documents, 10 documents, 15 documents, and 20 documents. The following Table 2 shows these averages precisions obtained:

Table 2. Precision values of the query expansion method

Metric	without query expansion	with query expansion
P5	0.77	0.82
P10	0.66	0.76
P15	0.69	0.72
P20	0.695	0.87

These results show that the query expansion gives better results than the use of the query without expansion.

5   Conclusion and Future Work

In this paper we have defined a new social user profile based on the folksonomy structure. We have also defined a new approach using the social user profile for personalizing and improving the result merging process in a distributed information retrieval. The results obtained shows that the integration of the social profile, in result merging process, improved the relevance of the retrieval result. In addition, the second evaluation indicates that the use of the query expansion process with the social profile gave good results for the merging process in a distributed information system. In our future work we plan to integrate the social user profile for personalizing and improving the source selection process in a distributed information system.

References

1. Lawrence, S., Giles, C.L.: Accessibility of information on the web. Nature 400, 107–107 (1999)
2. Saito, K., Kimura, M., Ohara, K., Motoda, H.: Selecting information diffusion models over social networks for behavioral analysis. In: Balcázar, J.L., Bonchi, F., Gionis, A., Sebag, M. (eds.) ECML PKDD 2010, Part III. LNCS, vol. 6323, pp. 180–195. Springer, Heidelberg (2010)
3. Hotho, A., Jäschke, R., Schmitz, C., Stumme, G.: Information retrieval in folksonomies: Search and ranking. In: Sure, Y., Domingue, J. (eds.) ESWC 2006. LNCS, vol. 4011, pp. 411–426. Springer, Heidelberg (2006)
4. Vallet, D., Cantador, I., Jose, J.M.: Personalizing web search with folksonomy-based user and document profiles. In: Gurrin, C., He, Y., Kazai, G., Kruschwitz, U., Little, S., Roelleke, T., Rüger, S., van Rijsbergen, K. (eds.) ECIR 2010. LNCS, vol. 5993, pp. 420–431. Springer, Heidelberg (2010)
5. Schenkel, R., Crecelius, T., Kacimi, M., Neumann, T., Parreira, J.X., Spaniol, M., Weikum, G.: Social wisdom for search and recommendation. IEEE Data Eng. Bull. 31, 40–49 (2008)
6. Bender, M., Crecelius, T., Kacimi, M., Michel, S., Neumann, T., Parreira, J.X., Schenkel, R., Weikum, G.: Exploiting social relations for query expansion and result ranking. In: IEEE 24th International Conference on Data Engineering Workshop, ICDEW 2008, pp. 501–506. IEEE (2008)
7. Jeon, H., Kim, T., Choi, J.: Personalized information retrieval by using adaptive user profiling and collaborative filtering. AISS 2, 134–142 (2010)
8. Fox, E.A., Shaw, J.A.: Combination of multiple searches, pp. 243–243. NIST Special Publication SP (1994)
9. Yuwono, B., Lee, D.L.: Server ranking for distributed text retrieval systems on the internet. In: DASFAA, vol. 97, pp. 41–49. Citeseer (1997)
10. Si, L., Callan, J.: A semisupervised learning method to merge search engine results. ACM Transactions on Information Systems (TOIS) 21, 457–491 (2003)
11. Si, L., Callan, J.: Modeling search engine effectiveness for federated search. In: Proceedings of the 28th Annual International ACM SIGIR Conference on Research and Development in Information Retrieval, pp. 83–90. ACM (2005)
12. Kechid, S., Drias, H.: Personalizing the source selection and the result merging process. International Journal on Artificial Intelligence Tools 18, 331–354 (2009)
13. Kechid, S., Drias, H.: Personalised distributed information retrieval-based agents. International Journal of Intelligent Systems Technologies and Applications 9, 49–74 (2010)
14. Smyth, B.: Web search: Social & collaborative. In: COnférence en Recherche d'Infomations et Applications - CORIA 2011, 8th French Information Retrieval Conference, Avignon, France, March 16-18, p. 3 (2011)
15. Rocchio, J.J.: Relevance feedback in information retrieval (1971)

Learning to Rank for Personalised Fashion Recommender Systems via Implicit Feedback

Hai Thanh Nguyen[1], Thomas Almenningen[2], Martin Havig[2],
Herman Schistad[2], Anders Kofod-Petersen[1,2],
Helge Langseth[2], and Heri Ramampiaro[2]

[1] Telenor Research, 7052 Trondheim, Norway
{HaiThanh.Nguyen,Anders.Kofod-Petersen}@telenor.com
[2] Department of Computer and Information Science (IDI),
Norwegian University of Science and Technology (NTNU), 7491 Trondheim, Norway
thomas_almenningen@hotmail.com, {mcmhav,herman.schistad}@gmail.com,
{anderpe,helgel,heri}@idi.ntnu.no

Abstract. Fashion e-commerce is a fast growing area in online shopping. The fashion domain has several interesting properties, which make personalised recommendations more difficult than in more traditional domains. To avoid potential bias when using explicit user ratings, which are also expensive to obtain, this work approaches fashion recommendations by analysing implicit feedback from users in an app. A user's actual behaviour, such as *Clicks*, *Wants* and *Purchases*, is used to infer her implicit preference score of an item she has interacted with. This score is then blended with information about the item's *price* and *popularity* as well as the *recentness* of the user's action wrt. the item. Based on these implicit preference scores, we infer the user's ranking of other fashion items by applying different recommendation algorithms. Experimental results show that the proposed method outperforms the most popular baseline approach, thus demonstrating its effectiveness and viability.

1 Introduction

The fashion domain has several interesting properties and behaves differently from most other domains. The domain is characterised by being related to clothes, popularity, time and cultural grouping. The main drivers of fashion is the need for belonging, for individuals to share a common thought or opinion.

Hanf et al. [1] argue that customers are rational wrt. price and quality. Further, the forming of a subculture happens through individuals seeking out other individuals with similar tastes in a variety of aspects [2]. Finally, brands also greatly affect what a the consumer purchases. A study done on the behaviour of the consumer [3] showed that knowing the brand of two almost identical products made the consumer crowd shift towards the more well known brand.

Our work is related to a fashion recommender for an online fashion portal in Europe. A user can interact with the portal via an app in several ways: When starting the app, she is met with a welcome-screen with streams of information containing news-items, such as sales and new collections available, and editorial contents (see Figure 1). The feeds, or the built-in search functionality, will

R. Prasath et al. (Eds.): MIKE 2014, LNAI 8891, pp. 51–61, 2014.

Fig. 1. Screenshots from the application. From the left to right: editorial information, product details, and a "love list".

lead her to storefronts, brands, or specific items. Users can indicate interest in products by looking at them (Clicks), marking them as "loved"(Wants), or as "intended for purchase"(Purchases).

The goal of this research is to extend the app with a separate personalised stream, which will present the user with the items she is most likely to buy. A design criteria is that the system shall be completely unobtrusive. Thus, the recommendations cannot rely on users explicitly rating items, but rather be guided by the rich history of interaction between the user and the app. The system therefore relies solely on *implicit feedback*, that is, the user's preference are to be automatically inferred from her behaviour [4,5]. Note that the system will not only use the user's purchase history, which can be sparse, but also consider which items the user has clicked, and which have been "loved".

The use of implicit feedback systems, while immensely popular, e.g., in news recommendation systems, also raise some challenges, see, e.g., [6,7]: Most notably are: *i*) every interaction a user has with an item is a sign of interest, and the system therefore never receives negative feedback; *ii*) feedback related to an item is multi-faceted (e.g., it can be both clicked and "loved"), and the different types of feedback will have to be combined into a single numerical value, as defined an implicit preference score for a recommendation algorithms; and *iii*) it is difficult to evaluate such a system compared to explicit-rating-systems, because the system does not have a target rating to compare its predictions to. The success of an implicit feedback system therefore relies on a well-defined strategy for inferring user preferences from implicit feedback data, combining event types into implicit scores, and evaluating these scores and recommendations by using a suitable metric.

2   Related Work

Different recommender systems for fashion exist, each using different mechanisms for acquiring information to uncover customers' preferences. One example is the fashion coordinates recommender system by Iwata et al. [8], which analyses photographs from fashion magazines. Full body images are segmented into their top and bottom parts, and used together with collected visual features of the products, the system learned which top matches to which bottom.

Shen et al. [9] proposed a system, where users can register clothes, add brands, type, material and descriptions (e.g., "I use these at home"). The system makes recommendations based on the situation for which the user needs help finding clothes. Yu-Chu et al. [10] and Ying et al. [11] described similar systems.

Liu et al. [12] combined the approaches from Iwata et al. [8], Ying et al. [11] and Shen et al. [9], suggesting a system that recommends clothes both based on the photographs and the occasion the clothes are to be worn.

While designing *SuitUp!*, Kao [13] did a survey among the system's potential users. An interesting finding was that many enjoyed a *Hot-or-Not* feature, which gives the user a set of items she can like or dislike. Not only did this make the users more engaged in the system, but it also produced ratings that could be used for recommendation of new items.

Common to all of these systems is that they employ explicit feedbacks from their users. However, a crucial design requirement from the owner of the fashion app in this paper is that recommendations must be unobtrusive, leaving harvesting of explicit feedbacks unattainable. Unfortunately, there is not much research on using implicit feedback for fashion recommendation systems available, but some work has been carried out in other domains. Among these are Xu et. al. [14], who proposed a recommendation system for documents, images and video online. It uses the user's attention time spent consuming different items to infer her interests. In music recommendation, Yang et. al. [15] suggested to use implicit feedback collected from a user during a short time period to extract her local preference, which is to represent her taste during the next few minutes. It has been shown that implicit feedback from users can be used for news recommendation [7], and Parra et al. [16] showed a strong relation between users' implicit feedback and explicit ratings.

3 Design and Implementation

3.1 Generating Implicit Scores

The first step towards building a recommender system is to translate data capturing users' actions into a number that can be understood as the "score" the users would give to particular items. The most important factor when creating such numbers is to understand the data available and their implications for user preferences. Once the data is analysed, suitable generalisations can be chosen.

Often, counting events can be useful; in our case a natural assumption is that a high number of clicks correlates with a preference for that item. Also, scores are dynamic in the sense that an item viewed a long time ago will probably be less relevant today than if it had been clicked yesterday. Therefore, the time when a user last viewed items and in which order should affect the score calculation.

It is important to recognise that when scores are not explicitly given by users, calculated implicit scores become the system's equivalent of a ground truth. The subsequent processing depends on the assumption that generated implicit scores represent users' preferences well. Unfortunately, this assumption is not verifiable, thus it is very important to generate scores using relevant features and methods.

Our point of departure is the work by Ghosh [17]. However, we extend this work by basing our weights of different events on the statistical properties of the dataset instead of manually defining the scores for each event. Furthermore, we create scores on a continuous scale between a minimum and a maximum value.

We focus on three events, *Clicks*, *Wants* and *Purchases*. Notice that the events are naturally ordered, i.e., wanting an item is a stronger indicator than clicking it, and intending purchase is stronger than wanting the item. We decided to let a user's interaction with an item be defined through the strongest event, so we for instance, disregard all click events when an item is wanted. Among the three events, 61 % of the registrations in our data set are of the type *Clicks*, 35 % are *Wants*, and the remaining 4 % are *Purchases*.

This distribution is used to generate the ranges of scores in Table 1. Note that the intervals for each event ensures natural ordering, and the split at, e.g., 96 is due to 96 % of the events in our dataset being *Wants* or weaker.

Table 1. Score Mapping

Event type	*Clicks*	*Wants*	*Purchases*
Score range	[0, 61]	[61, 96]	[96, 100]

Important properties in the fashion domain that must be captured by the system include seasons and trends, price sensitivity and popularity. Based on this, we adjust the relevance of a user's action wrt. an action's *recentness* as well as an item's *price* and overall *popularity*. We will do so using a *penalisation function*, giving each event a specific value inside the range of possible scores.

In general, when a user u triggers an event e, e.g. *Clicks*, we have a range of possible scores to give this event. We use S_e to denote this score, and let m_e and M_e denote the minimum and maximum score possible for event e, respectively. We then use a penalisation function $p_u(x)$ taking a feature value x (e.g., related to an item's price), and returns a number between zero and one to adjust the score inside the possible range. Specifically, we will use (1):

$$S_e = M_e - (M_e - m_e) \cdot p_u(x) \tag{1}$$

This formalises a procedure for finding S_e, the score given to event e after penalisation, which was originally assumed to be in the interval $[m_e, M_e]$. Note that penalisation also implicitly add negative feedback. This is an important aspect to keep in mind when working with implicit feedback, as modern recommender engines work better when based on both positive and negative feedback.

3.2   Considering Recentness

As mentioned, fashion is about timing and following recent trends. Thus, *recentness* is a natural feature to determine how well an item is liked. In our system we penalise items the user has not considered recently. To do so, we look at the number of days since the user did the event in question (denoted by x), and compare this to the oldest event this user has in the database, denoted by F_u.

We then enforce a *linear penalisation*, letting $p_u(x) = x/F_u$. Thus, an event happening today ($x = 0$) will be penalised with $p_u(0) = 0$, whilst the oldest event in the data base will be penalised with $p_u(0) = 1$. Some examples of the use of this penalisation (assuming $F_u = 14$) are given in Table 2.

Table 2. Example of penalisation

Event type	x	$p_u(\cdot)$	S_e
Purchases	3	0.21	99.2
Wants	7	0.50	78.5
Clicks	0	0.00	62.0
Clicks	14	1.00	0

A linear penalty function does not fit well with the idea of seasons. As an example, a store may be selling warm clothes from November to March, but wants to focus its recommendations at summer-clothes when the season changes. In order to mimic this behaviour we introduce a sigmoid function, and we chose to work with a parametrisation where we can fine tune both the steepness (s) and shift (c) of the "S"-shape (see Equation 2) to fit well with the data.

$$p_u(x) = \frac{1}{1 + \exp\left(-s \cdot x/F_u + c\right)} \tag{2}$$

Considering recentness in this way could obscure the preferences of users who have been absent from the app for some time, because all these users' events would be heavily penalised. In order to mitigate this problem, we adjust our features to consider the *ordering* of events instead of timing. Assume the user has triggered N events in total, now let x be the number of events triggered after the event e in question. In this case it seems appropriate to use a linear penalisation, with $p_u(x) = x/(N-1)$. This definition ensures that the difference in penalisation between the two most recent items is equal to the difference between the two oldest items. In total, this gives us two different ways of incorporating recency into the score of an event e. We discuss how these are combined in Section 3.5.

3.3   Considering Price

Users differ in the price range of items they frequent, and as people tend to be price sensitive, the prices of the items should also come into the score calculation.

Using the user's "typical" price-range, we can create personalised scores, penalising items that are not in the price range preferred by the user. This procedure is done in two steps: first we find the average price a_u of all items related to user u. Secondly, we calculate the difference between the price of item i triggering event e (denoted by π_i) and a_u. We define the penalisation function as:

$$p_u(\pi_i) = \begin{cases} \min\left(1, \frac{\pi_i - a_u}{F}\right) & \text{if } \pi_i \geq a_u \\ \min\left(1, \frac{a_u - \pi_i}{2F}\right) & \text{if } \pi_i < a_u, \end{cases}$$

where F is a constant controlling the price elasticity of the user-group. Note that we make the assumption that items being cheaper than the user's average (i.e., $\pi_i < a_u$) should also be penalised, but less strictly, making cheaper items half as sensitive to penalisation, compared to more expensive items ($\pi_i \geq a_u$).

3.4 Considering Popularity

Finally, as fashion is strongly related to the (global) popularity of items, we will consider popularity as a feature. By comparing a user's behaviour to the rest of the population, we can tell if the user's activities conforms to common standards or if her taste is more unique; giving significant clues about items to recommend.

The goal is to classify to what degree a user likes popular items, then define a penalty function that reduces the score of items that are too popular (if the user is not under peer-pressure) or too obscure (if the user prefers popular items). Assume we have ordered all items wrt. decreasing popularity. We define t_i, the popularity of item i, as the fraction of items that are more popular than item i.

Thus, t_i is zero if item i is the most popular, one if it is the least popular. The average popularity of the items user u looks at (denoted a_u) can now be calculated by the average of the relevant t_i values. We build the penalty function using two linear functions: one when the popularity of item i is below a_u and the one for when the popularity is equal or higher. We also introduce a constant c, being the penalisation given to the most popular item overall (when $t_i = 0$):

$$p_u(t_i) = \begin{cases} c \cdot (a_u - t_i)/a_u & \text{if } t_i < a_e \\ (t_i - a_u)/(1 - a_u) & \text{if } t_i \geq a_u. \end{cases}$$

3.5 Combining the Different Penalisations

The are four penalisation functions, each capturing a specific aspect of the domain and defining sets of implicit scores: *price*, *popularity* and two different uses of *recentness*. To infer a user's implicit score for an item, we *blend* these features.

When linearly combining M models $m_1 \ldots m_M$, we choose M weights ω_j, where each weight represents the importance we give the corresponding model in the final blend. For a given combination of a user and an item, each model m_j proposes calculates a score S_j, and the blended results is a sum over all models:

$$S_e = \sum_{j=1}^{M} \omega_j \cdot S_j$$

This requires setting weights for M different factors. Optimally we would like to run one blend, compare the result to a gold standard and adjust the weights accordingly. This is is done in simple blending schemes using basic approaches like linear regression, or in advanced schemes, like binned linear regression, bagged gradient boosted decision trees, and kernel Ridge regression blending [18].

However, as discussed, we lack a ground truth for making implicit scores, and cannot use supervised learning techniques to optimise the weights. In fact, we make several assumptions on how implicit features contribute to higher preference and thus higher scores, yet we lack the means to confirm these assumptions. Then for M different models we assume that they all contribute an equal amount to the final results, thus having the weights $\omega_j = 1/M$.

4   Test and Evaluation

4.1   Experimental Data

The data in our data set originates from users performing actions in the fashion app, thereby triggering events. These actions can range from accessing a storefront, scrolling the page or purchasing a product. Table 3 gives some key figures of the data set collected over a period of 6 months. We note that the app was in beta-release during this time period, and we expect a dramatic increase in the number users and events when the app goes public. Still, the behaviour of the beta testers is assumed to be indicative also of future use.

Table 3. Overview of the key figures in the data set

Users	Items	Storefronts	Brands	Product events	Clicks	Wants	Purchases
1 532	5 688	144	22	35 324	21 400	12 436	1 488

As seen from Table 3, the average amount of *Purchases* per user is below one. Using purchases alone would therefore in most cases render the recommendations incomplete. *Clicks* and *Wants* on the other hand have a much higher occurrence. Together, these three events average to more than 20 observations per user, and even if the data still is sparse, give a richer description of user behaviour than purchases alone.

4.2   Evaluation Metrics

We selected two ranking measurements as evaluation metrics for our experiment, namely the area under curve (AUC) and the mean average precision

(MAP@k) [19]. The AUC tells us the probability that, when two items are chosen randomly, the system is able to rank the most relevant item before the least relevant one. On the other hand, MAP@k counts how many relevant items we are able to retrieve among the top k elements presented, and also account for how high on the ranked list these elements are. Our app presents the user with a feed containing 20 items, and we therefore used $k = 20$ in our experiments.

4.3 Recommendation Algorithms and Settings

We selected different recommendation algorithms for our experiments, including both binary and non-binary approaches. These algorithms are then compared with a non-personalised baseline, which recommends the most popular items overall without personalisation.

Most Popular. selects the most-popular items, then uses dithering to randomize the recommendations given to the users. Dithering adds noise to the results by permuting them so that the top results have a significant probability for remaining highly ranked, while the degree of mixing increases for the lower ranked items. Dithering enhances the user's experience of the system, because she is not presented with the same list every time [20].

Binary Recommenders. only use binary data of users' interactions. We apply two algorithms, namely the k-nearest neighbour item-based collaborative filtering (kNN-item) and Bayesian personalized ranking (BPR) [21]. We used the kNN-item implementation in Mahout [22]. The BPR algorithm, which learns the latent factors of a matrix factorization using binary implicit feedback, has been implemented in MyMediaLite [23].

Non-binary Recommenders. base their recommendations on the inferred implicit scores. We use the alternating least-squares with weighted λ-regularization (ALS-WR), which was initially developed for the Netflix competition [24]. Using cross-validation, we found the following parameters to be reasonable: No. latent features: 100, λ: $100 - 150$, number of iterations: $10 - 15$, and α: 10. Additionally, the k-nearest neighbour user-based collaborative filtering with cosine similarity measure (kNN-user) was also included in our experiments.

4.4 Experimental Results

Random Split. For the first experiment we divided the data set randomly into training and test sets, without looking at the ordering of the events. Thus, we chose to only factor in popularity and price, and did not use the recentness feature. The results are given in Table 4.

The non-binary recommenders dominated the binary methods as well as the baseline ("Most-Popular") approach with respect to all metrics. In particular, the ALS-WR with blend 1, which combines price and popularity, provided the better results compared to the ALS-WR with using each feature alone and also

Table 4. Random split results 90:10. The results are averaged over 10 runs. Blend 1 combines price and popularity.

Model	AUC	MAP@20$_{\text{Clicks}}$	MAP@20$_{\text{Wants}}$	MAP@20$_{\text{Purchases}}$
Most-Popular	0.751516	0.014772	0.011147	0.009899
kNN-item, $k = 200$	0.735779	0.004435	0.009703	0.006054
BPR	0.728643	0.012015	0.005894	0.013378
ALS-WR (popularity)	0.835715	0.019811	0.019074	0.012906
ALS-WR (price)	0.825122	0.031473	0.039653	0.024824
ALS-WR (blend 1)	0.830996	0.026687	0.031277	0.019471
kNN-user, $k = 200$	0.796258	0.026697	0.039213	0.028266

Table 5. Time-based split results. Results are averaged over 10 runs. Blend 1 combines price and popularity, blend 2 combines price, popularity and recentness.

Model	AUC	MAP@20$_{\text{Clicks}}$	MAP@20$_{\text{Wants}}$	MAP@20$_{\text{Purchases}}$
Most-Popular	0.61321	0.00462	0.00304	0.00000
kNN-item, $k = 200$	0.46432	0.00257	0.00062	0.00277
BPR	0.60168	0.00457	0.00285	0.00476
ALS-WR (popularity)	0.63819	0.00043	0.00577	0.00108
ALS-WR (price)	0.64873	0.00289	0.00892	0.02174
ALS-WR (recentness)	0.66477	0.00370	0.01782	0.04796
ALS-WR (blend 1)	0.65104	0.00375	0.00740	0.02700
ALS-WR (blend 2)	0.65446	0.00482	0.01230	0.01691
kNN-user, $k = 200$	0.58544	0.00292	0.00788	0.05324

compared to the kNN-user. Overall, the ALS-WR with blended features worked best. This is somewhat surprising, as algorithms optimized by minimizing the root mean square error typically do not perform well when asked to give top-k recommendations [25]. We plan to investigate this finding further as more data from our domain becomes available.

Time-Based Split. By splitting the dataset wrt. time, we make the evaluation more realistic and also get the opportunity to factor in and measure the effect of recency. It has been particularly few purchase events in the last couple of months, only 0.07% of the last 10% of all events have been purchases. We ended up splitting the dataset on time, training the model on the first 80% of the ratings in chronological order, making predictions for the last 20%. In this period 0.6% of all events are purchases. The results are shown in Table 5.

The low AUC scores can be attributed to the fact that 143 new users and 870 previously unrated items were introduced during the test period. To our surprise, none of the binary recommenders outperformed the baseline in terms of AUC. The ALS-WR systems performed best overall, with the exception of the approach that only penalised based on popularity. The models that factored in recentness gave the best performance, confirming the importance of this feature in the domain.

5   Conclusion

In this paper, we have presented a framework for a personalised fashion recommender systems, and discussed the challenges connected to the domain. Particular focus was given to the fact that we do not have access to explicit ratings, that are commonly used in more traditional domains. Rather, we have analysed users' behaviour from a fashion app to infer their implicit preference scores, which are later used for recommendations. To test our approach, we have applied different recommendation algorithms, and compared them to a baseline that always recommends the globally most popular items. The experimental results demonstrates the effectiveness and viability of our method. Nevertheless, there are still areas of this work that are worth further exploring. First, we want to run an online experiment, where we ask users to explicitly rate items and combine this with their implicit feedbacks. Second, we plan to investigate learning-to-rank algorithms further, aiming to develop an even more effective method for personalised fashion recommendations.

Acknowledgements. The authors' would like to thank the following people, without whom this work would not have been possible: Dr. Humberto Castejon Martinez, Dr. Cyril Banino-Rokkones and Dr. Rana Juwel at Telenor Research; Markus Krüger and Matias Holte from Telenor Digital.

References

1. Hanf, C.H., Wersebe, B.: Price, quality, and consumers' behaviour. Journal of Consumer Policy 17(3), 335–348 (1994)
2. Vignali, G., Vignali, C.: Fashion marketing & theory. Access Press (2009)
3. Li, Y., Hu, J., Zhai, C., Chen, Y.: Improving one-class collaborative filtering by incorporating rich user information. In: Proc. of CIKM 2010, pp. 959–968. ACM (2010)
4. Nichols, D.M.: Implicit rating and filtering. In: The DELOS Workshop on Filtering and Collaborative Filtering, pp. 31–36 (1997)
5. Oard, D., Kim, J.: Implicit feedback for recommender systems. In: Proc. of the AAAI Workshop on Recommender Systems, pp. 81–83 (1998)
6. Hu, Y., Koren, Y., Volinsky, C.: Collaborative filtering for implicit feedback datasets. In: Proc. of IEEE ICDM 2008, pp. 263–272. IEEE CS (2008)
7. Ilievski, I., Roy, S.: Personalized news recommendation based on implicit feedback. In: Proc. of the 2013 International News Recommender Systems Workshop and Challenge, pp. 10–15. ACM (2013)
8. Iwata, T., Watanabe, S., Sawada, H.: Fashion coordinates recommender system using photographs from fashion magazines. In: Proc. of AAAI, pp. 2262–2267 (2011)
9. Shen, E., Lieberman, H., Lam, F.: What am i gonna wear?: Scenario-oriented recommendation. In: Proc. of the 12th International Conference on Intelligent User Interfaces, pp. 365–368. ACM (2007)
10. Yu-Chu, L., Kawakita, Y., Suzuki, E., Ichikawa, H.: Personalized clothing-recommendation system based on a modified bayesian network. In: Proc. of the 12th IEEE/IPSJ Inter. Symp. on Applications and the Internet, pp. 414–417 (2012)

11. Ying Zhao, K.A.: What to wear in different situations? a content-based recommendation system for fashion coordination. In: Proc. of the Japanese Forum on Information Technology (FIT2011) (2011)
12. Liu, S., Feng, J., Song, Z., Zhang, T., Lu, H., Xu, C., Yan, S.: Hi, magic closet, tell me what to wear! In: Proc. of ACM Multimedia 2012, pp. 619–628. ACM (2012)
13. Kao, K.: SuitUp! (2010) Social space centered around clothing items, http://bit.ly/YYyWoE (Online; accessed August 29, 2014)
14. Xu, S., Jiang, H., Lau, F.C.: Personalized online document, image and video recommendation via commodity eye-tracking. In: Proc. of RecSys, pp. 83–90 (2008)
15. Yang, D., Chen, T., Zhang, W., Lu, Q., Yu, Y.: Local implicit feedback mining for music recommendation. In: Proc. of ACM RecSys 2012, pp. 91–98. ACM (2012)
16. Parra, D., Amatriain, X.: Walk the talk: Analyzing the relation between implicit and explicit feedback for preference elicitation. In: Proc. of the 19th Inter. Conf. on User Modeling, Adaption, and Personalization, pp. 255–268 (2011)
17. Ghosh, P.: From explicit user engagement to implicit product rating (2014), http://bit.ly/1lidNQn (last accesed August 11, 2014)
18. Jahrer, M., Töscher, A., Legenstein, R.: Combining predictions for accurate recommender systems. In: Proc. of ACM SIGKDD 2010, pp. 693–702. ACM (2010)
19. Shani, G., Gunawardana, A.: Evaluating recommender systems. Technical Report MSR-TR-2009-159, Microsoft Research (2009)
20. Ellen Friedman, T.D.: Practical Machine Learning: Innovations in Recommendation. O'Reilly (2014)
21. Rendle, S., Freudenthaler, C., Gantner, Z., Schmidt-Thieme, L.: BPR: Bayesian personalized ranking from implicit feedback. In: Proc. of the 25th Conference on Uncertainty in Artificial Intelligence, pp. 452–461. AUAI Press (2009)
22. Apache Software Foundation: Apache Mahout: Scalable machine-learning and data-mining library, http://mahout.apache.org
23. Gantner, Z., Rendle, S., Freudenthaler, C., Schmidt-Thieme, L.: MyMediaLite: A free recommender system library. In: Proc. of ACM RecSys 2011. (2011)
24. Bennett, J., Lanning, S.: The Netflix prize. In: KDD Cup and Workshop in conjunction with KDD (2007)
25. Cremonesi, P., Koren, Y., Turrin, R.: Performance of recommender algorithms on top-n recommendation tasks. In: Proceedings of the Fourth ACM Conference on Recommender Systems, RecSys 2010, pp. 39–46 (2010)

Convergence Problem in GMM Related Robot Learning from Demonstration

Fenglu Ge, Wayne Moore, and Michael Antolovich

Charles Sturt University, Bathurst, 2795, Australia
fge@csu.edu.au

Abstract. Convergence problems can occur in some practical situations when using Gaussian Mixture Model (GMM) based robot Learning from Demonstration (LfD). Theoretically, Expectation Maximization (EM) is a good technique for the estimation of parameters for GMM, but can suffer problems when used in a practical situation. The contribution of this paper is a more complete analysis of the theoretical problem which arise in a particular experiment. The research question that is answered in this paper is how can a partial solution be found for such practical problem. Simulation results and practical results for laboratory experiments verify the theoretical analysis. The two issues covered are repeated sampling on other models and the influence of outliers (abnormal data) on the policy/kernel generation in GMM LfD. Moreover, an analysis of the impact of repeated samples to the CHMM, and experimental results are also presented.

1 Introduction

Learning from Demonstration (LfD) is a method that can train a robot to learn a task [1]. The authors used a GMM based LfD method to teach a small mobile robot to traverse a simulated mining tunnel (See Figure 1). The trajectory course that the robot was to take consisted of a number of actions labelled "Move", "Turn" and "Stop" in Figure 1 and Table 1. These actions were to be taught as a whole.

The basic training strategy can be summarized as follows [1]:

Step 1 A teleoperator controlled a P3-DX robot to run through a training environment in Figure 1, teaching the robot to finish a task. The robot used its built-in sonar array to record data at the same time.

Step 2 The robot used recorded training dataset via algorithms to generate a policy.

Step 3 The robot tried to finish the task independently. At every time point, it recorded environment data as an input. The policy processed the input to find an output with the highest probability, where the output was an action the robot needed to take.

Step 4 If the robot didn't work properly, the task would be interrupted by a supervisor. Extra training was performed, and new training data was added to the previous training data. Step 2 and Step 3 repeated until the robot could finish the task independently.

R. Prasath et al. (Eds.): MIKE 2014, LNAI 8891, pp. 62–71, 2014.
© Springer International Publishing Switzerland 2014

A number of findings using this method occurred during the training sessions:

1. Extra trainings of "Turn" can increase the learning efficiency,
2. Outlier samples made significantly negative contributions on policy generation within LfD.

A theoretical analysis related to the above findings has been presented in [2].
The authors of [2] have made a number of assumptions among which is the
assumption that the covariance matrix is approximaty to zero. This among other
assumptions is addressed in Section 2 below. Furthermore, potential solutions
are also presented in this paper.

A	Move	Moving Forward
B	Turn	90° Turning
C	Stop	Stop before the muck pile

Table 1. Training actions

Fig. 1. Route of the robot during training
period

2   Theoretical Analysis

2.1   The Influence of Repeated Samples on Other Models

Table 2. Parameters used in GMM

k	the index of a cluster/component
R	the number of repeated samples
X_{rep}	the sample repeated
$p(k|X_i)$	the probability of k givenX_i
$p(k)$	the probability of k
N	the number of samples
μ_k^t	the mean value of the cluster k at the t-th iteration.
Σ_k^t	the variance of the cluster k at at the t-th iteration.

Table 2 shows parameters used in GMM based method. Assume the compo-
nent k is the only one that has R repeated X_{rep}, fully consisting of X_{rep} in a
certain trajectory. If the EM [3] algorithm nearly converges, the component k
has higher probability to generate X_{rep} than $X_{i \neq rep}$, which means:

$$p(k|X_{i=rep}) - p(k|X_{i \neq rep}) > 0 \qquad (1)$$

The left part of Equation 1 can be expanded as

$$\frac{p^t(X_{rep}|k)p^t(k)}{p^t(X_{rep})} - \frac{p^t(X_{i\neq rep}|k)p^t(k)}{p^t(X_{i\neq rep})}$$

$$= \frac{p^t(X_{rep}|k)p^t(k)p^t(X_{i\neq rep}) - p^t(X_{i\neq rep}|k)p^t(k)p^t(X_{rep})}{p^t(X_{rep})p^t(X_{i\neq rep})}$$

$$= \frac{p^t(X_{rep}|k)p^t(X_{i\neq rep}) - p^t(X_{i\neq rep}|k)p^t(X_{rep})}{p^t(X_{rep})p^t(X_{i\neq rep})}p^t(k) \tag{2}$$

where $p^t(X_{rep})$ and $p^t(X_{i\neq rep})$ are prior probabilities recognized as constants, and

$$p^{t+1}(k) = \frac{1}{N}(\sum_{i\neq rep}^{N} p^t(k|X_i) + \sum_{rep} p^t(k|X_{rep}))$$

$$= \frac{1}{N}(\sum_{i\neq rep}^{N} p^t(k|X_i) + R * p^t(k|X_{rep})) \tag{3}$$

Because the EM algorithm converges to a local maximum, before it converges, it can be concluded that:

1.

$$p^{t+1}(X_{rep}|k) > p^t(X_{rep}|k) \tag{4}$$

2.

$$p^{t+1}(X_{i\neq rep}|k) < p^t(X_{i\neq rep}|k) \tag{5}$$

based on Equations 3, 4 and 5, if R is large enough, it can be concluded:

$$p^{t+1}(k) > p^t(k) \tag{6}$$

The mean value μ [3] can be rewritten as:

$$\mu_k^{t+1} = \frac{Rp^t(k|X_{rep})X_{rep} + \sum_{i\neq rep}^{N-R} p^t(k|X_{i\neq rep})X_i}{\sum_{i=1}^{N} p^t(k|X_i)} \tag{7}$$

By combining Equations 1, 2 and 6, it can be concluded that the difference between $p(k|X_{rep})$ and $p(k|X_{i\neq rep})$ increases with iterations. Since $p(k)$ will converge to a bounded value with the increase of iterations, $p(k|X_{i\neq rep})$ will become a tiny value. Then the $X_{i\neq rep}$ part in Equation 7 can be eliminated, and Equation 7 can generate the following result:

$$\mu_k^{t+1} \approx X_{rep} \tag{8}$$

i.e., the mean value of the component k is just the repeated element X_{rep}. Then the covariance [3] can be rewritten as :

$$
\begin{aligned}
\Sigma_k^{t+1} &= \frac{\sum_{i=1}^{N} p^t(k|X_i)(X_i - X_{rep})(X_i - X_{rep})^T}{\sum_{i=1}^{N} p^t(k|X_i)} \\
&= \frac{\sum_{i\neq rep}^{N} p^t(k|X_i)(X_i - X_{rep})(X_i - X_{rep})^T}{\sum_{i\neq rep}^{N} p^t(k|X_i) + Rp^t(k|X_{rep})}
\end{aligned}
\tag{9}
$$

which can be recognized as a constant close to zero, which addressed the convariance assumption in [2]. Also

$$
\begin{aligned}
p^{final}(k) &= \frac{1}{N}\Big(\sum_{i\neq rep}^{N} p^{final-1}(k|X_i) + \sum_{rep} p^{final-1}(k|X_{rep}) \Big) \\
&= \frac{1}{N}\Big(\sum_{i\neq rep}^{N} p^{final-1}(k|X_i) + R * p^{final-1}(k|X_{rep}) \Big) \\
&\approx \frac{R}{N}
\end{aligned}
\tag{10}
$$

Therefore, the expectation of component k will converge at the next iteration. It also means that the parameters related to component k are all close to constants.

1. Based on [3],

$$
Q(\Theta, \Theta^g) = \mathbf{Constant} + \sum_{l=1}^{K}\sum_{i=1}^{N}(-\frac{1}{2}log(|\Sigma_l|)-
\tag{11}
$$
$$
\frac{1}{2}(X_i - \mu_l)^T \Sigma_l^{-1}(X_i - \mu_l)p(k|X_i, \Theta^g)
$$

where Q is the difference between previous expectation and current expectation, Θ and Θ^g are current Gaussian Parameters Model and previous Gaussian Parameters Model. Because the Σ_k for component k is very close to zero, $-log(|\Sigma_k|)$ and Σ_k^{-1} become very large, making the largest contribution for Q, meaning that other contributions can be ignored. Therefore, the EM algorithm cannot converge easily or correctly. Regularization [4] or whitening [5] is required to fix the problem.

2. The data around the edge of models occupy a small percentage of ordinary models. For a model full of repeated data, all data are edge data when it converges. And these edge data will have a higher percentage in neighbouring models than normal. The percentage can be high enough to collapse the convergence. Therefore, extra related data should be added to decrease the influence of the model with repeated samples.

2.2 The Influence of Outliers on Policy Generation

During the experiments, the robot also acquired one or two outliers in the turning section. For example, a real distance to the forward wall was 1.7 metres, while the acquired value was more than 5 metres. The errors were deleted by a built-in program. The theoretical influence on the performance of the GMM based algorithm has been investigated. At the start of the EM algorithm, it's reasonable that an outlier X_{out} is close to one component, in the name of k, though the X_{out} is very far away from the rest of samples. Recall the gaussian function in [3], then we have

$$p^t(X_{out}|k) = \frac{1}{\sqrt{(2\pi)^d |\Sigma(k)|}} e^{-\frac{1}{2}(X_{out}-\mu_k)' \Sigma_k^{-1}(X_{out}-\mu_k)} \tag{12}$$

Based on the Bayes Rule and Equation 12:

$$
\begin{aligned}
\frac{p^t(k|X_i)}{p^t(k|X_{out})} &= \frac{p(X_i|k)p(X_{out})}{p(X_i)p(X_{out}|k)} \\
&= \frac{p(X_{out})}{p(X_i)} exp(-\frac{\| X_{out} - \mu_k \|^2 - \| X_i - \mu_k \|^2}{2\sigma^2})
\end{aligned}
\tag{13}
$$

where $i \neq X_{out}$. Then the following can be deduced:

1. Since X_{out} is far away from other samples, it means it is also far away from the center of other components. Then the value of $X_{out} - \mu_k$ is very large, and the result of Equation 13 is very close to zero.
2. It also means the numerator of Equation 13 is very close to 0. Therefore, it can be concluded that $p^t(k|X_{out})$ will be very close to 1 based on:

$$\sum_{i=1}^{N} p(k|X_i) = 1 \tag{14}$$

3. Therefore, μ_k is very close to X_{out}.
4. The new μ_k can be substituted into Equation 13. The result of Equation 13 is still 0 since $X_{i \neq out}$ is very far from μ_k (X_{out}).

Applying the same strategy to the E-Step and M-step of EM algorithm, It can also be concluded that the covariance is close to a constant related to X_{out} and $p^{t+1}(k)$ is also a constant related to X_{out}. Therefore, the parameters of component k have no relation to the iteration of the EM algorithm. Until now, it can be concluded that the component k finally converges to the abnormal data point X_{out} , which is not the reasonable center of the component k. Hence, it can be seen that the outlier makes the algorithm fail.

3   Experimental Results

3.1   Results Based on Simulation Data

The axes in Figure 2 to 6 represent the x,y coordinates of the samples in metres. Figure 2 shows the contours of two GMM models. Figures 3 and 4 show the contours with **50 and 100** samples replaced by 50 repeated samples and 100 repeated samples separately, where the repeated sample chosen from the top

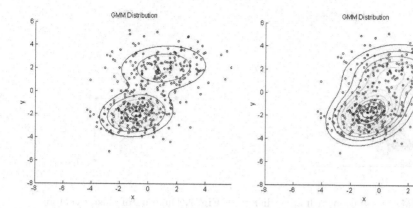

Fig. 2. GMM distribution with two clusters. Each cluster has 200 samples.

Fig. 3. GMM distribution with unequal samples taken from Figure 2 (200 in lower cluster and 150 in upper cluster). A sample taken from the upper cluster in Figure 2 has been replicated 50 times and added to the upper cluster.

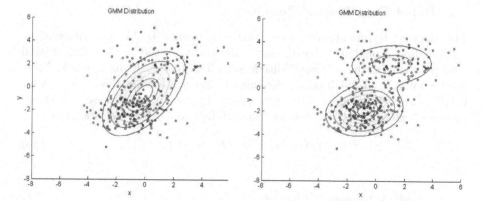

Fig. 4. GMM distribution with unequal samples taken from Figure 2 (200 in lower cluster and 100 in upper cluster). A sample taken from the upper cluster in Figure 2 has been replicated 100 times and added to the upper cluster.

Fig. 5. GMM Distribution with 49 repeated samples removed in the Figure 3.

right model of Figure 2. It shows the two models can not be recognized easily with the increase of repeated samples in a model. Based on Figure 3, Figure 5 shows the result after 49 repeated samples of 50 were removed. It shows the clusters can be recognized correctly.

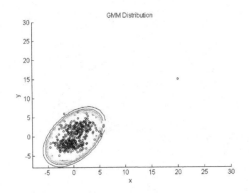

Fig. 6. GMM Distribution with an outlier   **Fig. 7.** Diagram of cluster centers

Figure 6 shows the GMM distribution with a sample replaced by $[20, 15]$. Even this sample is not very far away from the two models, only one model can be recognized easily, showing the outlier producing a negative contribution for model recognition.

3.2   Robot Experimental Results

The following results are from robot LfD experiments [1]. The experimental result is 5 dimensional and very difficult to illustrate as easily as the 2 dimensional data presented in the 3.1. For Multiple sampling, the results as shown in Table 3 were obtained. Table 3 shows the sum of distances among cluster centres for GMM as shown in Figure 7 (refer to Figure 1) and Table 1. These centres are the mean values of each model. The sum of distances can be calculated as:

$$Sum\ of\ distances\ =\ AB\ +\ AC\ +\ BC \tag{15}$$

Table 3. The sum of distances among centres

Original	3.8814
One extra turnings	5.2371
Two extra turnings	5.2642
Three extra turnings	5.2512
Four extra turnings	5.6472
Five extra turnings	6.1935

Table 4. Influence of outliers

without any outlier	1.27	1.89	0.71
with an outlier	4.14	4.19	0.99

It shows with extra rotation("B") trainings, the sum of distances increases from 3.88 to around 6.19, meaning the three models can be separated more easily, verifying the theoretical analysis in section 2. Outliers also occurred during "B" training. Table 4 shows the influence of an outlier. Each row presents the distances among cluster centres. It shows the outlier is far away from others, distorting the performance of GMM by generating the wrong centres. The outlier was removed by internal filter for abnormal data. Since it is a practical simulation, it was relatively easy to remove sensor readings that seemed improbable. For example, very large distance measurement from the sonar sensors. This may not be easy in other situations.

4   Further Analysis of Influence of the Repeated Samples on Continuous Hidden Markov Model

Table 5. Simulation sum of distances among states

Original	1.3879
50 more repeated samples	1.2656
100 more repeated samples	1.1534

Table 6. Actual sum of distances among states

Original	4.4242
20 more repeated samples	4.0957
40 more repeated samples	3.4386

The problem of repeated samples also emerges in the Continuous Hidden Markov Model as shown in Table 5 (2 state, 1 cluster in each state) and Table 6 (3 states, 1 cluster in each state). In Table 5, 50 and 100 repeated samples were used by the method explained in the GMM section. with extra repeated samples, the sum of distances among state decreased. This used simulated data generated using Matlab. In Table 6, 20 and 40 extra repeated samples were used by the same strategy as the GMM part. It shows that the more Action stops ("C") were used, the smaller the sum of distances among states became. These results were from actual data.

4.1   Theoretical Analysis

Parameters of CHMM [6] are presented in Talbe 7:

Assume there are N repeated observations o_{rep} at cluster k of state i given a sequence l

$$\mu_{ik}^l = \frac{\sum_{t=1}^{T_l} P_{\Theta}(q_t = i, m_t = k|o^l)o_t^l}{\sum_{t=1}^{T_l} P_{\Theta}(q_t = i|o^l)}$$
$$= \frac{NP_{\Theta}(i,k)o_{rep} + \sum P_{\Theta}(i,k)o_{\neq rep}}{\sum_{t=1}^{T_l} P_{\Theta}(q_t = i)}$$

$$(16)$$

Because $k = 1$ in this experiment, it shows

$$\frac{\partial \mu_i^l}{\partial o_{rep}} = \frac{NP_{\Theta}(q_t = i) + MP_{\Theta}(q_t \neq i)}{\sum_{t=1}^{T_l} P_{\Theta}(q_t = i)}$$

$$(17)$$

Table 7. Parameters used in CHMM

π	initial state probability
l	index of a training sequence
q_i	the state at time i
a_{ik}	state transition probability between state i and state j
$b_i(o_t)$	emission probability
c_{ik}	emission probability of cluster k at state i
μ_{ikn}	mean value of variable n in cluster k at state i
μ_{ik}	mean value of cluster k at state i
Σ_{ikn}	variance of variable n in cluster k at state i
Σ_{ik}	variance of cluster k at state i
Φ	Gaussian parameters model
Θ	CHMM parameters model
o^l	the l-th training sequence
n	index of a P-dimensional observation

and

$$c_i = 1 \tag{18}$$

if N is close to T_l, then

$$\frac{\partial \mu_i^l}{\partial o_{rep}} \approx 1 \tag{19}$$

Then,

$$\mu_i^l \approx O_{rep} + \textbf{constant} \tag{20}$$

and

$$\Sigma_i^l \approx \textbf{constant} \tag{21}$$

Based on the above assumption that N is close to T_l, it shows:

$$\Phi(O_{rep}, \mu_i, \Sigma_i) \approx 1 \tag{22}$$

Go back to the Q function:

$$Q(\Theta, \Theta^g) = \sum_q P_{\Theta(q|o)} log(\pi_{q_1}|\Theta_g) +$$

$$\sum_t \sum_q P_{\Theta(q|o)} log(a_{q_t, q_{t+1}}|\Theta^g) +$$

$$\sum_t \sum_q P_{\Theta(q|o)} log(c_{q_t}|\Theta^g) +$$

$$\sum_t \sum_q P_{\Theta(q|o)} log(\Phi(O_t, \mu_{q_t}, \Sigma_t)|\Phi^g) \tag{23}$$

Because $c_i = 1$,

$$\sum_t \sum_q P_{\Theta(q|o)} log(c_{q_t}|\Theta^g) = 0 \tag{24}$$

Basing on the Equation 22

$$\sum_t \sum_q P_{\Theta(q|o)} log(\Phi(O_t, \mu_{q_t}, \Sigma_t)|\Phi^g) \approx \infty \qquad (25)$$

Therefore, it can be concluded that enough o_{rep} can make the EM algorithm collapse, which means the CHMM parameters generated can not easily recognize every state. The experimental results also verify this analysis.

5   Conclusion

This paper presents issues on local features of the training data, which can affect the performance of the GMM based algorithm. It also investigates the internal mechanism contributing to the performance of robot learning, exposing how the problems were detected and fixed, giving potential solutions for such issues in a practical scenario. Furthermore, it also delivers an analysis on the influence of repeated samples to the CHMM. Experimental results verify the analysis.

References

1. Ge, F., Moore, W., Antolovich, M., Gao, J.: Application of learning from demonstration to a mining tunnel inspection robot. In: 2011 First International Conference on Robot, Vision and Signal Processing (RVSP), pp. 32–35. IEEE (2011)
2. Archambeau, C., Lee, J.A., Verleysen, M.: On convergence problems of the em algorithm for finite gaussian mixtures. In: Proc. 11th European Symposium on Artificial Neural Networks, pp. 99–106 (2003)
3. Bilmes, J.: A gentle tutorial of the em algorithm and its application to parameter estimation for gaussian mixture and hidden markov models. Technical report (1998)
4. Neumaier, A.: Solving ill-conditioned and singular linear systems: A tutorial on regularization. SIAM Review 40, 636–666 (1998)
5. O'Dowd, R.J.: Ill-conditioning and pre-whitening in seismic deconvolution. Geophysical Journal International 101(2), 489–491 (1990)
6. Movellan, J.R.: Tutorial on Hidden Markov Models (2003),
   http://mplab.ucsd.edu/tutorials/hmm.pdf

Hybridization of Ensemble Kalman Filter and Non-linear Auto-regressive Neural Network for Financial Forecasting

Said Jadid Abdulkadir[1], Suet-Peng Yong[1], Maran Marimuthu[2], and Fong-Woon Lai[2]

[1] Department of Computer and Information Sciences
[2] Department of Management and Humanities,
Universiti Teknologi Petronas,
31750, Tronoh, Malaysia
said_g02156@utp.edu.my,
{yongsuetpeng,maran.marimuthu,laifongwoon}@petronas.com.my

Abstract. Financial data is characterized as non-linear, chaotic in nature and volatile thus making the process of forecasting cumbersome. Therefore, a successful forecasting model must be able to capture long-term dependencies from the past chaotic data. In this study, a novel hybrid model, called UKF-NARX, consists of unscented kalman filter and non-linear auto-regressive network with exogenous input trained with bayesian regulation algorithm is modelled for chaotic financial forecasting. The proposed hybrid model is compared with commonly used Elman-NARX and static forecasting model employed by financial analysts. Experimental results on Bursa Malaysia KLCI data show that the proposed hybrid model outperforms the other two commonly used models.

Keywords: chaotic time-series, ensemble model, non-linear autoregressive network, financial forecasting.

1 Introduction

Forecasting is a dynamic process and perplexing task in the financial division. It helps financial market analysts to evade stock trading losses and obtain huge profits by coming up with promising business policies. Hence, financial companies can make precise forecasts by planning some required interventions to meet their business performance targets [14]. Furthermore, stock trading companies are usually scrutinized by short and long term investors while concerning the expectations from shareholders. Stockholders may also like to analyze their investments by comparing the analysis of forecasting companies.

An example of financial time-series forecasting is the stock prices in the share market which are characterized by non-linearity, noisy, chaotic in nature and volatile thus making the process of forecasting cumbersome. The goal of financial

R. Prasath et al. (Eds.): MIKE 2014, LNAI 8891, pp. 72–81, 2014.

forecasters is to innovate numerous techniques that can forecast effectively by following legal trade strategies and avoiding losses. The general idea of successful stock market prediction to achieve best results is by using minimum required input data and the least complex stock market model [4]. The intricate nature of stock market forecasting has led to the need for further improvements in the use of intelligent forecasting techniques to drastically decrease the dangers of inaccurate decision making.

Financial controllers who adhere to the ideas of an efficient market hypothesis and random walk theories disbelieve that stock market can be predicted [15]. Nevertheless, fanatics of technical and fundamental analysis have shown numerous ways to counter the claim by adherents of random walk theory and efficient market hypothesis. Therefore, numerous approaches for tackling the chaotic nature of forecasting have been suggested. However, new improvements in the area of soft computing through the use of computational intelligence have offered new ideas in forecasting chaotic data in stock market and also modelling its non-linearity.

In this paper, deriving from computational intelligence method, a hybrid neural network model consisting of unscented kalman filter and parallel non-linear autoregressive neural network is developed to enhance the performance of financial forecasting based on Kuala Lumpur Composite Index (KLCI) data. The remainder of the paper is organized as follows: section 2 outline some related work in financial forecasting and its development; in section 3, the proposed hybrid neural network model is discussed; while section 4 addresses the experimental setup and performance analysis followed by the conclusion in the last section.

2   Related Works

Financial analysts have employed the use of static forecasting models which are a sequence of one-step ahead forecasts made at different points in time. Static forecasting uses actual rather than forecast value for the lagged variable and which can be done only if there are actual data available [5].

Computational Intelligence forecasting techniques such as fuzzy logic, genetic algorithms (GA) and artificial neural networks (ANN) are the most famous used techniques to cope with problems that have not been solved by complex mathematical systems. ANN applications have been widely used for forecasting in a variety of areas [9], [11], [13], [16], [21]. ANN was used for the solution of numerous financial problems [9]. It is also used in forecasting of financial markets, particularly forecasting of stock market indexes which are considered to be a barometer of the markets in many countries [21,19]. However, the problem of over-fitting [7] arises when a model describes noise instead of the underlying relationship, hence affecting the accuracy of forecasting. The forecast of Kuala Lumpur Composite Index (KLCI) has been investigated using ANN [21], fuzzy logic [2] and artificial neural fuzzy inference system (ANFIS) [23]. However, ANFIS has strong computational complexity restrictions and translates prior knowledge into network topology hence being sensitive to the number of input variables.

The advances of ANNs over the last few years is its ability of easily allowing more than one model to be combined with itself with multiple training, which is also referred to as hybrid modelling [7]. This technique has huge advantages because each part of the model performs and captures patterns within the data applied hence increasing the forecasting ability of each model inside the when being hybridized. A number of studies have employed the use of hybrid modelling in financial forecasting, for instance, a hybrid model consisting of neural networks and support vector machines (SVM) [10], a radial basis function neural network model hybridized with SVM [18], and another hybrid forecasting model developed from the integration of generalized linear auto-regression (GLAR) and neural networks (ANN) [22]. Apart from that, a hybrid neural network and fuzzy regression model has also been used in foreign exchange rate forecasting [8].

Elman and NARX network [3] have been hybridized for chaotic forecasting too, the hybrid model minimized the problem of vanishing gradients in recurrent networks, but did not consider the over fitting problem. In this paper, a novel hybrid model is proposed for financial forecasting by addressing the main issues of vanishing gradient [12] and over fitting [7] in recurrent neural networks. Its performance is compared to the recently developed Elman-NARX model.

3   Proposed Model

A novel hybrid model consisting of unscented Kalman filter and non-linear auto-regressive network with exogenous input is proposed to enhance multi-step-ahead forecasting of chaotic financial data. The hybrid model addresses the problem of vanishing gradient [12] experienced in network training by employing the use of bayesian regulation in training of the hybrid model and also the problem of over-fitting [7] by filtering the chaotic data before forecasting. The function of UKF is to create a better forecasting model by filtering the chaotic data because tiny errors in noise form [1] will be amplified hence affecting the forecasting performance of non-linear auto-regressive input network.

3.1   Unscented Kalman Filter (UKF)

Unscented Kalman filter addresses the approximation issues of extended Kalman filter [17]. In UKF, a minimal number of sigma points are selected that captures the mean and covariance of the state distribution which are obtained using a Gaussian random variable.

The random variable undergoes the process of non-linear unscented transformation. Assuming X has mean $\bar{X}$ and covariance P_k, each sigma point is propagated through the non-linear process model:

$$X_k^{f,j} = f(X_{k-1}^j) \qquad (1)$$

The transformed points are used to compute the mean and covariance of the forecast value of X_k:

$$X_k^f = \sum_{j=0}^{2n} w^j X_k^{f,j} \tag{2}$$

$$P_k^f = \sum_{j=0}^{2n} w^j (X_k^{f,j} - X_k^f)(X_k^{f,j} - X_k^f)^T + Q_{k-1} \tag{3}$$

We propagate then the sigma points through the non-linear observation model:

$$Z_{k-1}^{f,j} = h(X_{k-1}^j) \tag{4}$$

With the resulted transformed observations, their mean and covariance (innovation covariance) is computed:

$$Z_{k-1}^f = \sum_{j=0}^{2n} w^j Z_{k-1}^{f,j} \tag{5}$$

$$Cov(\tilde{Z}_{k-1}^f) = \sum_{j=0}^{2n} w^j (Z_k^{f,j} - Z_k^f)(Z_{k-1}^{f,j} - Z_{k-1}^f)^T + R_k \tag{6}$$

The cross covariance between $\tilde{X}_k^f$ and $\tilde{Z}_{k-1}^f$ is:

$$Cov(\tilde{X}_{k-1}^f, \tilde{Z}_{k-1}^f) = \sum_{j=0}^{2n} w^j (X_k^{f,j} - Z_k^f)(Z_{k-1}^{f,j} - Z_{k-1}^f)^T + R_k \tag{7}$$

The information obtained form the time update step is combined with the measurement step Z_k. The gain K_k is given by:

$$K_k = Cov(\tilde{X}_k^f, \tilde{Z}_{k-1}^f)Cov^{-1}(\tilde{Z}_{k-1}^f) \tag{8}$$

The posterior covariance is updated from the following formula:

$$P_k = P_k^f - K_k Cov(\tilde{Z}_{k-1}^f)K_k^T \tag{9}$$

In our work for this paper, Q_k and R_k are the process and measurement noise which are set up as 1 and 0.001 respectively.

3.2   Recurrent Network

Non-linear auto-regressive network with exogenous input (NARX) can be easily applied for prediction of time series data with the embedded input reconstruction of the network [20]. Hence, the filtered dataset is then fed into the non-linear auto-regressive with exogenous input model in parallel mode as shown in Figure 1, which is mathematically expressed as:

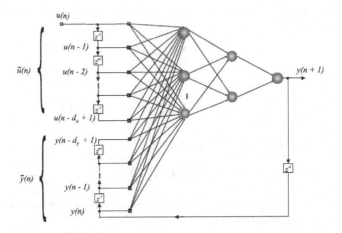

Fig. 1. Parallel-NARX recurrent network architecture (z^{-1} = unit time delay)

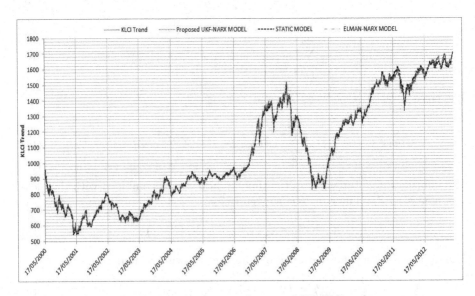

Fig. 2. Graph showing 12 years forecasted KLCI Trend output

$$y(n+1) = f[y(n), ..., y(n-d_y+1); u(n-k), u(n-k+1), ..., u(n-d_u-k+1)] \quad (10)$$

where $u(n)\epsilon\mathbb{R}$ and $y(n)\epsilon\mathbb{R}$ denote, respectively, the input and output of the model at discrete time step n, while $d_u \geq 1$ and $d_y \geq 1$, $d_u \leq d_y$, are the input-memory and output-memory orders, respectively. The parameter $k(\geq 0)$ is a delay term assumed to be zero hence referred to as the process dead-time.

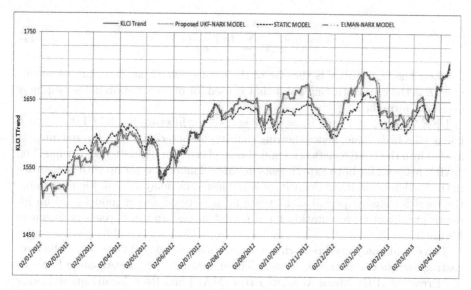

Fig. 3. Graph showing 1.5 years forecasted sample from the total forecasted period of 12 years

When gradient algorithms are used for training, the value decreases to zero as the number of time steps increases. This problematic outcome is commonly referred to as the vanishing gradient problem that results in reduced network performance on standard neural network models [12]. In the proposed model, Bayesian regulation is used as a training algorithm to adjust the parameters of the network so as to move the equilibrium in a way that will result in an output that is close as possible to the target output [6].

4   Results and Discussions

In this paper, Kuala Lumpur Composite Index (KLCI) traded in FTSE Bursa Malaysia that serves as an example of daily financial indexes, is used as real life financial time series dataset for our case study. The transaction date was from 12 April, 1988 to 12 April, 2013 with a total of 6524 daily samples over a period of 25 years. 3156 daily data points were used as training data and the remaining 3368 points are used for testing of the proposed hybrid model which translates to around 12 years of forecasting.

Determining chaos in time series analysis is a very crucial step to differentiate between chaotic and random time series. Largest Lyapunov Exponent (LLE)[3] is used to determine if KLCI data is chaotic or not, a negative LLE means the time series is not chaotic and a positive LLE shows the existence of chaos in the tested time series. The following equation is used to obtain the lyapunov exponent:

$$\lambda = \lim_{N \to \infty} \frac{1}{N} \sum_{t=0}^{N-1} ln \mid \frac{dR}{dx}(x(t)) \mid \qquad (11)$$

where the range of N is the size of the dataset which is 6524 samples in this case, R is the initial starting point that is used for differentiation in the lyapunov exponent formula.

UKF with $Q_k = 1$ and $R_k = 0.001$ is used for filtering the KLCI data which had a +3.8 LLE value, hence showing chaotic characteristics. The filtered outputs are fed into the NARX network trained using Bayesian regulation algorithm. The NARX network for both experiments in parallel mode was set up with 10 neurons in the hidden layer, input delay $d_y = 3$ and feedback delay $d_n = 4$. Figure 2 shows the total forecasted period of 12 years using three different models and an in-depth graphical representation for a period of one and half years is shown in Figure 3.

Three commonly used performance metrics are employed to evaluate the forecasting accuracy in different aspects. Those metrics are Mean Absolute Percentage Error (MAPE), Mean Absolute Error (MAE) and Root Mean Squared Error (RMSE). The obtained forecasting errors from the three models are shown in Table 1. The results show that the proposed UKF-NARX model outperforms other commonly used forecasting models by having the least value of error i.e. 0.5641, 5.7331, 8.3878 for MAPE, MAE and RMSE respectively. Figure 4 shows the histograms of the forecasting errors from each model respectively.

Table 1. Model comparison for forecasted KLCI trend

FORECASTING MODEL	MAPE	MAE	RMSE
Static Financial Model	0.8769	9.0408	11.4886
ELMAN-NARX Model	0.7654	6.3147	9.9556
UKF-NARX Model	**0.5641**	**5.7331**	**8.3878**

The regression (R) value is obtained by measuring the correlation between the forecasted and target output in the testing phase, the importance of this value is to check the model success in forecasting the dependant variable within the KLCI sample. The proposed UKF-NARX model had a regression value of 0.9985 depicted from Figure 5a; Elman-NARX model had a value of 0.9911 as shown in Figures 5b while the static model had a regression value of 0.982 which is shown in Figure 5c. All the three models had an accepted regression value which is closer to the value of 1 translating to a close model relationship with almost a perfect fit.

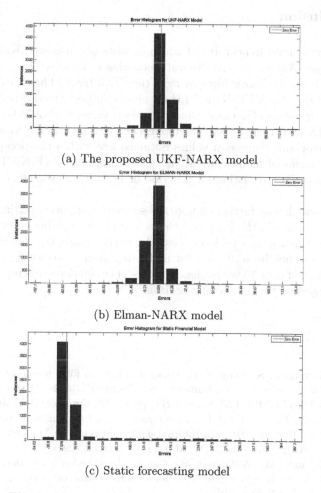

(a) The proposed UKF-NARX model

(b) Elman-NARX model

(c) Static forecasting model

Fig. 4. Error histograms for the three comparing models

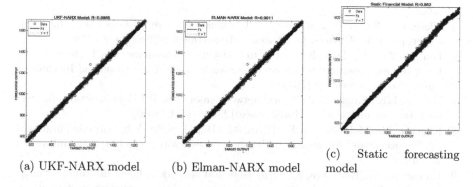

(a) UKF-NARX model   (b) Elman-NARX model   (c) Static forecasting model

Fig. 5. Regression analysis for the comparing models

5   Conclusion

In this study, a novel hybrid model which consists of Unscented Kalman Filter and Non-linear Auto-regressive Neural Networkis (UKF-NARX) was proposed for multi-step-ahead chaotic forecasting of the KLCI trend. The experimental results showed that the UKF-NARX hybrid model outperformed other commonly used models in terms of accuracy and regression value. It should be noted that Elman-NARX and static forecasting models can also be used for forecasting because the error and regression values obtained are within the accepted range. However, in terms of model enhancement, the proposed UKF-NARX model is better for financial forecasting for a period of 12 years, as shown is our investigations.

Future research may further explore the selection parameter settings for input and feedback delays in the proposed model and the forecasting effect of increasing the period to a longer period if possible. Furthermore, the daily KLCI time series trend has not been applied for multi-step-ahead forecasting with a forecasting horizon of over 10 years, hence no model comparison was reported in the literature based on the forecasting horizon.

References

1. Jadid Abdulkadir, S., Yong, S.-P.: Unscented kalman filter for noisy multivariate financial time-series data. In: Ramanna, S., Lingras, P., Sombattheera, C., Krishna, A. (eds.) MIWAI 2013. LNCS, vol. 8271, pp. 87–96. Springer, Heidelberg (2013)
2. Abdullah, L., Ling, C.Y.: A fuzzy time series model for kuala lumpur composite index forecasting. In: 2011 4th International Conference on Modeling, Simulation and Applied Optimization (ICMSAO), pp. 1–5. IEEE (2011)
3. Ardalani-Farsa, M., Zolfaghari, S.: Chaotic time series prediction with residual analysis method using hybrid elman–narx neural networks. Neurocomputing 73(13), 2540–2553 (2010)
4. Atsalakis, G.S., Valavanis, K.P.: Surveying stock market forecasting techniques–part ii: Soft computing methods. Expert Systems with Applications 36(3), 5932–5941 (2009)
5. Klose, C., Pircher, M., Sharma, S.: Univariate time series forecasting. In: Term Paper, pp. 1–46. University of Vienna - Department of Economics (2004)
6. Fan, S., Yan, T.: Two-phase air–water slug flow measurement in horizontal pipe using conductance probes and neural network. IEEE Transactions on Instrumentation and Measurement (2014)
7. Hansen, L.K., Salamon, P.: Neural network ensembles. IEEE Transactions on Pattern Analysis and Machine Intelligence 12, 993–1001 (1990)
8. Khashei, M., Mokhatab, R.F., Bijari, M., Hejazi, S.R.: A hybrid computational intelligence model for foreign exchange rate forecasting. J. Ind. Eng. Int 7(15), 15–29 (2011)
9. Kumar, K., Bhattacharya, S.: Artificial neural network vs linear discriminant analysis in credit ratings forecast: a comparative study of prediction performances. Review of Accounting and Finance 5(3), 216–227 (2006)

10. Lai, K.K., Yu, L., Wang, S.-Y., Wei, H.: A novel nonlinear neural network ensemble model for financial time series forecasting. In: Alexandrov, V.N., van Albada, G.D., Sloot, P.M.A., Dongarra, J. (eds.) ICCS 2006. LNCS, vol. 3991, pp. 790–793. Springer, Heidelberg (2006)

11. Ludwig, R.S., Piovoso, M.J.: A comparison of machine-learning classifiers for selecting money managers. Intelligent Systems in Accounting, Finance and Management 13(3), 151–164 (2005)

12. Mahmoud, S., Lotfi, A., Langensiepen, C.: Behavioural pattern identification and prediction in intelligent environments. Applied Soft Computing 13(4), 1813–1822 (2013)

13. Rana, M., Koprinska, I., Khosravi, A., Agelidis, V.G.: Prediction intervals for electricity load forecasting using neural networks. In: The 2013 International Joint Conference on Neural Networks (IJCNN), pp. 1–8. IEEE (2013)

14. Swanson, R.A.: Analysis for improving performance: Tools for diagnosing organizations and documenting workplace expertise. Berrett-Koehler Publishers (2007)

15. Timmermann, A., Granger, C.W.: Efficient market hypothesis and forecasting. International Journal of Forecasting 20(1), 15–27 (2004)

16. Ukil, A., Jordaan, J.A.: A new approach to load forecasting: Using semi-parametric method and neural networks. In: King, I., Wang, J., Chan, L.-W., Wang, D. (eds.) ICONIP 2006. LNCS, vol. 4233, pp. 974–983. Springer, Heidelberg (2006)

17. Wan, E.A., Van Der Merwe, R.: The unscented kalman filter. Kalman Filtering and Neural Networks, 221–280 (2001)

18. Wang, D., Li, Y.: A novel nonlinear rbf neural network ensemble model for financial time series forecasting. In: 2010 Third International Workshop on Advanced Computational Intelligence (IWACI), pp. 86–90. IEEE (2010)

19. Ward, S., Sherald, M.: The neural network financial wizards. Technical Analysis of Stocks and Commodities, 50–55 (1995)

20. Xie, H., Tang, H., Liao, Y.H.: Time series prediction based on narx neural networks: An advanced approach. In: International Conference on Machine Learning and Cybernetics, vol. 3, pp. 1275–1279. IEEE (2009)

21. Yao, J., Tan, C.L., Poh, H.L.: Neural networks for technical analysis: a study on klci. International journal of theoretical and applied finance 2(02), 221–241 (1999)

22. Yu, L., Wang, S., Lai, K.K.: A novel nonlinear ensemble forecasting model incorporating glar and ann for foreign exchange rates. Computers & Operations Research 32(10), 2523–2541 (2005)

23. Yunos, Z.M., Shamsuddin, S.M., Sallehuddin, R.: Data modeling for kuala lumpur composite index with anfis. In: Second Asia International Conference on Modeling & Simulation, AICMS 2008, pp. 609–614. IEEE (2008)

Forecast of Traffic Accidents Based on Components Extraction and an Autoregressive Neural Network with Levenberg-Marquardt

Lida Barba[1,2] and Nibaldo Rodríguez[2]

[1] Engineering Faculty, Universidad Nacional de Chimborazo,
33730880 Riobamba, Ecuador
lbarba@unach.edu.ec
[2] School of Informatics Engineering, Pontificia Universidad Católica de Valparaíso,
2362807 Valparaíso, Chile

Abstract. In this paper is proposed an improved one-step-ahead strategy for traffic accidents and injured forecast in Concepción, Chile, from year 2000 to 2012 with a weekly sample period. This strategy is based on the extraction and estimation of components of a time series, the Hankel matrix is used to map the time series, the Singular Value Decomposition(SVD) extracts the singular values and the orthogonal matrix, and the components are forecasted with an Autoregressive Neural Network (ANN) based on Levenberg-Marquardt (LM) algorithm. The forecast accuracy of this proposed strategy are compared with the conventional process, SVD-ANN-LM achieved a MAPE of 1.9% for the time series Accidents, and a MAPE of 2.8% for the time series Injured, in front of 14.3% and 21.1% that were obtained with the conventional process.

Keywords: Autoregressive Neural Network, Levenberg-Marquardt, Singular Value Decomposition.

1 Introduction

Analysis and classification of traffic accidents have been mainly developed [1], [2], [3], [4]. Some techniques applied use explanatory variables [5], [6], [7], they could be internal or external variables, disaggregated data to simplify the process were used by [8] and [9].

The Autoregressive Neural Networks is a nonlinear technique used to predict time series in diverse fields of the engineering [10,11,12,13,14]. The forecast by means of the ANNs however not always can model the time series due to their high variability.

In this work is applied an strategy based on the Singular Value Decomposition, it supports the extraction of components of low and high frequency of a time series. SVD traditionally has been applied for de-noising [15,16] and the dimensionality reduction [17,18,19], in various problems [20,21,22].

The strategy improve the forecast of traffic accidents occurred in Concepción, Chile. It involves SVD, the Hankel matrix, and an Autoregressive Neural Network based on Levenberg-Marquardt. The paper is structured as follows. Section 2 describes the improved forecast strategy. Section 3 presents the forecast accuracy metrics. Section 4 presents the Results and discussion. Section 5 shows the conclusions.

R. Prasath et al. (Eds.): MIKE 2014, LNAI 8891, pp. 82–90, 2014.

2   Improved Forecast Strategy

2.1   Singular Value Decomposition of the Hankel Matrix

The SVD is the technique applied to preprocess the time series. The first step consist in to map the original values of the observed time series in a Hankel matrix, after SVD extracts the singular values and the orthogonal matrix from the Hankel matrix, and with these elements are extracted the components of low and high frequency of the time series. In Figure 1 is shown the SVD process, the time series is represented with x, H is the mapped matrix that contains the time series, C_L and C_H are the obtained components, of low and high frequency respectively.

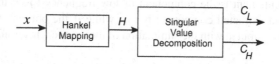

Fig. 1. Singular Value Decomposition of the Hankel matrix

Hankel Mapping: The raw time series is mapped in a trajectory matrix, named Hankel matrix, which elements $x_{ij} = x_{ji}$. The mapping process is illustrated as follows:

$$H_{m \times n} = \begin{bmatrix} x_1 & x_2 & \dots & x_n \\ x_2 & x_3 & \dots & x_{n+1} \\ \vdots & \vdots & \vdots & \vdots \\ x_m & x_{m+1} & \dots & x_N \end{bmatrix} \tag{1}$$

where H is a matrix of order $m \times n$, $x_1 \dots x_N$, are the raw values of the time series, of length N. The value of n is computed with $n = N - m - 1$.

Singular Value Decomposition: Given a real matrix H of order $m \times n$, the matrix product $U \Sigma V^T$ is a Singular Value Decomposition of H, for $m < n$, Σ is a matrix of order $m \times n$ that contains the singular values of H, U is an orthogonal matrix of order $m \times m$ that contains the left singular vectors of H, and V is an orthogonal matrix of order $n \times n$ that contains the right singular vectors of H [23].

$$H = U \times \Sigma_\times V^T \tag{2}$$

The main diag of Σ is composed by the numbers $\sigma_1, \sigma_2, \dots, \sigma_m$, they are called singular values of H, and with U, and V are used during the components extraction process, with

$$A_i = \sigma(i) \times U(:,i) \times V(:,i)^T \tag{3}$$

The matrix A_i contains the i-th component, for $i = 1 \dots m$, the extraction process gives each vector C_i that represent each i-th component. The elements of C_i are located in the first row and in the last column of A_i:

$$C_i = \left[A_i(1,:) \ A_i(2,n:m)^T \right] \tag{4}$$

The optimal number of components m is determined with the computation of the differential energy ΔE of the singular values, as follows:

$$\Delta E_i = E_i - E_{i+1} \tag{5}$$

where E_i is the energy of the i-th singular value, and $i = 1, \ldots, m - 1$. The energy is computed with

$$E_i = \frac{\sigma_i^2}{\left(\sum_{i=1}^{m} \sigma_i^2\right)} \tag{6}$$

where σ_i is the i-th singular value of the Hankel matrix obtained before. The optimal number of components is found through the maximum differential energy peak. The first extracted component is the component of low frequency C_L, if the optimal was $m = 2$, the second extracted component is C_H, while if the optimal was $m > 2$, the component C_H is computed with the summation of the components from 2 to $m - th$, as follows:

$$C_H = \sum_{i=2}^{m} C_i \tag{7}$$

2.2 Components Estimation Process

The process of estimation of components is developed separately using an Autoregressive Neural Network based on the Leveberg-Marquardt algorithm, and it is illustrated in Figure 2.

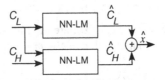

Fig. 2. Components Estimation Process

Autoregressive Neural Network structure: The ANN has a common structure of three layers [24], at the input layer the ANN receives the lagged terms of C_L and C_H, at the hidden layer is applied the sigmoid transfer function, and at the output is obtained the estimated component. The output is

$$\hat{x}(n) = \sum_{j=1}^{N_h} b_j \varphi_j \tag{8a}$$

$$\varphi_j = f\left[\sum_{i=1}^{k} w_{ji} Z_i\right] \tag{8b}$$

where $\hat{x}$ is the estimated value, n is the time instant, N_h is the number of hidden nodes, b_j and w_{ji} are the linear and nonlinear weights of the ANN, and Z_i is the regressor vector that contains the lagged components. The sigmoid transfer function is

$$f(x) = \frac{1}{1+e^{-x}} \tag{9}$$

The weights of the ANN connections, b_j and w_{ji} are adjusted with the LM algorithm.

Levenberg-Marquardt Algorithm: The LM algorithm is a variation of the Newton method, where the parameter u determines its behaviour, if u is increased, it approaches the steepest descent algorithm with small learning rate, and if u is decreased to zero, the algorithm becomes Gauss-Newton [25]. The ANN weights updating is computed with

$$W_{n+1} = W_n + \Delta(W_n) \tag{10a}$$

$$\Delta(W_n) = -[J^T(W_n) \times J(W_n) + u_n \times I]^{-1} J^T(W_n) \times V(W_n) \tag{10b}$$

where W_n is the vector of weights in the time instant n, $\Delta(W_n)$ is the increment for the weights vector, J^T is the transposed Jacobian matrix, u is the scalar parameter, I is the identity matrix, and V is the error vector.

The computation of the Jacobian matrix J is based on the first order partial derivatives of the fitness function (MSE metric, described in section 3 respect to the ANN weights with:

$$J(W_{ji}) = \left[\frac{\partial V(W_{ji})}{\partial W_{ji}} \right] \tag{11}$$

The matrix J has order $N \times k$, where N is the sample size and k is the number of ANN connections, $\partial V(W_{ji})$ is the partial derivative of the error and ∂W_{ji} is the partial derivative of W_{ji}.

3   Forecast Accuracy Metrics

The forecast accuracy is computed with the metrics Root Mean Squared Error (RMSE), Mean Absolute Percentage Error (MAPE), and Relative Error (RE).

$$RMSE = \sqrt{\frac{1}{N_v} \sum_{i=1}^{N_v} (x_i - \hat{x}_i)^2} \tag{12}$$

$$MAPE = \left[\frac{1}{N_v} \sum_{i=1}^{N_v} |(x_i - \hat{x}_i)/x_i| \right] \times 100 \tag{13}$$

$$RE = \sum_{i=1}^{N_v} (x_i - \hat{x}_i)/x_i \tag{14}$$

where N_v is the validation sample size, x_i is the i-th observed value, and $\hat{x}_i$ is the i-th estimated value.

4   Results and Discussion

The proposed time series forecast strategy is evaluated with two time series of traffic accidents occurred in Concepción - Chile, available from the CONASET web site[26]. Two time series were used, the number of accidents, and the number of injured people, each series contains 680 records from year 2000 to 2012. The training set contains the 70% of the sample, consequently the validation set contains the 30% of the sample.

4.1   Components Extraction

The time series is mapped in the Hankel matrix, the rows number of the H matrix (m) is computed with $N/2$ (N sample length). The differential energy ΔE of each singular value of H is computed to find the optimal m. The Figures 3a and 3b show the ΔE of the components for each series, the maximum peak represents the optimal m, therefore for the time series Number of Accidents was found $m = 6$, and for the time series Injured people the optimal was $m = 4$. The first component extracted is the component of low frequency C_L, while the rest components are summated to find the component of high frequency C_H.

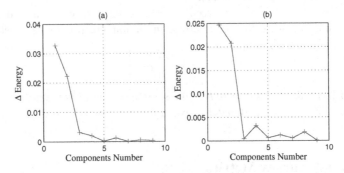

Fig. 3. Differential energy (a) Number of Accidents, (b) Injured people

4.2   Forecast

The optimal number of lagged values used by the ANN-LM method was determined with the computation of the RMSE metric, and the optimal lags for the time series Number of Accidents was $k = 11$, while for Injured $k = 10$, as shown in Figure 4. The number of hidden nodes was computed with the natural logarithm of the training sample (normally computed in our experiments), then the optimal configurations were ANN-LM(11,6,1) for Accidents and ANN-LM(10,6,1) for Injured.

The parameters of the algorithm LM were set in $u = 0.001$, u−increment of 10, u−decrement of 0.1, minimum performance gradient $1e - 10$, and a maximum number of 500 iterations.

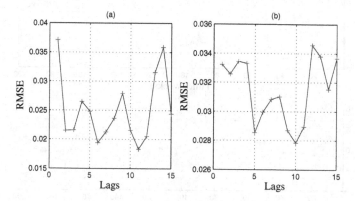

Fig. 4. Lags calibration (a) Number of accidents (b) Injured people

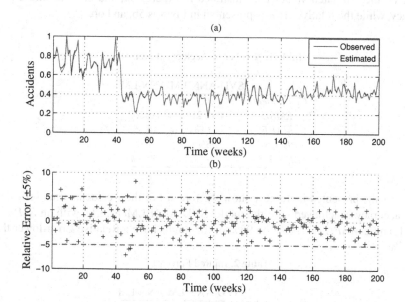

Fig. 5. Number of Accidents (a) SVD-ANN-LM(11,6,1) (b) Relative Error

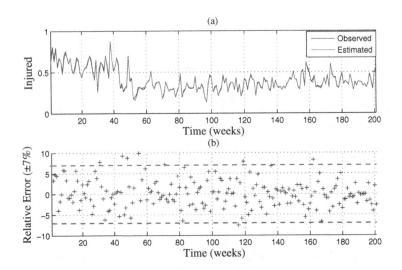

Fig. 6. Injured (a) SVD-ANN-LM(10,6,1) (b) Relative Error

The validation process was developed with each testing data set of Accidents and Injured. The best results are shown in Figures 5, 6, Table 1, and Table 2. The observed values vs. the estimated values are illustrated in Figures 5a, and 6a, reaching a good accuracy, while the relative error is presented in Figures 5b, and 6b.

Table 1. Number of Accidents Forecast

	SVD-ANN-LM	ANN-LM
Components	6	–
RMSE	0.0121	0.0862
MAPE	1.9%	14.3%
$RE \pm 5\%$	95%	–
$RE \pm 10\%$	–	50.5%

The results presented in Table 1 show the forecast of the Accidents time series, the major accuracy is achieved with the model SVD-ANN-LM(11,6,1), with an *RMSE* of 0.0121, and a *MAPE* of 1.9%, the 95% of the points have a Relative Error lower than the ±5%.

Table 2. Injured Forecast

	SVD-ANN-LM	ANN-LM
Components	4	–
RMSE	0.0146	0.0981
MAPE	2.8%	21.1%
$RE \pm 7\%$	94%	–
$RE \pm 10\%$	–	34.3%

The results presented in Table 2 show the forecast of the Injured time series, the major accuracy is achieved with the model SVD-ANN-LM(10,6,1), with an *RMSE* of 0.0146, and a *MAPE* of 2.8%, the 94% of the points have a Relative error lower than the ±7%.

5 Conclusions

The proposed one-step-ahead forecast strategy shows an improved accuracy, the data preprocessing is implemented with the Singular Value Decomposition of the Hankel matrix. The components of low and high frequency of the time series were extracted with the singular values and the orthogonal matrix. The estimation process is developed by an Autoregressive Neural Network based on the Levenberg-Marquardt algorithm, the forecast is obtained with the summation of the two estimated components.

The time series of traffic accidents occurred in Concepción, Chile from year 2000 to 2012 were forecasted, for the time series Number of Accidents, SVD-ANN-LM achieves an RMSE of 0.0121, and a MAPE of 1.9%, in front of the conventional ANN-LM that presents an RMSE of 0.0862 and a MAPE of 14.3. For the time series Injured were achieved an RMSE of 0.0146 and a MAPE of 2.8%, in front of the conventional ANN-LM that presents an RMSE of 0.0981 and a MAPE of 21.1.

In the future this strategy will be evaluated with data of traffic accidents of other regions of Chile, other countries, and with time series of other engineering fields.

References

1. Abellán, J., López, G., de Oña, J.: Analysis of traffic accident severity using Decision Rules via Decision Trees. Expert Systems with Applications 40(15), 6047–6054 (2013)
2. de Oña, J., López, G., Mujalli, R., Calvo, F.: Analysis of traffic accidents on rural highways using Latent Class Clustering and Bayesian Networks. Accident Analysis & Prevention 51, 1–10 (2013)
3. Chang, L., Chien, J.: Analysis of driver injury severity in truck-involved accidents using a non-parametric classification tree model. Safety Science 51(1), 17–22 (2013)
4. Fogue, M., Garrido, P., Martinez, F., Cano, J., Calafate, C., Manzoni, P.: A novel approach for traffic accidents sanitary resource allocation based on multi-objective genetic algorithms. Expert Systems with Applications 40(1), 323–336 (2013)
5. Commandeur, J., Bijleveld, F., Bergel-Hayat, R., Antoniou, C., Yannis, G., Papadimitriou, E.: On statistical inference in time series analysis of the evolution of road safety. Accident Analysis & Prevention 60, 424–434 (2013)
6. Weijermars, W., Wesemann, P.: Road safety forecasting and ex-ante evaluation of policy in the Netherlands. Transportation Research Part A: Policy and Practice 52, 64–72 (2013)
7. Antoniou, C., Yannis, G.: State-space based analysis and forecasting of macroscopic road safety trends in Greece. Accident Analysis & Prevention 60, 268–276 (2013)
8. García-Ferrer, A., de Juan, A., Poncela, P.: Forecasting traffic accidents using disaggregated data. International Journal of Forecasting 22(2), 203–222 (2006)
9. Quddus, M.: Time series count data models: An empirical application to traffic accidents. Accident Analysis & Prevention 40(5), 1732–1741 (2008)
10. Rojas, I., Pomares, H., Bernier, J.L., Ortega, J., Pino, B., Pelayo, F.J., Prieto, A.: Time series analysis using normalized PG-RBF network with regression weights. Neurocomputing 42(1-4), 267–285 (2002)

11. Palmer, A., Montaño, J., Sesé, A.: Designing an artificial neural network for forecasting tourism time series. Tourism Management 27(5), 781–790 (2006)
12. Liu, F., Ng, G.S., Quek, C.: RLDDE: A novel reinforcement learning-based dimension and delay estimator for neural networks in time series prediction. Neurocomputing 70(7-9), 1331–1341 (2007)
13. Roh, S.B., Oh, S.K., Pedrycz, W.: Design of fuzzy radial basis function-based polynomial neural networks. Fuzzy Sets and Systems 185(1), 15–37 (2011)
14. Gheyas, I.A., Smith, L.S.: A novel neural network ensemble architecture for time series forecasting. Neurocomputing 74(18), 3855–3864 (2011)
15. Yang, W., Tse, P.: Development of an advanced noise reduction method for vibration analysis based on singular value decomposition. NDT & E International 36(6), 419–432 (2003)
16. Reninger, P., Martelet, G., Deparis, J., Perrin, J., Chen, Y.: Singular value decomposition as a denoising tool for airborne time domain electromagnetic data. Journal of Applied Geophysics 75(2), 264–276 (2011)
17. Li, C., Park, S.: An efficient document classification model using an improved back propagation neural network and singular value decomposition. Expert Systems with Applications 36(2), 3208–3215 (2009)
18. Shih, Y., Chien, C., Chuang, C.: An adaptive parameterized block-based singular value decomposition for image de-noising and compression. Applied Mathematics and Computation 218(21), 10370–10385 (2012)
19. Kavaklioglu, K.: Robust electricity consumption modeling of Turkey using Singular Value Decomposition. International Journal of Electrical Power & Energy Systems 54, 268–276 (2014)
20. Al-Zaben, A., Al-Smadi, A.: Extraction of foetal ECG by combination of singular value decomposition and neuro-fuzzy inference system. Physics in Medicine and Biology 51(1), 137 (2006)
21. Yin, H., Zhu, A., Ding, F.: Model order determination using the Hankel matrix of impulse responses. Applied Mathematics Letters 24(5), 797–802 (2011)
22. Cong, F., Chen, J., Dong, G., Zhao, F.: Short-time matrix series based singular value decomposition for rolling bearing fault diagnosis. Mechanical Systems and Signal Processing 34(1-2), 218–230 (2013)
23. Shores, T.S.: Applied Linear Algebra and Matrix Analysis, pp. 291–293. Springer, Heidelberg (2007)
24. Freeman, J.A., Skapura, D.M.: Neural Networks, Algorithms, Applications, and Programming Techniques. Addison-Wesley, California (1991)
25. Hagan, M.T., Demuth, H.B., Bealetitle, M.: Neural Network Design. Hagan Publishing, 12.19–12.27 (2002)
26. National Commission of Transit Security, http://www.conaset.cl/

Top-k Parametrized Boost

Turki Turki[1,4,*], Muhammad Ihsan[2,*], Nouf Turki[3],
Jie Zhang[4], Usman Roshan[4,*], and Zhi Wei[4,*]

[1] Computer Science Department, King Abdulaziz University
P.O. Box 80221, Jeddah 21589, Saudi Arabia
tturki@kau.edu.sa
[2] Department of Electrical Engineering, Stanford University
350 Serra Mall, Stanford, CA 94305, United States
aihsan@stanford.edu
[3] King Abdulaziz University
P.O. Box 80200, Jeddah 21589, Saudi Arabia
nouf.turkey@aol.com
[4] Department of Computer Science, New Jersey Institute of Technology
University Heights, Newark, NJ 07102
{ttt2,jz266,zhiwei}@njit.edu, usman@cs.njit.edu

Abstract. Ensemble methods such as AdaBoost are popular machine learning methods that create highly accurate classifier by combining the predictions from several classifiers. We present a parametrized method of AdaBoost that we call Top-k Parametrized Boost. We evaluate our and other popular ensemble methods from a classification perspective on several real datasets. Our empirical study shows that our method gives the minimum average error with statistical significance on the datasets.

Keywords: Ensemble methods, AdaBoost, statistical significance.

1 Introduction

Ensemble methods are popular machine learning methods that produce a single highly accurate classifier by combining the predictions from several classifiers [1]. Among many such methods the AdaBoost(AB) [2] is a very popular choice. AB outputs a single classifier by combining T weighted classifiers and prediction [3] is given by

$$h^*(x_j) = \arg\max_y \sum_{i=1}^{T} \alpha_i I(h_i(x_j) = y) \qquad (1)$$

where $x_j \in R^d$ for $j = 1...m$, $y \in \{-1, +1\}$, h_i the ith classifier that maps the instance x_j to y , α_i the weight of h_i, and $I(.)$ is an indicator function that

---

* Corresponding author.

R. Prasath et al. (Eds.): MIKE 2014, LNAI 8891, pp. 91–98, 2014.
© Springer International Publishing Switzerland 2014

outputs 1 if its argument is true and 0 otherwise. In this paper we consider a parametric version of Equation 1 that we call the Parametrized AdaBoost(P-AdaBoost) given by

$$h^*(x_j) = \arg\max_y \sum_{i=1}^{T} \beta\alpha_i I(h_i(x_j) = y), \qquad (2)$$

where $\beta = \sum_{i=1}^{k} \beta_i \leq 1$ and $0 < \beta_i \leq 1$. In addition to P-AdaBoost, we present a method that we call Top-k Boost which uses P-AdaBoost to search for the top-k parameter values in β that achieve the best classification results. We combine the k parameter values to produce optimal classifier. For given dataset we obtain this optimal classifier and β by cross validation. Both P-AdaBoost with Top-k Boost yield our method which we call Top-k Parametrized Boost. Compared to other popular ensemble methods our empirical study on 25 datasets shows that Top-k Parametrized Boost yields the minimum average error with statistical significance.

The rest of this paper is organized as follows. In section 2 we review related work. In section 3 we present our method Top-k Parametrized Boost. Following that we present empirical study and discussion before concluding.

2   Related Work

AdaBoost [2] combines sequentially classifiers(e.g. decision trees) to produce highly accurate classifier. In detail, the AdaBoost algorithm [4] which is outlined in Algorithm 1 works as follows. Line 1 receives as an input a set of m labeled training examples $S = \{(x_1, y_1), ..., (x_m, y_m)\}$ where the label associated to the instance $x_m \in R^d$ is $y_m \in \{-1, +1\}$ drawn i.i.d to a distribution used for both validation and training. Lines 2-4 initialize F, which will hold T weighted classifiers [3]. Lines 5-7 assign equal weight distribution to all training examples. The for loop of lines 8-16 update weight distribution and combine T weighted classifiers. In line 9 the learning algorithm will find classifier $h_t \in H$ using weight distribution D_t to map instances in x to y with small error. Line 10 incurs the loss $Pr_{D_t}[y_i \neq h_t(x_i)] = \sum_{i=1}^{m} D_t I[y_i \neq h_t(x_i)]$. Line 11 calculates α_t the weight of the classifier h_t. Line 12 stores the weighted classifier $\alpha_t h_t$. Lines 13-15 update weight distribution [3] by upweighting of examples which are incorrectly classified to focus on and decreasing the weights of correctly classified examples. Line 17 gives the final weighted classifier. Line 18 yields the prediction by taking the sign of the sum of T weighted classifiers (Equation 1).

Algorithm 1. AdaBoost algorithm

1:   **AdaBoost**$(S = \{(x_1, y_1), ..., (x_m, y_m)\})$
2:     **for** $i = 1$ **to** m **do**
3:         $F_0(i) \leftarrow 0$
4:     **end for**
5:     **for** $i = 1$ **to** m **do**
6:         $D_1(i) \leftarrow \frac{1}{m}$
7:     **end for**
8:     **for** $t = 1$ **to** T **do**
9:         $h_t \leftarrow fit\ classifier\ h_t \in H\ with\ D_t$
10:       $\epsilon_t \leftarrow Pr_{D_t}[y_i \neq h_t(x_i)]$
11:       $\alpha_t \leftarrow \frac{1}{2} \log \frac{1-\epsilon_t}{\epsilon_t}$
12:       $F_t \leftarrow F_{t-1} + \alpha_t h_t$
13:       **for** $i = 1$ **to** m **do**
14:           $D_{t+1}(i) \leftarrow \frac{D_t(i)exp(-\alpha_t h_t(x_i)y_i)}{\sum\limits_{i=1}^{m} D_t(i)}$
15:       **end for**
16:     **end for**
17:     $F \leftarrow \sum\limits_{t=1}^{T} \alpha_t h_t$
18:     $h^* = sgn(F)$

---

3   Top-k Parametrized Boost

As shown above the AdaBoost algorithm [3,4] receives the labeled training sample S as an input(line 1) and outputs the final weighted classifier as an output (line 17). Our algorithm outlined in Algorithm 3 uses as a subroutine a parametrized version of the AdaBoost algorithm [3,4] which is outlined in Algorithm 2. The P-AdaBoost algorithm receives a fixed parameter $\beta \in (0, 1]$ and a training sample $S = \{(x_1, y_1), ..., (x_m, y_m)\}$ where the label associated to the instance $x_m \in R^d$ is $y_m \in \{-1, +1\}$(line 1). Lines 2-11 are the same as AdaBoost algorithm outlined in Algorithm 1. Line 12 stores the parameterized weighted classifiers. Line 17 gives the final parametrized weighted classifier. Line 18 yields the prediction by taking the sign of the sum of T parametrized weighted classi-fiers(Equation 2). Line 19 incurs the loss $Pr[y \neq h^*(x)] = \sum\limits_{i=1}^{m} I[y_i \neq h^*(x_i)]$. Line 20 returns (F, E) the final classifier and the corresponding error respectively.

    We can now use the P-AdaBoost algorithm as a subroutine in the Top-k Boost algorithm which is outlined in Algorithm 3. The Top-k Boost algorithm outputs the final parametrized weighted classifier(line 33 in Algorithm 3) that outperforms the AdaBoost's final classifier(line 17 in Algorithm 1). To output the final parametrized weighted classifier(line 33 in Algorithm 3), we make the initial call Top-k Boost(β, S), where $\beta =< \beta[1], ..., \beta[n] >$ for fixed number of parameter values n to be precisely specified in section 4.1 such that $\beta_i \in (0, 1]$ and

$S = \{(x_1, y_1), ..., (x_m, y_m)\}$ where $x_m \in R^d$ and $y_m \in \{-1, +1\}$. The for loop of lines 2-5 iterates n times to store the parameter values in β and the errors of the corresponding classifiers returned by P-AdaBoost algorithm in (P, E) respectively. Lines 7-12 store the error of classifier h_l with $\beta_l = 1$ in $error$ variable(line 9), where $F \leftarrow \sum_{t=1}^{T} \beta \alpha_t h_t$(line 17 in Algorithm 2) $= F \leftarrow \sum_{t=1}^{T} \alpha_t h_t$(line 17 in Algorithm 1) when $\beta = 1$. The $error$ variable(line 9) corresponds to the same error incurred by AdaBoost algorithm outlined in Algorithm 1. The for loop of lines 14-19 searches for parameter values in β already stored in P that achieve the same or better performance than one by $\beta_l = 1$. The Min subroutine in line 22 finds the ith smallest error e_i to return the corresponding ith parameter b_i which is stored in P_i. The for loop of lines 26-32 adds the top-k parameter values in P, where $\sum_{i=1}^{k} P_i \leq 1$. In line 33 we call P-AdaBoost by passing the top-k parameter values in β and training sample S to obtain the final classifier as shown in line 33. Line 34 yields the prediction by taking the sign of the sum of T parametrized weighted classifiers(Equation 2).

Algorithm 2. P-AdaBoost algorithm

1: **P-AdaBoost**$(\beta, S = \{(x_1, y_1), ..., (x_m, y_m)\})$
2:      **for** $i = 1$ **to** m **do**
3:          $F_0(i) \leftarrow 0$
4:      **end for**
5:      **for** $i = 1$ **to** m **do**
6:          $D_1(i) \leftarrow \frac{1}{m}$
7:      **end for**
8:      **for** $t = 1$ **to** T **do**
9:          $h_t \leftarrow fit\ classifier\ h_t \in H\ with\ D_t$
10:         $\epsilon_t \leftarrow Pr_{D_t}[y_i \neq h_t(x_i)]$
11:         $\alpha_t \leftarrow \frac{1}{2} \log \frac{1-\epsilon_t}{\epsilon_t}$
12:         $F_t \leftarrow F_{t-1} + \beta \alpha_t h_t$
13:         **for** $i = 1$ **to** m **do**
14:             $D_{t+1}(i) \leftarrow \frac{D_t(i) exp(-\alpha_t h_t(x_i) y_i)}{\sum_{i=1}^{m} D_t(i)}$
15:         **end for**
16:     **end for**
17:     $F \leftarrow \sum_{t=1}^{T} \beta \alpha_t h_t$
18:     $h^* = sgn(F)$
19:     $E \leftarrow Pr[y \neq h^*(x)]$
20:     **return** (F, E)

Algorithm 3. Top-k Boost algorithm

1:   **Top-k Boost**$(\beta, S = \{(x_1, y_1), ..., (x_m, y_m)\})$
2:       **for** $l = 1$ **to** $length(\beta)$ **do**
3:           $(h_l, \epsilon_l) \leftarrow P - AdaBoost(\beta_l, S)$
4:           $(P, E) \leftarrow (\beta_l, \epsilon_l)$
5:       **end for**
6:       $(error, index) \leftarrow (0, 0)$
7:       **for** $l = 1$ **to** $length(\beta)$ **do**
8:           **if** $P_l = 1$ **then**
9:               $(error, index) \leftarrow (E_l, l)$
10:              *break*
11:          **end if**
12:      **end for**
13:      $k \leftarrow 1$
14:      **for** $l = 1$ **to** $length(\beta)$ **do**
15:          **if** $l \neq index$ *and error* $\geq E_l$ **then**
16:              $(b_k, e_k) \leftarrow (P_l, E_l)$
17:              $k \leftarrow k + 1$
18:          **end if**
19:      **end for**
20:      $P \leftarrow 0$
21:      **for** $i = 1$ **to** k **do**
22:          $b_i \leftarrow Min(e, i)$
23:          $P_i \leftarrow b_i$
24:      **end for**
25:      $\beta \leftarrow P_1$
26:      **for** $i = 2$ **to** k **do**
27:          **if** $\beta + P_i \leq 1$ **then**
28:              $\beta \leftarrow \beta + P_i$
29:          **else**
30:              *break*
31:          **end if**
32:      **end for**
33:      $(F, E) \leftarrow P - AdaBoost(\beta, S)$
34:      $h^* = sgn(F)$

---

4   Empirical Study

To evaluate the performance of our method, an empirical study from a classification perspective [5] is performed on 25 real datasets shown in Table 1 from the UCI Machine Learning Repository [6]. This section describes the experimental methodology, then presents the experimental results.

4.1   Experimental Methodology

We compare three ensemble classification algorithms: Top-k Parametrized Boost (T-K PB), AdaBoost(AB), Random forests(RF) [7]. For each dataset we use the

Table 1. Datasets from the UCI Machine Learning repository that we used in our empirical study [6]

Code	Dataset	Classes	Dimension	Instances
1	Haberman's Survival	2	3	306
2	Skin Segmentation	2	3	245057
3	Blood Transfusion Service Center	2	4	748
4	Liver-disorders	2	6	345
5	Diabetes	2	8	768
6	Breast Cancer	2	10	683
7	MAGIC Gamma Telescope	2	10	19020
8	Planning Relax	2	12	182
9	Heart	2	13	270
10	Australian Credit Approval	2	14	690
11	Climate	2	18	540
12	Two norm	2	20	7400
13	Statlog German credit card	2	24	1000
14	Breast cancer	2	30	569
15	Ionosphere	2	34	351
16	Qsar	2	41	1055
17	SPECTF heart	2	44	267
18	Spambase	2	57	4597
19	Sonar	2	60	208
20	Digits	2	63	762
21	Ozone	2	72	1847
22	Insurance company coil2000	2	85	5822
23	Hill valley	2	100	606
24	BCI	2	117	400
25	Musk	2	166	476

10-fold cross-validation with the same splits for each algorithm. For ensemble method we find the best parameter values with further cross-validation on the training set.

In T-K PB we let β range from $\{1,.9,.8,.7,.6,.5,.4,.375,.35,.3,.25,.2,.15,.1,0.05\}$. Recall that T-K PB is given by $h^*(x_j) = \underset{y}{\operatorname{argmax}} \sum_{i=1}^{T} \beta \alpha_i I(h_i(x_j) = y)$. For each parameter value we select the parameter value that gives the minimum error on the training. Thus we obtain the top-k values of parameter β with cross-validation on the training set. We then apply the top-k parameter values of β to the validation set. We wrote our code in R.

4.2   Experimental Results on Twenty Five Datasets

We measure the error as the number of the validation instances incorrectly predicted divided by the number of validation instances. We compute the error of our classification tasks for each training-validation split in the cross-validation and take the the average result of ten folds to be the average cross-validation error. In Table 2 we show the average cross-validation error on each dataset. Over the 25 datasets T-K PB gives the minimum average error of 10.452% and has the minimum error in 22 out of 25 datasets. The second best is RF that gives an average error of 15.24% and has the minimum error in 2 out of 25 datasets. AB gives higher average error of 15.548% and has the minimum error in 1 out of 25 datasets.

We measure the statistical significance of the methods with Wilcoxon rank test [5, 8]. This test is a standard test to measure the statistical significance between two methods in many datasets. It shows that if one method outperforms the other in many datasets then it is considered statistically significant. In Table 3 the p-values show that our method T-K PB outperforms the other two methods in the 25 datasets with statistical significance.

Table 2. Average cross-validation error of different ensemble algorithms on each of the 25 real datasets from the UCI machine learning repository. The method with the minimum error is shown in bold.

Code	Dataset	T-K PB	AB	RF
1	Haberman's Survival	0.31154	0.35769	**0.26923**
2	Skin Segmentation	**0.00005**	0.0005	0.00044
3	Blood Transfusion Service Center	0.21470	**0.21176**	0.22647
4	Liver-disorders	**0.236**	0.3	0.244
5	Diabetes	**0.21618**	0.27353	0.24412
6	Breast Cancer	**0.00794**	0.02381	0.02222
7	MAGIC Gamma Telescope	0.16142	0.16395	**0.13205**
8	Planning Relax	**0.3**	0.39167	0.375
9	Heart	**0.14**	0.185	0.16
10	Australian Credit Approval	**0.06167**	0.14167	0.14833
11	Climate	**0.026**	0.044	0.06
12	Two norm	**0.00795**	0.02534	0.02356
13	Statlog German credit card	**0.24222**	0.26333	0.25889
14	Breast cancer	**0.00408**	0.03061	0.03878
15	Ionosphere	**0.00323**	0.03871	0.04194
16	Qsar	**0.10211**	0.13368	0.12737
17	SPECTF heart	**0.11765**	0.22941	0.26471
18	Spambase	**0.03392**	0.04573	0.04508
19	Sonar	**0.03889**	0.11667	0.11111
20	Digits	**0.00278**	0.01528	0.0125
21	Ozone	**0.04463**	0.05593	0.05876
22	Insurance company coil2000	**0.05682**	0.06731	0.06469
23	Hill valley	**0.04245**	0.41509	0.43868
24	BCI	**0.18**	0.28667	0.33333
25	Musk	**0.06087**	0.06957	0.10870
	Average error	0.10452	0.15548	0.1524

Table 3. P-values of Wilcox rank test(two-tailed test) between all pairs of methods

	AB	RF
T-K PB	0.00002	0.00044
AB		0.77948

5 Discussion

T-K PB finds the top-k parameter values to the given dataset. We add these top-k values such that its sum is not greater than one, then we take the sign of the sum of T parametrized weighted classifiers to make prediction. This approach is better than AB and RF(results shown in Table 2 and Table 3).

In this study we varied β for T-K PB in the cross validation to obtain the top-k parameter values. We chose the standard decision trees as classifiers in AB and our method T-K PB. We combined T decision trees to construct the final classifier for T-K PB and AB where $T = 100$. In the current experiments T-K PB is the slowest method but still computationally tractable for large datasets. However, T-K PB achieves better accuracy than the other two methods most of the time.

We chose RF [7] to be compared with our method due to its stability and its popularity as a powerful method. We used the standard package for RF in R [9]

6 Conclusion

We introduce a parametrized method of AdaBoost and optimize it with cross-validation for the classification. Our method outperforms the other popular methods by giving the minimum average error with statistical significance on many real datasets selected from UCI machine learning repository.

References

1. Maclin, R., Opitz, D.: Popular ensemble methods: An empirical study. arXiv preprint arXiv:1106.0257 (2011)
2. Freund, Y., Schapire, R.E.: A desicion-theoretic generalization of on-line learning and an application to boosting. In: Vitányi, P.M.B. (ed.) EuroCOLT 1995. LNCS, vol. 904, pp. 23–37. Springer, Heidelberg (1995)
3. Tan, P.N., Steinbach, M., Kumar, V.: Introduction to Data Mining, 1st edn. Addison-Wesley Longman Publishing Co., Inc., Boston (2005)
4. Mohri, M., Rostamizadeh, A., Talwalkar, A.: Foundations of machine learning. MIT press (2012)
5. Japkowicz, N., Shah, M.: Evaluating Learning Algorithms. Cambridge University Press (2011)
6. Asuncion, A., Newman, D.J.: UCI machine learning repository (2007)
7. Breiman, L.: Random forests. Machine Learning 45, 5–32 (2001)
8. Kanji, G.K.: 100 statistical tests. Sage (2006)
9. Liaw, A., Wiener, M.: Classification and regression by randomforest. R News 2, 18–22 (2002)

Unsupervised Feature Learning for Human Activity Recognition Using Smartphone Sensors

Yongmou Li, Dianxi Shi, Bo Ding, and Dongbo Liu

National Key Laboratory for Parallel and Distributed Processing,
College of Computer, National University of Defense Technology, Changsha, China
liyongmou@gmail.com

Abstract. Feature representation has a significant impact on human activity recognition. While the common used hand-crafted features rely heavily on the specific domain knowledge and may suffer from non-adaptability to the particular dataset. To alleviate the problems of hand-crafted features, we present a feature extraction framework which exploits different unsupervised feature learning techniques to learning useful feature representation from accelerometer and gyroscope sensor data for human activity recognition. The unsupervised learning techniques we investigate include sparse auto-encoder, denoising auto-encoder and PCA. We evaluate the performance on a public human activity recognition dataset and also compare our method with traditional features and another way of unsupervised feature learning. The results show that the learned features of our framework outperform the other two methods. The sparse auto-encoder achieves better results than other two techniques within our framework.

Keywords: human activity recognition, unsupervised feature learning, sparse auto-encoder, denoising auto-encoder.

1 Introduction

Extensive work has been done in the area of human activity recognition (HAR) [1]. HAR has a great potential for applications for human health and wellness, for example we can promote a healthier lifestyle to who lacks adequate physical activity by monitoring and analysis the daily activities of him. Research on HAR mainly involves the use of different sensing technologies. Especially, the widespread use of smartphone equipped with a rich set of sensors make it a popular platform for HAR. The most commonly used sensors are accelerometer and gyroscope in HAR using smartphone.

Human activity recognition (HAR) is a classical time series or sequence classification problem, for which the task is to determine those contiguous portions of sensor data streams are generated by which one of activities of interest. The processing pipeline of HAR includes three stages: segmenting sensor data streams, extracting features and classification. For data segmentation, a widely adopted approach is the sliding window technique, where a fixed length window moving

R. Prasath et al. (Eds.): MIKE 2014, LNAI 8891, pp. 99–107, 2014.

along the stream with some overlapping separates data flow to frames. As for any pattern recognition task, the keys to successful HAR are: appropriate feature representations of the sensor data and suitable classification algorithm [2]. Because we can test a couple of state-of-the-art classification algorithms and select the best one by cross validation, the most difficult problem lies in designing good feature representation of the sensor data.

The features used in most of researches on HAR are selected by hand. These hand-crafted features rely on the specific domain knowledge, and maybe result in loss of information after extracting features. The Fourier and Wavelet representations, which are commonly used in HAR, suffer from non-adaptability to the particular dataset [3]. Our solution for these problems is using the unsupervised feature learning techniques to automatically learn feature representation from massive data, which is data adaptive.

In this paper, we present a feature extraction framework which exploits different unsupervised feature learning techniques to learning useful feature representation from accelerometer and gyroscope sensor data for HAR. Our framework can also reduce the complexity the used unsupervised learning model. we conducted a number of experiments on a public HAR dataset. The results shows that our method is better than the other two methods.

2   Related Work

Extensive work has been done in the area of HAR using smartphone sensors, which has been summarized in [1]. However, to the best of our knowledge, there are only three works [2,4,5] applying feature learning techniques in activity recognition. The authors in [2] firstly introduce feature learning methods to the area of activity recognition, they used Deep Belief Networks (DBN) and PCA to learn features for activity recognition in ubiquitous computing. The authors in [4] following the work of [2] applied shift-invariant sparse coding technique. The authors in [5] also used sparse coding to learn features, while they evaluated their method on a dataset collected by smartphone sensors.

Our work is different from the previous three works in certain aspects. In the previous three works, they concatenated the data of all channels in an analysis window as the input of the feature learning algorithms, we see this way as the 'monolithic' way. We adopt the channel-wise way, which learns feature from each channel then concatenates learned features for all channels as the feature representation. We explore different auto-encoders in HAR, which have not been studied. The channel-wise method can reduce the complexity of learning model and, hence, does not need so much data to fit the model well.

3   Proposed Methods

In this section, we present a channel-wise feature extraction framework for accelerometer data and gyroscope sensor data based on unsupervised feature learning, which is integrated into a general HAR procedure (Fig. 1). As show in Fig. 1, unsupervised feature learning technique is used to learn a transformation from raw sensor data to features used in classification. The accelerometer and gyroscope in smartphone have three axes (x, y, z), which is consistent with the 3D space we live in. As the coordinate system of smartphone is based on the screen, the sensor data of three axes easily influenced by the place of screen. We generate a resultant $r = \sqrt{x^2 + y^2 + z^2}$ to avoid the influence. We view the x, y, z and r as four channels. In the framework, we firstly learn features in each channel, then concatenate the features of all channels as the final feature representation.

The channel-wise way of unsupervised feature learning has the following advantages:

- reduce the complexity of feature learning model and hence faster
- more flexible, different channels can use different hyper-parameters even different models

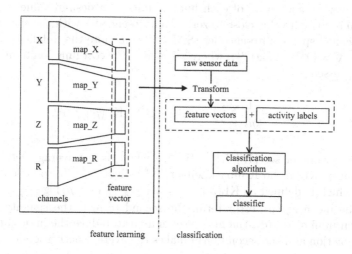

Fig. 1. Unsupervised feature learning for HAR - overview

In this paper, we focus on the following three unsupervised feature leaning techniques: sparse auto-encoder (SAE), denoising auto-encoder (DAE) and principal component analysis (PCA).

3.1 SAE Based Unsupervised Feature Learning

An auto-encoder is a neural network with a single hidden layer, where the target values are set equal to the inputs during training. Formally, let N and K represent the number of input units and the number of hidden units, the parameters of an auto-encoder can be described as follow: $W^{(1)} \in \mathbb{R}^{(K \times N)}$ and $W^{(2)} \in \mathbb{R}^{(N \times K)}$ represent the weights from the input units to the hidden units and from the hidden units to the output units, $b^{(1)} \in \mathbb{R}^K$ and $b^{(2)} \in \mathbb{R}^N$ represent the biases of the hidden units and the output units. An auto-encoder can be divided into two parts, one part is the encoder which transforms the input $x \in \mathbb{R}^N$ to an hidden representation $h(x) \in \mathbb{R}^K$, the other part is the decoder which reconstructs the input from the hidden representation. The hidden representation $h(x)$ and the reconstruction of x denoted as $\hat{x}$ are calculated as:

$$h(x) = \sigma(W^{(1)}x + b^{(1)})$$

$$\hat{x} = \sigma(W^{(2)}h(x) + b^{(2)})$$

where $\sigma(\cdot)$ is the activation function, typically the sigmoid function. The goal of the auto-encoder model is to learn the hidden representation $h(x)$, which also can be seen as a feature mapping function. We can stack auto-encoders to learn more complicate feature mapping.

In order to discover interesting structure from input data, we can impose a sparsity constraint on the hidden units [6]. The sparsity constraint means that limit the average activation of each hidden units to a desired value $\rho \in (0,1)$, typically ρ is a small value close to zero. An auto-encoder with the sparsity constraint is called sparse auto-encoder (SAE), if we denote the unlabeled training dataset as $X = \{x^{(1)}, \ldots, x^{(m)}\}$, where $x^{(i)} \in \mathbb{R}^N$, the cost function to minimized can be expressed as:

$$J(\theta) = \frac{1}{m} \sum_{x \in X} \|\hat{x} - x\|^2 + \beta \sum_{j=1}^{K} \mathrm{KL}(\rho \parallel \hat{\rho}_j) \tag{1}$$

where θ is parameter set $\{W^{(1)}, b^{(1)}, W^{(2)}, b^{(2)}\}$, $\hat{\rho}_j$ is the average activation of hidden unit j, KL is the KullbackLeibler (KL) divergence (also called relative entropy) which is defined as $\mathrm{KL}(\rho \parallel \hat{\rho}_j) = \rho \log \frac{\rho}{\hat{\rho}_j} + (1 - \rho) \log \frac{1-\rho}{1-\hat{\rho}_j}$. The first term is the mean square error term, the second term is the sparsity penalty term, when minimize $J(\theta)$, the first term is used to reduce the information loss of reconstruction and the second term makes the average activation at a desired level.

There are several hyper-parameters in a SAE: the number of hidden units K, the sparsity parameter ρ and the sparsity coefficient β. The commonly used method of choosing them is by cross validation accuracy, while this method depends on the labeled training set and the classification algorithm. In order to avoid these limitations, we found that we can choose the parameters by the reconstruction error.

3.2   DAE Based Unsupervised Feature Learning

The denoising auto-encoder (DAE) is an extension of a ordinary auto-encoder and it was introduced as a building block for deep networks in [7]. In order to learn more robust features, the DAE try to reconstruct the clean (denoised) input from a corrupted version of it, which is generated by adding noise. There are two types of noise commonly used in DAE, following [8]:

- Gaussian noise: $GN(x) = x + \mathcal{N}(0, \sigma^2 I)$, where $\sigma^2 I$ is the covariance matrix of the Gaussian noise added.
- Masking noise: for each input x, randomly choose a percentage p of elements and set them to be 0.

The hyper-parameters of DAE are the number of hidden units K and the noise. We can choose them by cross-validation.

3.3 PCA Based Unsupervised Feature Learning

Principal Components Analysis (PCA) is a dimensionality reduction method that uses an orthogonal transformation to convert a number of (possibly) correlated variables into a (smaller) number of uncorrelated variables, while the loss of information is minimized. PCA is also an approach of unsupervised feature learning since it automatically discovers compact and meaningful representation of raw data without relying on domain knowledge [2]. Essentially, PCA learns a linear transformation $f(x) = W^T x$ to map raw data to features, which is the same as an auto-encoder with a linear activation function.

4 Experiments and Results

To evaluate the proposed methods for HAR we conducted a number of experiments on a public HAR dataset [9] from the UCI repository [10]. We also compared our method with traditional features and the monolithic way of unsupervised feature learning which combines the data of all channels in an analysis window as the input of a unsupervised feature learning method. To make them more comparable, we carefully selected the hyper-parameters of unsupervised feature learning models and the number of traditional features. After obtaining the features we train the classifier in the training set and report the classification accuracy of the test set as the results.

To fully exploit the potential of features, we use well-known and state-of-the-art Support Vector Machine (SVM) for classification. The SVM we used is extended by one-versus-all approach for multi-class classification and adopts a standard radial Gaussian kernel. To avoid the influence of the hyper-parameters of SVM on the results we select the soft margin parameter C and the Gaussian kernel parameter γ by grid search using 10-fold cross validation.

4.1 Dataset and Preprocessing

The HAR dataset [9] was created by a group of 30 volunteers with ages ranging from 19 to 48 years. Each person performed six activities, including *standing, sitting, laying, walking, walking upstairs* and *walking downstairs*, while wearing a Samsung Galaxy S II smartphone on the waist. The signals of 3-axial linear acceleration and 3-axial angular velocity were captured at a fixed rate of 50Hz using the smartphone inertial accelerometer and gyroscope. The sensor signals were preprocessed by using noise filter and then segmented by applying fixed-width sliding windows of 2.56 sec and 50% overlap (128 readings per window). A feature vector of 561 elements was extracted by using many complicated signal processing techniques from three windows of accelerometer signals and three windows of gyroscope signals. The dataset were divided into the training set and the test set by randomly selecting 70% volunteers, the training set includes 7352 examples and the test set 2947 examples.

We did not use the original features of the HAR dataset, only used the raw sensor data. We view the three axes data as three channels (i.e. x, y, z), and add the resultant of three axes as another channel (r). For the sake of outliers and using sigmoid function in auto-encoders, we truncated the value of each channel to the middle 99.5% of it and then scaled it to [0.1, 0.9].

4.2 Features

We not only leaned features using unsupervised learning techniques, including SAE, DAE and PCA, but also extracted statistical features and features from FFT coefficients as comparisons. For each unsupervised learning technique, we adopted two ways of learning features: channel-wise and monolithic. To allow comparison of different feature representations we ensured that each of them had approximately the same dimensions.

SAE. For the channel-wise method, we stacked two SAEs with the sigmoid function as the activation in each channel, the first SAE has 64 hidden units, the second SAE using the hidden layer output of the first SAE as the input has 16 hidden units. The feature mapping is through a 128-64-16 fully connected neural network in each channel. There are 8 channels, so finally 128 features were extracted in an example. For the monolithic way, we also stacked two SAEs using the same activation function, the structure of the first SAE is 1024-512-1024, the second 512-128-512. Eventually we also extracted 128 features.

The remaining hyper-parameters of SAE are the sparsity parameter ρ and the sparsity coefficient β. We performed a grid search for $\rho \in [0.05, 0.4]$ in steps of 0.05 and $\beta \in \{1.0, 3.0, 9.0\}$ to find the a combination of ρ and β that gave the lowest reconstruction error for all sensor data. In grid search, we trained all SAEs for 100 epochs using batch gradient descent algorithm implemented in Pylearn2 [11]. After selecting ρ and β, we trained SAEs for 200 epochs to get the last feature mapping network.

DAE. For the two ways of feature learning, we used all the same configuration including the activation function, architecture and training method in DAEs as in SAEs, except the hyper-parameters and the selection criteria in grid search. We performed a grid search for noise type and its parameter: the standard deviation σ of Gaussian Noise was set to 0.1 and the percentage p of Masking Noise [0.5, 0.20] in steps of 0.05. As in DAEs the input is corrupted, it is not appropriate to select hyper-parameters by the lowest reconstruction error. When performing the grid search, the noise type and its parameter were chosen by 10-fold cross validation in the training set.

PCA. We also extracted 128 features for the two ways of unsupervised feature leaning. For the channel-wise method we applied PCA to reduce 128 dimensions of sensor data to 16 dimensions in each channel. For another way, we used PCA learned a transformation from the 1024 dimensions data of all channels to the 128 dimensions feature vector.

Statistical Metrics. Statistical measures are probably the most common feature representation used in HAR. We computed mean, standard deviation, energy, and entropy for each channel. Together with six correlation coefficients of four channels accelerometer data and six of gyroscope data this yielded a 44 dimensions feature vector.

FFT Coefficients. Another commonly used features are extracted from Fast Fourier Transform coefficients . For each channel we performed FFT to get FFT coefficients then found 8 dominant frequencies from the first half of coefficients spectrum as the spectrum is symmetric. A dominant frequency needs record the value and the magnitude, so we also extracted 16 features per channel. After concatenating the 16 features of all channels, a 128 dimensions feature representation was generated.

4.3   Results

After extracting the features, we performed classification experiments using library LibSVM (version 3.18) [12]. We firstly applied a grid search on $C \in \{2^{-5}, 2^{-4}, \dots, 2^5\}$ and $\gamma \in \{2^{-4}, 2^{-3}, \dots, 2^4\}$ using 10-fold cross validation, and then used the optimal hyper-parameters to train SVM in the training set. We finally reported the accuracy in the test set as the classification performance. We also report the cross validation (CV) accuracy, because the difference between CV accuracy and test accuracy can also reflect some degree of over-fitting.

Table 1. Results

Features	CV Accuracy	Test Accuracy
SAEs-c	95.24%	**92.16%**
SAEs-m	94.91%	83.81%
DAEs-c	94.86%	90.50%
DAEs-m	94.21%	82.78%
PCA-c	95.42%	91.82%
PCA-m	94.71%	89.79%
Stat	97.55%	88.56%
FFT	92.86%	85.14%

In table 1, we present the results of CV accuracy and the test accuracy for different features used in this paper. We notes the channel-wise way of unsupervised features learning using SAEs as SAEs-c and the monolithic way of unsupervised feature learning using SAEs as SAEs-m in table 1, similarly for DAEs and PCA.

The results clearly show that the learned features of our framework outperform traditional features including statistical measures (Stat) and FFT. The channel-wise ways of unsupervised learning has a obvious advantages comparing with the monolithic way, this demonstrates the effectiveness of the proposed methods. Interestingly, SAEs only have slightly better than PCA through the channel-wise way. We reason that this is mainly due to the nonlinear nature of SAE, which is more flexible and needs more data to fit the model. The differences between CV accuracies and test accuracies show that traditional features suffer from serious over-fitting, while learned features relatively small.

5   Conclusion

In this paper we present a channel-wise feature extraction framework which exploits different unsupervised feature learning techniques to learning useful feature representation from accelerometer and gyroscope sensor data for HAR. We compare our methods to traditional features and to the monolithic way of feature learning. The evaluation was performed on a public HAR dataset. The channel-wise feature extraction using SAEs outperform other methods implemented in this paper. For SAEs we selected hyper-parameters by the reconstruction error, which is proved a effective method. Thus, our framework combining with unsupervised feature learning techniques especially SAE, has great potential for application in human activity recognition.

Acknowledgments. This work is partially supported by National Science Foundation of China (no. 61202117, 91118008).

References

1. Incel, O.D., Kose, M., Ersoy, C.: A review and taxonomy of activity recognition on mobile phones. BioNanoScience 3(2), 145–171 (2013)
2. Plötz, T., Hammerla, N.Y., Olivier, P.: Feature learning for activity recognition in ubiquitous computing. In: Proceedings of the Twenty-second International Joint Conference on Artificial Intelligence, vol. 2, pp. 1729–1734. AAAI Press (2011)
3. Hoyer, P.O.: Non-negative sparse coding. In: Proceedings of the 2002 12th IEEE Workshop on Neural Networks for Signal Processing, pp. 557–565. IEEE (2002)
4. Vollmer, C., Gross, H.-M., Eggert, J.P.: Learning features for activity recognition with shift-invariant sparse coding. In: Mladenov, V., Koprinkova-Hristova, P., Palm, G., Villa, A.E.P., Appollini, B., Kasabov, N. (eds.) ICANN 2013. LNCS, vol. 8131, pp. 367–374. Springer, Heidelberg (2013)
5. Bhattacharya, S., Nurmi, P., Hammerla, N., Plötz, T.: Using unlabeled data in a sparse-coding framework for human activity recognition. arXiv preprint arXiv:1312.6995 (2013)
6. Ng, A.: Sparse autoencoder. CS294A Lecture notes (2011),
   http://web.stanford.edu/class/cs294a/sparseAutoencoder2011.pdf
7. Vincent, P., Larochelle, H., Bengio, Y., Manzagol, P.A.: Extracting and composing robust features with denoising autoencoders. In: Proceedings of the 25th International Conference on Machine Learning, pp. 1096–1103. ACM (2008)

8. Vincent, P., Larochelle, H., Lajoie, I., Bengio, Y., Manzagol, P.A.: Stacked denoising autoencoders: Learning useful representations in a deep network with a local denoising criterion. The Journal of Machine Learning Research 11, 3371–3408 (2010)

9. Anguita, D., Ghio, A., Oneto, L., Parra, X., Reyes-Ortiz, J.L.: A public domain dataset for human activity recognition using smartphones. In: European Symposium on Artificial Neural Networks, Computational Intelligence and Machine Learning, ESANN (2013)

10. Bache, K., Lichman, M.: UCI machine learning repository (2013), http://archive.ics.uci.edu/ml

11. Goodfellow, I.J., Warde-Farley, D., Lamblin, P., Dumoulin, V., Mirza, M., Pascanu, R., Bergstra, J., Bastien, F., Bengio, Y.: Pylearn2: a machine learning research library. arXiv preprint arXiv:1308.4214 (2013)

12. Chang, C.C., Lin, C.J.: Libsvm: a library for support vector machines. ACM Transactions on Intelligent Systems and Technology (TIST) 2(3), 27 (2011)

Influence of Weak Labels for Emotion Recognition of Tweets

Olivier Janssens, Steven Verstockt, Erik Mannens,
Sofie Van Hoecke, and Rik Van de Walle

Multimedia Lab – Ghent University – iMinds,
Gaston Crommenlaan 8, 9050 Ledeberg-Ghent, Belgium
{odjansse.janssens,steven.verstockt,erik.mannens
sofie.vanhoecke,rik.vandewalle}@ugent.be
http://www.mmlab.be/

Abstract. Research on emotion recognition of tweets focuses on feature engineering or algorithm design, while dataset labels are barely questioned. Datasets of tweets are often labelled manually or via crowdsourcing, which results in strong labels. These methods are time intensive and can be expensive. Alternatively, tweet hashtags can be used as free, inexpensive weak labels. This paper investigates the impact of using weak labels compared to strong labels. The study uses two label sets for a corpus of tweets. The weakly annotated label set is created employing the hashtags of the tweets, while the strong label set is created by the use of crowdsourcing. Both label sets are used separately as input for five classification algorithms to determine the classification performance of the weak labels. The results indicate only a 9.25% decrease in f1-score when using weak labels. This performance decrease does not outweigh the benefits of having free labels.

Keywords: Emotion recognition, Twitter, Annotation, Crowdsourcing.

1 Introduction

Emotions influence everything in our life, e.g. relationships and decision making and are therefore analysed in many research projects. The automatic detection of emotions in text allows for a broad range of applications, such as forecasting movie box office revenues during the opening weekend [14], 3D facial expression rendering based on recognized emotions in text [5], stock prediction [4], and helping to understand consumer views towards a product [3].

The main goals of emotion recognition is to get implicit feedback about certain events, people's actions, services and products. For example, to see if a student has trouble learning, or finds the course too easy, it is possible to monitor the user's emotions while working on an e-learning platform. It is also possible to monitor the user's emotion during a game, allowing the difficulty to be adapted automatically to the user. It is also possible to monitor the user's opinion about products via social media in order to enhance recommender engine user profiles, so that a better recommender engine can be built.

R. Prasath et al. (Eds.): MIKE 2014, LNAI 8891, pp. 108–118, 2014.

Emotion recognition is either done with supervised machine learning methods, where a labelled dataset is required, or unsupervised machine learning methods, where no labels for the data are provided. Research shows that supervised methods tend to outperform unsupervised methods [10]. In order to gather labelled data, several methods are employed:

A first method is a strong labelling method because it consists of manually labelling the collected corpus. An example of this can be found in the work of Roberts et al. [13], where 7,000 tweets are manually labelled. This method has several disadvantages. First of all, annotating a large dataset takes a lot of time and effort. Secondly, because annotating data manually takes a lot of time, often the dataset will be small. Finally, since there is little data, mistakes will have a bigger impact on the classifiers' performance.

To create bigger datasets crowdsourcing can be used, which is also a strong labelling method and is employed regularly in machine learning research. Crowdsourcing services allow for data to be manually labelled by a large group of people. Crowdsourcing is an attractive solution since it can provide an annotated dataset possibly cheap and easy. Different methods exist to crowdsource labels. One can build his own application and attract annotators, but it is also possible to use existing services such as Mechanical Turk or Crowdflower.

Another frequently used labelling method is a weak labelling method and consist of using the emotion hashtag, or an emotion linked to the hashtag, as label. This method requires the collection of tweets based on their emotion hashtags which are added by the author of the tweet. For example, in the work of Mohammad [10], the 6 basic emotions of Ekman [7] are used as hashtags (#anger, #disgust, #fear, #happy, #sadness, and #surprise) to query the twitter API to create a dataset. This method allows for a large dataset to be created without any costs in a short amount of time, though the assignment of these weak labels can be questioned [2]. For example, the hashtag can be used to indicate sarcasm, resulting in a hashtag that does not correspond with the true underlying emotion. Another problem when using a weak labelling method for tweets is that the hashtags need to be removed in order to create the ground truth. As a result the tweet might lose its emotional value. For example the tweet "*No modern family tonight #sad*" has no underlying emotion when the hashtag is removed. However, if the hashtags are kept in the dataset, data leakage [9] occurs which results in an unrealistically good classifier.

In this paper the classification performance impact of weakly labelled emotion labels for emotion recognition is investigated. The analysis is done on a corpus of tweets for which two label sets are constructed. The weak label set is annotated by the emotion linked with the author's emotional hashtags. The strong label set is constructed by the use of crowdsourcing using Crowdflower. Manual annotation is not included in this analysis because the corpus consists of 341,931 tweets, which is too large for manual annotation.

The remainder of this paper is as follows. Section 2 presents the related work in text based emotion recognition. Section 3 elaborates on the dataset creation, the preprocessing methods and the classification algorithms. Next, Section 4 shows the evaluation results for the different classification algorithms. Finally, section 5 lists the conclusions and the future work.

2　Related Work on Text Based Emotion Recognition

Current state-of-the-art emotion recognition techniques use lexicon based methods in order to classify text [10],[16]. Lexicon based techniques use dictionaries of words with pre-computed scores expressing the emotional value of the word. The advantage of this technique is that there is a known relation between the word and the emotion. However, the disadvantage of this method is that often no new features are extracted from the relevant dataset such as emoticons [15]. Opposed to lexicon based techniques there are learning based techniques that require the creation of a model by training a classifier with labelled examples [8]. This means that the vocabulary for the feature vectors are built based on the dataset, resulting in a vocabulary, which is more suitable for tweets.

3　Methodology

In order to study the impact of weakly labelled data on text based emotion recognition, first, a corpus together with two label sets is created. Second, the corpus is preprocessed so that it can be used by the classification algorithms. Finally, the corpus, together with one of the two label sets, is used as input for five classification algorithms for which several objective metrics are computed.

3.1　Dataset

The corpus consists of 341,931 English tweets, i.e. 171,485 tweets for the test set and 170,446 tweets for the training set. Every tweet is annotated with one of these emotions: 'anger', 'fear', 'joy', 'love', 'sadness', 'surprise' or 'thankfulness'. The dataset is a subset of the dataset collected by W. Wang et al. [17] where word lists for the emotions were constructed to query the Twitter Streaming API to collect tweets.

The strong label set for this corpus is created by Crowdflower. The job in Crowdflower is designed so that every tweet is labelled by at least 3 different people. Each person contributing to the job is given 10 test cases with known ground truth (which is not shown to the contributor) for training purposes before they could start the annotation job. Based on these test cases a trust score, which is the accuracy score achieved by a contributor on the set of test cases, is calculated. The trust score allows for Crowdflower to filter out bad annotators or automated programs to ensure that a high quality set of labels is created.

The weak label set for the corpus of tweets was collected by extracting the emotional hashtag from the tweet. For example the tweet "I hate when my mom compares me to my friends #annoying" is labelled as 'anger' since it contains the "#annoying" hashtag which can be found in the word list belonging to 'anger'. For more information on the dataset, such as the choice of emotions or how the dataset was split into a training set and test set, we refer to the work of W. Wang et al. [17]. Tweets with less than 5 words or URLs are discarded. Also the emotion hashtags themselves are removed in order to prevent data leakage.

Because the weak label set and strong label set come from different sources, the distribution of labels are different as can be seen in Figure 1a and 1b. Both label sets were split according to the training and test set of the corpus.

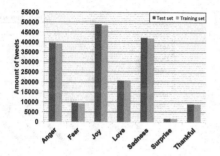 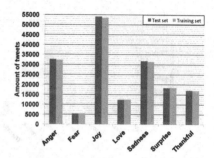

(a) Distribution of the weak labels constructed by the emotional hashtags

(b) Distribution of the strong labels constructed by crowdsourcing

Fig. 1. Label distributions

The training set is representative for the test set for both label sets as they have a similar distribution. However, there are some differences between the distribution of the strong labels and weak labels. For example, the weak label set has less tweets in the category 'surprise' compared to the strong label set. There are also more tweets associated with 'thankfulness' in the strong label set compared to the weak label set.

In order to improve the quality of the strong label set further, an additional filtering process is applied. This process uses the confidence score provided by Crowdflower. The confidence score describes the agreement between annotators and is calculated as follows: different users annotate the same tweet with one of the provided labels. In order to get a confidence score, all the trust scores of the users who voted for the same label for a tweet are added together. This number is then divided by the sum of all the trust scores of all the user who annotated that tweet. As a result the confidence score is always between 0 and 1.

Because of this confidence score, it is possible to filter out tweets and their respective labels which have no underlying emotion or are ambiguous. In the test set part of the corpus, only tweets are kept with a confidence score larger than 66%, which indicates that 2 of the 3 annotators agree on the emotion of a tweet. For the training set, tests with different ranges of confidences scores were done e.g.: >50%; >70%; >90%. The best result is achieved when tweets with a confidence score equal to 100% are kept in the training set because this results in less noise in the features as only the best features remain.

The resulting distributions of the reduced label sets can be found in Figures 2a and 2b. As result of the filter process based on the confidence score, the reduced training set (48,985 tweets) is smaller than the reduced test set (114,273 tweets) Overall it can still be stated that for both the reduced strong label set and the reduced weak label set, the test set distribution is similar to the training set distribution, although downscaled.

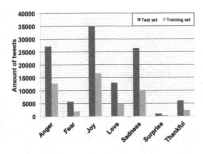
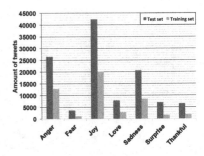

(a) Distribution of weak labels after confidence score based thresholding

(b) Distribution of weak labels after confidence score based thresholding

Fig. 2. Label distributions of reduced sets

Another insight is given by Table 1. Since there is one corpus and two sets of labels, it is possible to determine how the label sets differ. The table illustrates how the labels from the weak label set are redistributed in the strong label set. For example, from all the tweets labelled as 'joy' in the weak label set; 70.80% are labelled as 'joy' in the strong label set. Some other notable facts are:

Table 1. Redistribution of the weak label set to the strong label set

		From weak labels						
		Joy	Fear	Sadness	Thankfulness	Anger	Surprise	Love
To strong labels	Joy	70.80%	5.78%	3.53%	22.89%	1.62%	16.73%	19.63%
	Fear	3.53%	69.77%	2.94%	1.66%	1.37%	5.53%	1.32%
	Sadness	5.61%	12.97%	74.33%	4.72%	16.06%	40.84%	6.12%
	Thankfulness	4.31%	0.59%	0.56%	53.22%	0.20%	1.36%	6.59%
	Anger	2.74%	9.13%	15.06%	2.79%	79.28%	20.40%	2.21%
	Surprise	0.70%	0.21%	0.23%	0.35%	0.18%	8.12%	0.30%
	Love	12.31%	1.56%	3.34%	14.37%	1.30%	7.03%	63.83%
	Sum	100%	100%	100%	100%	100%	100%	100%

- A large part (12.31%) of tweets labelled as 'joy' according to the hashtag, are labelled as 'love' by crowdsourcing. Conversely, a large part (19.63%) of tweets labelled as 'love' according to the hashtag, are labelled as 'joy' by crowdsourcing.

- A large part (15.06%) of tweets labelled as 'sadness' according to the hashtag, are labelled as 'anger' by crowdsourcing. Conversely, a large part (16.06%) of tweets labelled as 'anger' according to the hashtag, are labelled as 'sadness' by crowdsourcing.
- A large part (12.97%) of tweets labelled as 'fear' according to the hashtag, are labelled as 'sadness' by crowdsourcing.
- Almost half of all tweets labelled as 'thankfulness' according to hashtag, are labelled as other emotions by crowdsourcing, with 'joy' being the largest one.
- Tweets annotated with the emotion 'surprise' according to the hashtag are distributed amongst the other emotions by crowdsourcing. Only 8.12% receives the 'surprise' label and a large part is labelled as 'sadness' (40.84%). Since there is only a 8.12% overlap in the label sets for the category 'surprise', it can be stated that the category 'surprise' in the weak label set contains of a lot of noise. Because the tweets were initially collected using words in lists linked to the categories [17], it is our belief that the noise stems from the chosen words in the word list for 'surprise'. For example the tweet "Got to see him today #unexpected" has the weak label 'surprise' since it was collected by the hashtag "#unexpected" which is listed in the 'surprise' word list. Nevertheless the tweet is labelled with the strong label 'joy' by crowdsourcing. Another example is the tweet "5-1 Edmonton over Blackhawks nearing end of the 1st Period #ASTONISHED". This tweet received the weak label 'surprise' because "#astonished" is present in the word list for 'surprise', nevertheless it is labelled as 'sadness' by crowdsourcing. Because of these observation it is our belief that the collection of the tweets based on certain hasthags present in the word list for the 'surprise' don't always return unambiguous tweets.

To summarize, 31.26% of the tweets are labelled differently by crowdsourcing compared to the weak label set. Generally, tweets in both label sets are labelled with the same valence, i.e., negative valence or positive valence. 'Joy', 'love', 'thankfulness' have a positive connotation, 'fear', 'sadness', 'anger' have a negative connotation and 'surprise' is neutral. 89.72% of the tweets with a positive connotation according to the weak label set also have a positive connotation in the strong label set. 93.83% of the tweets with a negative connotation according to the weak label set also have a negative connotation in the strong label set. These numbers indicate that only a small fraction of the labels differ when it comes to valence. The differences between the label sets will have an impact on the classification results, thus supporting the research question of this paper to investigate the impact of weak emotions labels on the classification results compared to strong emotion labels.

3.2   Preprocessing

In this paper a learning based technique is used to classify the tweets. The corpus is preprocessed first in order to transform it into feature vectors. Preprocessing consists of stemming, which reduces every word of every tweet to their stem if possible. An example: kneeling, kneeled, kneels are all reduced to kneel. As the feature space, when using learning based methods, can grow large very quickly, reducing conjugations of the same word to their stem reduces this feature space. In this paper, the Porter stemmer [18] is used because it outperforms other widely used stemmers such as the Paice-Husk [11] and Lovins [1] stemmers.

The next step transforms the tweets to feature vectors.

3.3   Feature Extraction Methods

The feature extraction methods transform words and sentences into a numerical representation which can be used by the classification algorithms. In this research, the combination of N-grams and TF-IDF feature extraction methods is used. As will be motivated below, this will preserve syntactic patterns in text and help solve class imbalance respectively.

N-gram Features: In the field of computational linguistics, an N-gram is a contiguous sequence of N items from a given sequence of text. An N-gram of size 1 is referred to as a "unigram", size 2 is a "bigram", size 3 is a "trigram". An N-gram can be any combination of letters, words or base pairs according to the application. In this paper, a combination of 1,2 and 3 grams of words is used and passed to the Term Frequency-Inverse Document Frequency algorithm.

Term Frequency-Inverse Document Frequency: One of the broadly used feature representation methods is the bag of words representation together with word occurrence [6], equalizing a feature value to the number of times it occurs. This method is fairly straightforward and easy, though it has a major downside as longer documents have higher average count values. In this paper, the tweets are transformed to feature vectors by using the bag of word representation together with word occurrence. The number of occurrences of each word are then normalized to create the so called Term Frequencies. Additionally, weights for words that occur in many documents are downscaled making them less dominant than those that occur only in a small portion of the corpus. The combination of term frequencies and downscaling weights of dominant words is called Term Frequency-Inverse Document Frequency (TF-IDF).

By composing the vocabulary for the feature vectors based on the dataset, the feature vectors will be very high dimensional. In this case the feature vectors consist of 48,036 features. However by combining TF-IDF and the N-grams method, a feature vector will be very sparse. This can be beneficial if the used algorithms can work with sparse data.

3.4   Classification Algorithms

At this point the tweets have been transformed to feature vectors. In this step the corpus and label sets are used as input for various machine learning algorithms. In this paper, five different classification algorithms are compared to ensure objective determination of the impact of the different label sets for the corpus.

The algorithms compared here are the ones frequently used in text classification as they deliver good results and work very fast. [12]

1. **SGD:** A linear classifier which uses stochastic gradient descent with the modified huber loss. SGD is a method where the gradient of the loss function is estimated each sample at a time. This means that not all the samples have to be loaded into the memory at once. This results in an algorithm that can be used on large datasets that normally do not fit in the memory. The choice of the loss function influences how the data will be handled. For example the modified huber loss reduces the effect of the outliers much more than the log loss function. Additionally, L2 regularization is added to reduce overfitting.

2. **SVM:** A linear support vector machine from the LIBLINEAR library. This SVM is optimised to work with datasets with a large number of samples.

3. **MNB:** Multinomial Naive Bayes, the Naive Bayes algorithm implemented for data where the features are assumed to have a multinomial distribution.

4. **NC:** Nearest Centroid classifier, which is known as the Rocchio classifier when using TF-IDF feature vectors for text classification. In a nearest centroid classifier, every class is represented by a centroid and every test sample is classified to the class with the nearest centroid.

5. **Ridge** A linear classifier that uses regularized least-squares.

Since linear classifiers are designed for binary classification problems, the one versus all method is used to combine them for the multi-class problem presented here. The classification results can be found in the section below.

4   Results

The results are subdivided in 2 subsections. The first subsection elaborates on the classification results by the comparison of different objective metrics. The second subsection discusses the confusion matrices of the best classifier.

4.1   Classification Metrics

For the comparison of the five classification algorithms, three metrics are compared, namely weighted precision, weighted recall and weighted f1-score. The most important metric here is the weighted f1-score because it incorporates the recall and the precision score, and a good classifier will maximize both. Since the classes are highly imbalanced in both label sets, the accuracy score is not included. For example, if the classifier labels all the samples as the majority class, a good accuracy score would still be achieved, though in fact the classifier was not able to learn the underlying pattern of the data.

The results can be found in Figure 3a, 3b and 3c respectively. For almost every algorithm and metric, the tests with the strong labels delivers better results. The best result for the weighted f1-metric is given by the SGD algorithm with a modified huber loss where a difference of 6.53% between the results using the weakly labelled label set and the strongly labelled label set is noticed. By taking

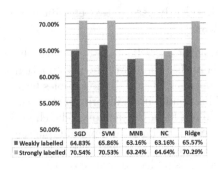

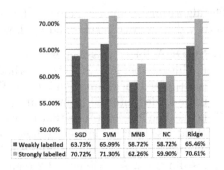

(a) Weighted precision   (b) Weighted recall

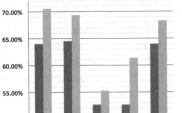

(c) Weighted f1-score

Fig. 3. Classification metrics comparison of the five classification algorithms

the strong labels as the ground truth of the dataset, a decrease of 9.25% of f1-score occurs when using weakly labelled data.

4.2   Confusion Matrices

To see if the algorithms confuse any emotions, which often is the case when there is a significant imbalance in the label sets, confusion matrices of the SGD classifier for both label sets tests are presented in Figure 4a and 4b.

Both classifiers display very little confusion, this can be attributed to the fact that tweets with possible ambiguity and/or which have no underlying emotion are filtered out using the confidence score.

In our previous work [8], a dataset of 498,885 tweets, annotated with the emotion hashtags provided by the authors of the tweets was used. In that case, no confidence score could be calculated to filter out tweets. The results showed that the classifiers confused some emotions, as it was possible that tweets were not annotated with the correct emotion, or did not have an underlying emotion at all, or were ambiguous. The classifiers of this paper do not show this shortcoming.

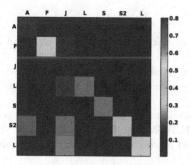 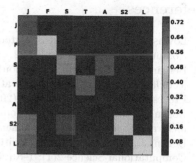

(a) SGD classifier with modified huber loss when using the stong labels. A='anger', F='fear', J='joy',
L='love', S='sadness', S2='surprise',
T='thankfulness'

(b) SGD classifier with modified huber loss when using the weak labels. A='anger', F='fear', J='joy',
L='love', S='sadness', S2='surprise',
T='thankfulness'

Fig. 4. Confusion matrices

5 Conclusion and Future Work

In this paper the impact of weak emotion labels on classification results is studied
by comparing the classification results of five classification algorithms using both
a weak label set and a strong label set. Both label sets are created for the same
corpus of tweets. However one set is created by using the emotion hashtag of
the tweet, and the other set is constructed by a crowdsourcing service. A filter
process is applied on the corpus to eliminate tweets and their corresponding
labels in both label sets which have a low confidence.

The results show that, when using weak labels, there is only a 9.25 % decrease
in f1-score compared to the results when using strong labels. This disadvantage
does not outweigh the benefits of weak labels, i.e. it is available free of charge
and requires almost no extra work to collect. Also, since hashtags and tweets
are published together, a continues stream of labelled data is created. This is
useful for online learning algorithms, such as SGD, since it gives access to a vast
amount of labelled data, which can result in improved classification results [17].

Also, it should be noted that emotion recognition of tweets remains a difficult
task. The best weighted f1-score for the strong labels is 70,56%, leaving room
for improvement.

Future work will consist of exploring new methods to create weak emotion
label sets that better approximate the results of a strong emotion label sets. A
possible improvement is to gathering tweets based on a combination of words
related to an emotion instead of just the hashtag.

References

1. Lovins, J.B.: Development of a stemming algorithm. Mechanical Translation and Computational Linguistics 11, 22–31 (1968)
2. Bergamo, A., Torresani, L.: Exploiting weakly-labeled web images to improve object classification: a domain adaptation approach. Advances in Neural Information Processing Systems 23, 181–189 (2010)
3. Bollen, J., Mao, H., Pepe, A.: Modeling public mood and emotion: Twitter sentiment and socio-economic phenomena. In: ICWSM, pp. 450–453 (2011)
4. Bollen, J., Mao, H., Zeng, X.: Twitter mood predicts the stock market. Journal of Computational Science 2(1), 1–8 (2011)
5. Calix, R.A., Mallepudi, S.A., Chen, B.C.B., Knapp, G.M.: Emotion Recognition in Text for 3-D Facial Expression Rendering. IEEE Transactions on Multimedia 12(6), 544–551 (2010)
6. Chaffar, S., Inkpen, D.: Using a heterogeneous dataset for emotion analysis in text. In: Butz, C., Lingras, P. (eds.) Canadian AI 2011. LNCS, vol. 6657, pp. 62–67. Springer, Heidelberg (2011)
7. Ekman, P.: Basic emotions. Handbook of Cognition and Emotion, vol. 98. John Wiley & Sons (1999)
8. Janssens, O., Slembrouck, M., Verstockt, S., Hoecke, S.V., Walle, R.V.D.: Real-time Emotion Classification of Tweets. In: IEEE/ACM International Conference on Advances in Social Networks Analysis and Mining, pp. 1430–1431 (2013)
9. Kaufman, S., Rosset, S.: Leakage in data mining: Formulation, detection, and avoidance. In: 17th ACM SIGKDD International Conference on Knowledge Discovery and Data Mining, pp. 556–563 (2012)
10. Mohammad, S.: #Emotional Tweets. In: First Joint Conference on Lexical and Computational Semantics, pp. 246–255. Association for Computational Linguistics, Montréal (2012)
11. Paice, C., Husk, G.: Another Stemmer. ACM SIGIR Forum 24, 56–61 (1990)
12. Pedregosa, F., Varoquaux, G., Gramfort, A., Michel, V., Thirion, B., Grisel, O., Blondel, M., Prettenhofer, P., Weiss, R., Dubourg, V., Vanderplas, J., Passos, A., Cournapeau, D., Brucher, M., Perrot, M., Duchesnay, E.: Scikit-learn: Machine learning in Python. Journal of Machine Learning Research 12, 2825–2830 (2011)
13. Roberts, K., Roach, M.A., Johnson, J.: EmpaTweet: Annotating and Detecting Emotions on Twitter. In: LREC, pp. 3806–3813 (2012)
14. Rui, H., Whinston, A.: Designing a social-broadcasting-based business intelligence system. ACM Transactions on Management Information Systems 2(4) (2011)
15. Suttles, J., Ide, N.: Distant supervision for emotion classification with discrete binary values. In: Gelbukh, A. (ed.) CICLing 2013, Part II. LNCS, vol. 7817, pp. 121–136. Springer, Heidelberg (2013)
16. Wang, A., Hoang, C., Kan, M.: Perspectives on crowdsourcing annotations for natural language processing. Language Resources and Evaluation 47(1), 9–31 (2013)
17. Wang, W., Chen, L., Thirunarayan, K., Sheth, A.P.: Harnessing Twitter "Big Data" for Automatic Emotion Identification. In: 2012 International Conference on Privacy, Security, Risk and Trust and 2012 International Confernece on Social Computing, pp. 587–592 (2012)
18. Willett, P.: The Porter stemming algorithm: then and now (2006), http://dx.doi.org/10.1108/00330330610681295

Focused Information Retrieval & English Language Instruction: A New Text Complexity Algorithm for Automatic Text Classification

Trisevgeni Liontou

Greek Ministry of Education
tliontou@enl.uoa.gr

Abstract. The purpose of the present study was to delineate a range of linguistic features that characterize the English reading texts used at the B2 (Independent User) and C1 (Advanced User) level of the Greek State Certificate of English Language Proficiency (KPG) exams in order to better define text complexity per level of competence. The main outcome of the research was the L.A.S.T. Text Difficulty Index that makes possible the automatic classification of B2 and C1 English reading texts based on four in-depth linguistic features, i.e. *lexical density, syntactic structure similarity, tokens per word family* and *academic vocabulary*. Given that the predictive accuracy of the formula has reached 80% on a new set of reading comprehension texts with 32 out of the 40 new texts assigned to similar levels by both raters, the practical usefulness of the index might extend to EFL testers and materials writers, who are in constant need of calibrated texts.

Keywords: Readability, Text complexity, Automatic text analysis, Text classification.

1 Introduction

The present study builds on earlier findings of research on reading assessment, according to which many text variables such as content, lexis and structure can have an impact on either the reading process or product and need to be taken into account during test design and validation (56). In fact, although a lot of research has been conducted in the field of second language acquisition with specific reference to ways of reading and text processing strategies, Alderson (2000: 104) stressed language testers' lack of success "to clearly define what sort of text a learner of a given level of language ability might be expected to be able to read or define text difficulty in terms of what level of language ability a reader must have in order to understand a particular text". Such information would be particularly useful in providing empirical justification for the kinds of reading texts test-takers sitting for various language exams are expected to process, which to date have been arrived at mainly intuitively by various exam systems (1, 3, 26, 56).

R. Prasath et al. (Eds.): MIKE 2014, LNAI 8891, pp. 119–134, 2014.
© Springer International Publishing Switzerland 2014

2   Background to the Study

Despite the considerable advances in exploring and understanding the various aspects of L2 reading performance, the available research has been rather unsuccessful in prioritizing those text features that have a direct impact on text complexity and need to be accounted for during the text selection and item design process. Weir (69) acknowledged that, although the Common European Framework of Reference for Languages (CEFR) attempted to describe language proficiency through a group of scales composed of ascending level descriptors, it failed to provide specific guidance as to the topics that might be more or less suitable at any level of language ability, or define text difficulty in terms of text length, content, lexical and syntactic complexity. In fact, according to Weir, the argument that the CEFR is intended to be applicable to a wide range of different languages" offers little comfort to the test writer, who has to select texts or activities uncertain as to the lexical breadth of knowledge required at a particular level within the CEFR" (ibid: 293). Alderson *et al.* (2) also stressed that many of the terms in the CEFR remain undefined and argued that difficulties arise in interpreting it because "it does not contain any guidance, even at a general level, of what might be simple in terms of structures, lexis or any other linguistic level". Therefore, according to Alderson *et al.*, the CEFR would need to be supplemented with lists of grammatical structures and specific lexical items for each language for item writers or item bank compilers to make more use of it.

Test designers' knowledge of the variables that can influence the reading process and product is, thus, in many respects linked to the validity of the reading tests; test designers need to focus on making their test items as relevant as possible to described levels of difficulty on an a priori basis, and further ensure that these are not biased against particular test-takers nor are they affected in an unexpected way by the readability of the text (37). By following such an approach, they will be able to provide evidence that the methods they employ to elicit data are appropriate for the intended purposes, that the procedures used provide stable and consistent data and, consequently, that the interpretations they make of the results are justified, since they are based on a valid and reliable exam system (21). To this end, a number of testing scholars have called for more research in order to enhance our knowledge about the factors contributing to text difficulty and more precisely define it in terms of actual performance on specific text types (c.f. 4, 6, 25, 55, 60, 61, 63, 68). With specific reference to the field of testing, Chalhoub-Deville and Turner (2000: 528-30) stressed the failure of well-established international exam systems to provide adequate documentation regarding how the level of text difficulty is determined, and which processes for text selection are applied, with a view to establishing internal consistency of their tests and equivalence across parallel test forms. According to them (ibid: 528-9) making such information available to the public is mandatory, in order to help all interested parties make informed evaluations of the quality of the tests and their ensuing scores.

3   Aim of the Study

The aim of the present study was to delineate a range of linguistic features that characterize the reading comprehension texts used at the B2 (Independent User) and C1 (Advance User) level of the Greek national exams in English for the State Certificate of Language Proficiency - known with their Greek acronym as KPG exams- in order to better define text readability per level of competence and create a statistical model for assigning levels to texts in accordance with the purposes of the specific exam battery. In order to explore these issues, the following research question was formed:

1.   Is there a specific set of text variables to better predict text difficulty variation between reading texts used at the B2 and C1 KPG English language exams?

KPG is a relatively new multi-level multilingual suite of national language examinations developed by teams of experts from the foreign language departments of the National and Kapodistrian University of Athens and the Aristotle University of Thessaloniki. The exams are administered by the Greek Ministry of Education, making use of the infrastructure available for the Panhellenic university entrance exams. Exams are administered twice a year and, since November 2003, more than 200,000 test-takers have taken part in the English exams. The level of the reading comprehension texts has been broadly defined in the Common Framework of the KPG examinations (2008: 16), according to which the B2 reading comprehension and language awareness paper is designed to test at an Independent User level the test-takers' abilities to understand the main ideas of texts of *average difficulty* on various topics, including abstract ideas or specialized information that requires some technical knowledge, whereas, at the C1 level, reading comprehension tasks are designed to test at an Advanced User level test-takers' abilities to understand texts relatively long and of a *higher level of difficulty*'. Nevertheless, till the time the present study was undertaken, it had not been possible to quantify the occurrence of in-depth lexicogrammatical features that could be more appropriate to the intended audience, i.e. prospective B2 or C1 KPG test-takers. By making use of advanced Computational Linguistics and Machine Learning systems, the current research was, thus, designed to fill this void and further add to our present state of knowledge on EFL text difficulty in general.

4   Methodology

The KPG English Reading Corpus: The text variables identified for analysis in the present research were chosen for both practical and theoretical reasons. First, from the practical standpoint of comparability, it was important to establish whether the presence of particular lexicogrammatical features in the KPG English language reading comprehension texts might introduce construct-relevant variance into test scores. At the same time, given the inherent complexity of the reading process and the fact that our ultimate purpose was the creation of a mathematical model capable of assigning levels to texts in a consistent and reliable way, in accord with the purposes of the KPG language exams in English, it was imperative to include a comprehensive list of text-based indices, in order to minimize the risk of omitting variables that might have contributed to the predictive capacity of the final model. Given that previous research had failed to produce a definite set of quantifiable text variables, no decision was a priori made in terms of their expected significance. All in all, thirty-four B2 reading comprehension texts used between November 2003 and November 2011 examination periods and twenty-nine C1 texts used between April 2005 and May 2012 examination periods were chosen for analysis with regard to 135 text variables (see Appendix 1 for a complete list of text variables). Finally, these two levels of competence were chosen for reasons of practicality since, when the research began, they were the only ones available and had attracted a great number of test-takers.

Automated Text Analysis Tools: Over the last ten years, advances in Computational Linguistics and Machine Learning systems have made it possible to go beyond surface text components, focusing on a wider range of *deep* text features that take into account semantic interpretation and the construction of mental models and can, thus, offer a principled means for test providers and test-takers alike to assess this aspect of test construct validity (27). In the present study *Coh-Metrix 2.1*, *Linguistic Inquiry and Word Count 2007* (LIWC), *VocabProfile 3.0*, *Computerized Language Analysis* (CLAN) suite of programs, *Computerized Propositional Idea Density Rater 3.0* (CPIDR), *Gramulator*, *Stylometrics* and *TextAnalyzer* were used to estimate 135 text variables. To begin with, a great extent of the present study was based on Coh-Metrix 2.1, a freely available web-based tool (available online at http://cohmetrix.memphis.edu/cohmetrixpr/index.html) developed at the Institute of Intelligent Systems of the University of Memphis, that uses lexicons, parts-of-speech classifiers and statistical representations of world knowledge to measure cohesion and text difficulty at deeper language levels (19, 18, 20, 27, 49). VocabProfile 3.0 and its updated BNC version 3.2, two freeware available web-based vocabulary profile programs (available online at http://www.lextutor.ca/vp.), were also used in order to estimate word frequency and obtain lists of word tokens (total number of running words), word types (different word forms) and word families (groups containing different forms of a word) for each text in the corpus (12, 13). Based on Nation's frequency lists, VocabProfile classifies the vocabulary of a text into frequency levels and outputs a profile that describes the lexical content of the text in terms of frequency bands, by showing how much coverage of the text each of the twenty lists accounts for (13, 50, 54).

Propositional idea density was estimated using the Computerized Propositional Idea Density Rater 3.0 (CPIDR), a computer program that determines the propositional idea density (P-density) of an English text on the basis of part-of-speech tags (8). Developed at the Institute for Artificial Intelligence of the University of Georgia, the Computerized Propositional Idea Density Rater is a user-friendly Windows application distributed as open-source freeware through http://www.ai.uga.edu/caspr. Following Kintsch's theory of comprehension and the representation of meaning in memory (1988: 165), CPIDR 3.0 functions based on the idea that propositions correspond roughly to verbs, adjectives, adverbs, prepositions and conjunctions (8, 15, 38, 64, 67). In order to assess lexical diversity, Malvern and Richards' (41) D-formula incorporated into the *vocd* command of the Computerized Language Analysis (CLAN) suite of programs (available online at http://childes.psy.cmu.edu) of the Child Language Data Exchange System (CHILDES) project was used (39, 41, 42). A minimum sample size of 50 words is required for *vocd* to compute a valid D, a measurement based on an analysis of the probability of new vocabulary being introduced into longer and longer samples of speech or writing. Texts with a high D value are expected to be comparatively more difficult to comprehend, because many unique words need to be encoded and integrated with the discourse context. On the other hand, low diversity scores indicate that words are repeated many times in a text, which should generally increase the ease and speed of text processing (39, 42, 43, 47). Malvern and Richards' measure was recently criticized by McCarthy and Jarvis (44) for being sensitive to text length variation. McCarthy and Jarvis concluded their research by questioning whether a single index has the capacity to encompass the whole construct of lexical diversity and urged researchers to select those measures that have proved more effective than others over particular lengths of texts (ibid:483). More recently, McCarthy and Jarvis (45) presented a new index of lexical diversity called the Measure of Textual Lexical Diversity (MTLD), which is calculated as the mean length of word strings that maintain a criterion level of lexical variation in a text and does not appear to vary as a function of text length. Once again, McCarthy and Jarvis concluded their study by advising researchers to consider using not one but three specific indices, namely the MTLD, vocd-D (or its hypergeometric distribution HD-D) and Maas index, when investigating lexical diversity, since each index appears to capture unique lexical information as to the construct under investigation (ibid: 391). Thus, in the present research, all three indices were estimated over available KPG reading comprehension texts, while their impact on text readability was further explored through correlational analyses. In addition to CLAN, the Linguistic Inquiry and Word Count 2007 (LIWC), a text analysis program developed by Francis and Pennebaker at the University of Texas and can count words in psychologically meaningful categories, based on a specially designed dictionary of 4,500 words divided into sub-sections according to the category they define, was used (59, 65). LIWC2007 made it possible to estimate the percent of word units that expressed a variety of psychological processes, such as cognitive, affective, social and perceptual ones, and further explore their impact on text readability. This part of the research should be best viewed as a springboard for an analysis of text features that may affect comprehension in a more subtle way and whose effect on exam performance has to date been largely ignored. LIWC has been validated in a number of health and personality studies and has been found to accurately identify complexity in language use (c.f. 5, 7, 33, 57, 58, 59). The Gramulator, a freely available textual analysis tool developed at the Institute of Intelligent Systems of the University of Memphis (https://umdrive.memphis.edu), was also used in order to identify not predefined constructs but differential linguistic features such as idioms and phrasal verbs within and across our two sets of texts (36, 46, 51, 62). The Gramulator has been used in a range of recent studies to analyze differentials in various corpora (c.f. 29, 31, 66). The Gramulator includes eight modules, two pre-processing and six post-processing ones. For the purposes of the present research, we used the Evaluator and Viewer modules to produce and analyze relevant text results, and the GPAT command to identify text genre per level of competence.

5   Text Analysis Findings

The ultimate purpose of the present research was the creation of a mathematical formula that could make possible the classification of English texts to two levels of language competence, that is, intermediate (B2) or advanced (C1), according to their level of linguistic complexity. In order to avoid contamination of results due to text length variation, the frequency counts of all indices were normalized to a text length of 100 words. IBM SPSS 20.0 statistical package data was used to compute descriptive statistics and perform reliability analyses, Pearson product moment correlations, T-tests, ANOVAs, multiple Linear and Binary Logistic Regressions.

In order to create our statistical model, regression analysis, a procedure commonly employed for predictive purposes, was used. The training set consisted of 63 KPG reading texts, 34 of which had already been used in past B2 level exams and 29 in C1 level exams. These texts had originally been chosen as appropriate for each level of competence based on the judgement of experienced KPG test designers and a series of piloting sessions. Most importantly, these texts had already been used in real KPG English language exams so, at least in principle, they could be taken to represent the best available evidence of texts classified according to their level of difficulty. Albeit empirical and subjective, such a classification was, nevertheless, practically useful, as it provided us with a set of rated texts on which to build and train our model. In other words, these levelled texts were taken to constitute the empirically established gold standards for analysis in the linear regression. Due to the high number of independent text variables and the two levels of text classification Binary Logistic Regression was carried out using the Forward method and the Wald criterion in IBM SPSS 20.0. Forward selection looks at each explanatory variable individually and selects the single explanatory variable that fits the data the best on its own as the first variable included in the model. Given the first variable, other variables are examined to see if they will add significantly to the overall fit of the model. Among the remaining variables, the one that adds the most is included. This latter step (examining remaining variables in light of those already in the model and adding those that add significantly to the overall fit) is repeated until none of the remaining variables would add significantly or there are no remaining variables (14).Following this method the program entered in the analysis one variable at a time, depending on whether it met a predefined set of statistical criteria (p<0.05), and also removed those that contributed less to the predictive power of the model. To avoid over-fitting our model, a minimum of 15 cases of data for each predictor variable was considered acceptable, with the final model containing 4 predictor variables. Thus, with a ratio of 15.75, the model followed the necessary statistical restrictions and distanced itself from over-fitting problems (24, 30). The Binary Logistic Regression Model that yielded the most accurate prediction of text classification regarding B2 and C1 KPG reading comprehension texts is the following:

Table 1. Regression Coefficients Table

	Beta (Standardized Coefficients)	Std. Error	Wald Test	df	Sig.	Exponent (Beta)
Constant	23.097	9.627	5.756	1	.016	107353273
V1. LexicalDensity	.202	.087	5.385	1	.020	1.224
V2. AcademicWordList	.814	.386	4.456	1	.035	2.258
V3. SyntacticStructure-Similarity	-97.184	46.835	4.306	1	.038	.000
V4. Tokens/Family	-60.381	20.239	8.901	1	.003	.000

The specific regression model succeeded in correctly predicting the level of 33 of the 34 B2 texts (97%) and 26 of the 29 C1 texts (90%). The total percentage of correct predictions on the training set was 94%. This result signifies that the combination of the four variables alone managed to correctly classify 59 of the 63 B2 and C1 KPG reading texts used in the analysis. In other words, for the training set, using these specific four variables, the model has managed to correctly predict the level of 94% of the total number of pre-classified passages. The Lexical Density (LD) variable refers to the proportion of content words to the total number of running words in a text estimated through LIWC2007, the Academic Word List variable relates to Coxhead's academic list and was estimated through VocabProfile, the Syntactic Structure Similarity variable for all sentences across paragraphs was calculated using Coh-Metrix 2.1, whereas the proportion of Tokens per Family was provided by VocabProfile. To determine whether the model adequately described the data, the Hosmer-Lemeshow test was used and the following contingency table created. According to the designers of SPSS 20.0, this statistic is the most reliable test of model fit for binary logistic regression, because it aggregates the observations into groups of "similar" cases. The statistic is, then, computed based upon these groups. As can be seen in the table below, the model describes the data of the training set in a rather satisfactory way given that expected and observed cases are almost identical.

Table 2. Contingency Table for Hosmer-Lemeshow Test

	Group=B2		Group=C1		Total
	Observed	Expected	Observed	Expected	
1	6	6.000	0	.000	6
2	6	6.000	0	.000	6
3	5	5.981	1	.019	6
4	6	5.738	0	.262	6
5	6	4.890	0	1.110	6
6	4	3.588	2	2.412	6
7	1	1.529	5	4.471	6
8	0	.236	6	5.764	6
9	0	.037	6	5.963	6
10	0	.002	9	8.998	9

Since in binary logistic regression models it is not possible to compute a single coefficient of determination (R^2) that summarizes the proportion of variance in the dependent variable associated with the predictor variables, with larger R^2 values indicating that more of the variation is explained by the model, two methods are used to estimate approximations (pseudo r-squared statistics) of the coefficient of determination, namely the Cox & Snell's R^2 and Nagelkerke's R^2. Cox & Snell's R^2 (16, 32) is based on the log likelihood for the model compared to the log likelihood for a baseline model. In other words, R^2 indicates the proportion of the variability in the dependent variable that is accounted for by the multiple regression equation and, since it includes a correction for shrinkage, it is an estimate for the general population, rather than the sample from which the data was obtained. However, with categorical outcomes, it has a theoretical maximum value of less than 1, even for a "perfect" model. Nagelkerke's R^2 (52) is an adjusted version of the Cox & Snell R^2 that adjusts the scale of the statistic to cover the full range from 0 to 1. Since what constitutes a "good" R^2 value varies between different areas of application, and in the field of text complexity there is no similar research so as to compare models and have a better indication of an acceptable R^2, we consider that our selected model which corresponds to step 4 and has the largest R^2 statistic (.841) is the "best" according to this measure.

Table 3. Model Summary

Step	-2 Log likelihood	Cox & Snell R Square	Nagelkerke R Square
1	48.880[a]	.453	.606
2	36.446[b]	.551	.737
3	30.786[c]	.590	.788
4	**24.357[c]**	**.630**	**.841**

Based on the model, the resulting regression equation or prediction formula, which may provisionally be called the **L.A.S.T. Text Difficulty Index** is as follows:

Predicted Text Level = 23.097 + (0.202***LexicalDensity**)
+ (0.814***AcademicWordList**)
                    + (-97.184***SyntacticSimilarityAll**)
                    + (-60.381***Tokens/Family**)

Thus, when adding specific numeric values to the predictor variables the binary logistic equation provides us with a score that ranges from 0 to 1. Given that in our analysis the categorical classification of the Dependent Variable (Text Level) was B2=0 and C1=1, scores higher than 0.5 designate the probability of a text been classified as an advanced one, whereas scores lower than 0.5 refer to the possibility of having an intermediate text. The formula does not only provide us with the predicted difficulty level of a given text (0≠1), but also makes possible the comparison of text scores within the same level of language competence. In other words, if we accept decimal places in the final score, we could achieve an even higher consistency in our selection of similar level texts by defining the range of values that is considered most appropriate for a specific exam condition. Finally, the formula can be implemented in any computer program that includes basic mathematical tools, such as Microsoft Excel 2007, for interested users to insert appropriate values and automatically obtain a rough text difficulty score in relation to these two levels, i.e. Intermediate or Advanced. Needless to say, the model performs reasonably well in making a distinction only between B2 and C1 reading texts and not the range of levels defined by the Common European Framework of References (A1/A2, B1/B2, C1/C2).

6   Model Validation Procedure

No matter the various r-squared statistics and additional tests used to check model variance, no regression model can be considered statistically reliable and practically useful, unless it proves itself successful in predicting similar features of a different sample from the one used to create it. In other words, for a model to be generalized, it should be capable of accurately predicting the same outcome variable from the same set of predictors in a different text group. Thus, if our model manages to classify a new set of texts into intermediate (B2) or advanced (C1) ones based on their text complexity features, we would be in a better position to argue that the newly created L.A.S.T. Text Difficulty Index is rather likely to successfully assign levels to a wider range of texts, besides the ones used in the KPG English language exams. To this end, in the present study, 40 new texts were selected to form the independent test set upon which the validity of the created model was checked (see Appendix 2 for a list of texts). To begin with, the model was used to select texts from a variety of sources, i.e. on-line and printed magazines, newspapers, practice and past exam papers from various examination batteries and EFL coursebooks, which represented a range of text genres and registers, such as articles, short stories, reports, biographies and film reviews and build our held-out test set. At the preliminary text selection phase, the number of collected texts reached 120, of which 40 were considered appropriate according to the predictive scores of the model (Ratio: 1/3). Each passage was a self-contained selection with a minimum of 450 and a maximum of 500 words. This word limit was set in order to avoid any effects on judges' ratings due to text length variations, that is, longer texts could be considered more difficult than shorter ones. Once the texts were selected, two expert judges were kindly requested to assess their level of difficulty and classify each one of them into one of the available categories, i.e. B2 or C1. To avoid contamination of human ratings due to text source effects no information was provided regarding the origin of the texts. The employment of expert judges at this stage of the research was considered as the only methodologically acceptable way for the practical usefulness and generalizability of the model to be adequately supported (11, 23, 28).

Both judges were considered highly-qualified for the task at hand. More specifically, the first judge was an English language university professor with a Ph.D. in English Linguistics and years of experience in text analysis, whereas the second one was a Doctor in English Language & Linguistics with years of experience in EFL teaching and testing. Both judges had taken part in various KPG research projects and were familiar with its specifications, but also with those of other examination systems. Most importantly, they were both acquainted with the Common European Framework of Reference (CEFR) and the distinction between intermediate and advanced levels of language competence. During an introductory session, judges were informed about the aims of the research and the importance of their contribution to the overall validity of the study. They were also provided with a specially designed handbook that included text selection guidelines and samples of rated texts followed by a short description of their textual characteristics. Judges confirmed lack of familiarity with the pre-selected texts and, working independently, rated them at their own premises. To avoid contamination of results, no personal communication was possible between the two judges. Once the text classification process was completed, individual ratings were compared and analysis of inter-rater reliability showed that consensus between the two judges reached 80% with 32 out of the 40 texts being assigned to similar levels by both raters. To solve the rating discrepancy in the remaining 8 texts, a third judge with a Ph.D. in English Language and years of experience as an EFL teacher and KPG researcher was kindly requested to assess their complexity and provide her own rating. This way, ratings for all 40 texts were collected and each text was classified as B2 or C1 based on inter-rater agreement. The final step of the validation process investigated the percentage of agreement between human ratings and the newly created automatic text classification formula. The comparison showed a 95% agreement between human judgement and automatic measurement (see Appendix 2 for the complete list of results). It is worth highlighting at this point that the two texts for which the model failed to reach agreement with human ratings were drawn from commercially available EFL coursebooks and the level predicted by our formula was in fact in accordance with the one assigned to them by their publishers.

7   Concluding Remarks

The preliminary results of the present study showed that the L.A.S.T. Text Difficulty Index could be used to draw a rough distinction between intermediate and advanced texts based on four linguistic features, i.e. lexical density, syntactic structure similarity, tokens per word family and technical vocabulary. This application might prove useful to test developers and other stakeholders interested in automatic text classification since texts calibrated to specific levels of language competence, can be fed into an electronic bank, from which test task writers of pen-and-paper or e-tests can more consistently choose source texts on the basis of specific text attributes. In addition, the proposed model might be of value in the context of classroom-based assessment, as well as for other exam batteries to better define what sort of text a reader of a prospective level of language ability should be able to process under real exam conditions. In a nutshell, the above mentioned results are considered promising not only because of the predictive accuracy of the model, but also because of the more in-depth lexical, semantic and syntactic variables included in it. Undoubtedly, the new formula could be best viewed as a springboard for assessing text difficulty based on a more sound theory of language. For now, it remains to be explored whether the L.A.S.T. Text Difficulty Index will perform as well with a wider range of reading texts from various disciplines.

8   Research Limitations

As with all studies, the implementation of the present one presented a number of challenges and limitations that we hope will be overcome in future research. To begin with, due to the fact that the KPG English language exam battery was still in its infancy, the number of available test source texts used for training our model was inevitably constrained to a total of sixty-three texts from the two levels of competence. Although this did not prevent the predictive power of the model from reaching 95% of accuracy in a new set of texts, it is strongly believed that a greater number of texts could strengthen the generalizability of present results. It is also worth pointing out that the proposed formula should by no means be treated as an absolute indicator of text difficulty that could replace expert human judges. On the contrary, the new index is intended to complement and increase the validity and reliability of human decisions and act as a supplementary tool during the text selection process of intermediate or advanced texts. In other words, the model can help test designers make an informed decision on the exact level of a bulk of preselected intermediate texts but cannot draw a distinction among a range of levels. In addition, (un)fortunately, despite the considerable advances in the field of Computational Linguistics and Machine Learning, no intelligent system has yet been developed to replicate human reasoning and, in the context of foreign language testing, decide on the appropriacy of a topic or judge how interesting and appealing it could be to prospective test-takers. It is also necessary to highlight that the inclusion of four specific variables in the created formula should not be taken to suggest that additional text features do not affect text complexity, but rather that the contribution or "loading" of the former to increased difficulty is sufficient for the model to reach its maximum predictive power, while eliminating the number of independent variables needed to this end.

References

[1] Alderson, C.: Assessing Reading. Cambridge University Press, Cambridge (2000)
[2] Alderson, C., Figueras, N., Kuijper, H., Nold, G., Takala, S., Tardieu, C.: The development of specifications for item development and classification within The Common European Framework of Reference for Languages: Learning, Teaching, Assessment: Reading and Listening: Final report of The Dutch CEF Construct Project. Unpublished Working Paper. Lancaster University, Lancaster (2004)
[3] Allen, D., Bernhardt, B., Berry, T., Demel, M.: Comprehension and text genre: an analysis of secondary school foreign language readers. The Modern Language Journal 72(2), 163–172 (1988)

[4] Bailin, A., Grafstein, A.: The linguistic assumptions underlying readability formulas: a critique. Language & Communication 21(3), 285–301 (2001)

[5] Beaudreau, S., Storandt, M., Strube, M.: A comparison of narratives told by younger and older adults. Experimental Aging Research 32(1), 105–117 (2005)

[6] Block, E.: See How They Read: Comprehension Monitoring of L1 and L2 Readers. TESOL Quarterly 26(2), 319–342 (1992)

[7] Bohanek, J., Fivush, R., Walker, E.: Memories of positive and negative emotional events. Applied Cognitive Psychology 19(1), 51–56 (2005)

[8] Brown, C., Snodgrass, T., Kemper, S., Herman, R., Covington, M.: Automatic measurement of propositional idea density from part-of-speech tagging. Behavior Research Methods 40(2), 540–545 (2008)

[9] Carr, N.: The factor structure of test task characteristics and examinee performance. Language Testing 23(3), 269–289 (2006)

[10] Chalhoub-Deville, M., Turner, C.: What to look for in ESL admission tests: Cambridge certificate exams, IELTS and TOEFL. System 28(4), 523–539 (2000)

[11] Chapelle, C., Jamieson, J., Hegelheimer, V.: Validation of a web-based ESL test. Language Testing 20(4), 409–439 (2003)

[12] Cobb, T.: Computing the vocabulary demands of L2 reading. Language Learning & Technology 11(3), 38–63 (2007)

[13] Cobb, T.: Learning about language and learners from computer programs. Reading in a Foreign Language 22(1), 181–200 (2010)

[14] Cook, P., Dixon, W., Duckworth, M., Kaiser, K., Koehler, W., Meeker, Stephenson, W.: Beyond Traditional Statistical Methods. Iowa State University Press, Iowa (2000)

[15] Covington, M.: CPIDR 3.0 User Manual. CASPR Research Report 2007-03. Artificial Intelligence Center, The University of Georgia (2007), http://www.ai.uga.edu/caspr

[16] Cox, D., Snell, E.: Analysis of Binary Data, 2nd edn. Chapman & Hall/CRC, New York (1989)

[17] Coxhead, A.: A new academic word list. TESOL Quarterly 34(2), 213–238 (2000)

[18] Crossley, S., Greenfield, J., McNamara, D.: Assessing Text Readability Using Cognitively Based Indices. TESOL Quarterly 42(3), 475–492 (2008)

[19] Crossley, S., Louwerse, M., McCarthy, P., McNamara, D.: A Linguistic Analysis of Simplified and Authentic Texts. The Modern Language Journal 91(1), 15–30 (2007)

[20] Crossley, S., Salsbury, T., McNamara, D., Jarvis, S.: Predicting lexical proficiency in language learner texts using computational indices. Language Testing 28(4), 561–580 (2011)

[21] Douglas, D.: Performance consistency in second language acquisition and language testing research: a conceptual gap. Second Language Research 17(4), 442–456 (2001)

[22] Durán, P., Malvern, D., Richards, B., Chipere, N.: Developmental trends in lexical diversity. Applied Linguistics 25(2), 220–242 (2004)

[23] Durán, N., McCarthy, P., Graesser, A., McNamara, D.: Using temporal cohesion to predict temporal coherence in narrative and expository texts. Behavior Research Methods 39(2), 212–223 (2007)

[24] Foster, J.: Data Analysis Using SPSS for Windows. Sage Publications Ltd, London (2001)

[25] Freedle, R., Kostin, I.: Does the text matter in a multiple-choice test of comprehension? The case for the construct validity of TOEFL's minitalks. Language Testing 16(1), 2–32 (1999)

[26] Fulcher, G.: Text difficulty and accessibility: Reading Formulas and expert judgment. System 25(4), 497–513 (1997)

[27] Graesser, A., McNamara, D., Louwerse, M., Cai, Z.: Coh-Metrix: Analysis of text on cohesion and language. Behavior Research Methods, Instruments & Computers 36(2), 193–202 (2004)

[28] Green, A., Ünaldi, A., Weir, C.: Empiricism versus connoisseurship: Establishing the appropriacy of texts in tests of academic reading. Language Testing 27(2), 191–211 (2010)

[29] Haertl, B., McCarthy, P.: Differential Linguistic Features in U.S. Immigration Newspaper Articles: A Contrastive Corpus Analysis Using the Gramulator. In: Murray, C., McCarthy, P. (eds.) Proceedings of the 24th International Florida Artificial Intelligence Research Society Conference, pp. 349–350. The AAAI Press, Menlo Park (2011)

[30] Hatch, E., Lazaraton, A.: The Research Manual: Design and Statistics for Applied Linguistics. Heinle & Heinle Publishers, Boston (1991)

[31] Hullender, A., McCarthy, P.: A Contrastive Corpus Analysis of Modern Art Criticism and Photography Criticism. In: Murray, C., McCarthy, P. (eds.) Proceedings of the 24th International Florida Artificial Intelligence Research Society Conference, pp. 351–352. The AAAI Press, Menlo Park (2011)

[32] Hutcheson, G.: Logistic Regression. In: Moutinho, L., Hutcheson, G. (eds.) The SAGE Dictionary of Quantitative Management Research, pp. 173–176. SAGE Publications Ltd., London (2011)

[33] Jarvis, S.: Short texts, best-fitting curves and new measures of lexical diversity. Language Testing 19(1), 57–84 (2002)

[34] Kahn, J., Tobin, R., Massey, A., Anderson, J.: Measuring Emotional Expression with the Linguistic Inquiry and Word Count. The American Journal of Psychology 120(2), 263–286 (2007)

[35] Kintsch, W.: The Role of Knowledge in Discourse Comprehension: A Construction Integration Model. Psychological Review 95(2), 163–182 (1988)

[36] Lamkin, T., McCarthy, P.: The Hierarchy of Detective Fiction: A Gramulator Analysis. In: Murray, C., McCarthy, P. (eds.) Proceedings of the 24th International Florida Artificial Intelligence Research Society Conference, pp. 257–262. The AAAI Press, Menlo Park (2011)

[37] Lee, J., Musumeci, D.: On Hierarchies of Reading Skills and Text Types. The Modern Language Journal 72(2), 173–187 (1988)

[38] Liu, H.: MontyLingua: An end-to-end natural language processor with common sense (Computer software and documentation) (2004),
http://web.media.mit.edu/~hugo/montylingua (retrieved March 23, 2012)

[39] MacWhinney, B.: The Childes Project: Tools for Analyzing Talk. Lawrence Erlbaum Associates, Mahwah (2000)

[40] MacWhinney, B., Snow, C.: The Child Language Data Exchange System: an update. Journal of Child Language 17(2), 457–472 (1990)

[41] Malvern, D., Richards, B.: A new measure of lexical diversity. In: Ryan, A., Wray, A. (eds.) Evolving Models of Language: Papers from the Annual Meeting of the British Association for Applied Linguistics Held at the University of Wales, pp. 58–71. Multilingual Matters, Clevedon (1996)

[42] Malvern, D., Richards, B.: Investigating accommodation in language proficiency interviews using a new measure of lexical diversity. Language Testing 19(1), 85–104 (2002)

[43] Malvern, D., Richards, B., Chipere, N., Durán, P.: Lexical diversity and language development: Quantification and Assessment. Palgrave Macmillan, Houndmills (2004)

[44] McCarthy, P., Jarvis, S.: vocd: A theoretical and empirical evaluation. Language Testing 24(4), 459–488 (2007)

[45] McCarthy, P., Jarvis, S.: MTLD, vocd-D, and HD-D: A validation study of sophisticated approaches to lexical diversity assessment. Behavior Research Methods 42(2), 381–392 (2010)

[46] McCarthy, P., Watanabe, S., Lamkin, T.: The Gramulator: A Tool to Identify Differential Linguistic Features of Correlative Text Types. In: McCarthy, P., Boonthum, C. (eds.) Applied natural language processing and content analysis: Identification, investigation, and resolution, pp. 312–333. IGI Global, Hershey (2012)

[47] McKee, G., Malvern, D., Richards, B.: Measuring vocabulary diversity using dedicated software. Literary and Linguistic Computing 15(3), 323–337 (2000)

[48] McNamara, D., Cai, Z., Louwerse, M.: Optimizing LSA measures of cohesion. In: Landauer, T., McNamara, D., Dennis, S., Kintsch, W. (eds.) Handbook of Latent Semantic Analysis, pp. 379–400. Routledge, New York (2011)

[49] McNamara, D., Louwerse, M., McCarthy, P., Graesser, A.: Coh-Metrix: Capturing Linguistic Features of Cohesion. Discourse Processes 47(4), 292–330 (2010)

[50] Meara, P.: Lexical Frequency Profiles: A Monte Carlo Analysis. Applied Linguistics 26(1), 32–47 (2005)

[51] Min, H., McCarthy, P.: Identifying Varietals in the Discourse of American and Korean Scientists: A Contrastive Corpus Analysis Using the Gramulator. In: Guesgen, H., Murray, C. (eds.) Proceedings of the 23rd International Florida Artificial Intelligence Research Society Conference, pp. 247–252. The AAAI Press, Menlo Park (2010)

[52] Nagelkerke, E.: A note on a general definition of the coefficient of determination. Biometrika 78(3), 691–692 (1991)

[53] Nation, P.: Using small corpora to investigate learner needs: two vocabulary research tools. In: Ghadessy, M., Henry, A., Roseberry, R. (eds.) Small Corpus Studies and ELT, pp. 31–45. John Benjamins, Amsterdam (2001)

[54] Nation, P.: How large a vocabulary is needed for reading and listening? The Canadian Modern Language Review 63(1), 59–82 (2006)

[55] Nevo, N.: Test-taking strategies on a multiple-choice test of reading comprehension. Language Testing 6(2), 199–215 (1989)

[56] Oakland, T., Lane, H.: Language, Reading, and Readability Formulas: Implications for Developing and Adapting Tests. International Journal of Testing 4(3), 239–252 (2004)

[57] Pasupathi, M.: Telling and the remembered self: Linguistic differences in memories for previously disclosed and previously undisclosed events. Memory 15(3), 258–270 (2007)

[58] Pennebaker, J., King, L.: Linguistic styles: Language use as an individual difference. Journal of Personality and Social Psychology 77(6), 1296–1312 (1999)

[59] Pennebaker, J., Booth, R., Francis, M.: Linguistic Inquiry and Word Count: LIWC 2007. LIWC.net, Austin (2007)

[60] Phakiti, A.: A Closer Look at Gender and Strategy Use in L2 Reading. Language Learning 53(4), 649–702 (2003)

[61] Purpura, J.: An analysis of the relationships between test takers' cognitive and metacognitive strategy use and second language test performance. Language Learning 47(2), 289–325 (1997)

[62] Rufenacht, R., McCarthy, P., Lamkin, T.: Fairy Tales and ESL Texts: An Analysis of Linguistic Features Using the Gramulator. In: Murray, C., McCarthy, P. (eds.) Proceedings of the 24th International Florida Artificial Intelligence Research Society Conference, pp. 287–292. The AAAI Press, Menlo Park (2011)

[63] Shokrpour, N.: Systemic Functional Grammar as a Basis for Assessing Text Difficulty. Indian Journal of Applied Linguistics 30(2), 5–26 (2004)

[64] Snowdon, D., Kemper, S., Mortimer, J., Greiner, L., Wekstein, D., Markesbery, W.: Linguistic ability in early life and cognitive function and Alzheimer's disease in late life: Findings from the Nun Study. The Journal of the American Medical Association 275(7), 528–532 (1996)

[65] Tausczik, J., Pennebaker, W.: The Psychological Meaning of Words: LIWC and Computerized Text Analysis Methods. Journal of Language and Social Psychology 29(1), 24–54 (2010)

[66] Terwilleger, B., McCarthy, P., Lamkin, T.: Bias in Hard News Articles from Fox News and MSNBC: An Empirical Assessment Using the Gramulator. In: Murray, C., McCarthy, P. (eds.) Proceedings of the 24th International Florida Artificial Intelligence Research Society Conference, pp. 361–362. The AAAI Press, Menlo Park (2011)

[67] Turner, A., Greene, E.: The construction and use of a propositional text base. Technical Report 63. Institute for the Study of Intellectual Behavior, University of Colorado (1977)

[68] Ungerleider, C.: Large-Scale Student Assessment: Guidelines for Policymakers. International Journal of Testing 3(2), 119–128 (2003)

[69] Weir, C.: Limitations of the Common European Framework for developing comparable examinations and tests. Language Testing 22(3), 281–300 (2005)

A   Appendix 1

Text Variables List

Basic Text Information
V1. words/ text
V2. Syllables/ word
V3. Words/ sentence
V4. Characters/ word
V5. Syllables/ 100 words
V6. No. of sentences
V7. Sentences/ paragraph
V8. Sentences/ 100 words
V9. Text Genre
V10. Sum 1 to 6-Letter Words
V11. Sum 7 to 14-Letter Words
V12. 1-Letter Word
V13. 2-Letter Word
V14. 3-Letter Word
V15. 4-Letter Word
V16. 5-Letter Word
V17. 6-Letter Word
V18. 7-Letter Word
V19. 8-Letter Word
V20. 9-Letter Word
V21. 10-Letter Word
V22. 11-Letter Word
V23. 12-Letter Word
V24. 13-Letter Word
V25. 14-Letter Word
Word Frequency Indices
V26. K1 Words
V27. K2 Words
V28. K3 Words
V29. K4 Words
V30. K5 Words
V31. Academic Word List
V32. Anglo-Sax Index
V33. Greco-Latin Cognates
V34. Log freq. content words
V35. Log min. freq. content words
Readability Indices
V36. Flesch
V37. Flesch-Kincaid
V38. Gunning-Fog
V39. Fry Graph
V40. Dale-Chall

Lexical Richness Indices
V41. vocd-D
V42. MTLD
V43. MAAS
V44. Apax Legomena
V45. Dis Legomena
V46. Entropy
V47. Relative Entropy
V48. Lexical Density
V49. Types per Text
V50. Tokens per Type
V51. Families / Text
V52. Tokens / Family
V53. Types / Family
V54. Verb Density
V55. Adverb Density
Text Abstractness Indices
V56. Concreteness content words
V57. Min. concreteness content words
V58. Noun hypernym
V59. Verb hypernym
Syntactic Complexity Indices
V60. Higher level constituents
V61. Noun Phrase incidence
V62. Modifiers per Noun Phrase
V63. Words before main verb
V64. Negations
V65. Passive sentences
V66. Syntactic structure similarity (adj. sentences)
V67. Sync. structure similarity (across paragraphs)
V68. Syntactic structure similarity (within paragraphs)
V69. Conditional operators
V70. All connectives
V71. Pos. additive connectives

V72. Pos. temporal connectives
V73. Pos. causal connectives
V74. Pos. logical connectives
V75. Neg. additive connectives
V76. Neg. temporal connectives
V77. Neg. causal connectives
V78. Neg. logical connectives
Cohesion & Coherence Indices
V79. Causal cohesion
V80. Causal content
V81. Intentional content
V82. Temporal cohesion
V83. Spatial cohesion
V84. Logical operators
Referential & Semantic Indices
V85. Anaphoric reference
V86. Adjacent anaphoric reference
V87. Argument overlap
V88. Adjacent argument overlap
V89. Stem overlap
V90. Adjacent stem overlap
V91. Content word overlap
V92. Pronoun ratio
V93. Personal pronouns
V94. Impersonal pronouns
V95. Propositional Idea Density
V96. LSA for adjacent sentences
V97. LSA for all sentences
V98. LSA for all paragraphs
Psycholinguistic Processes

V99. Affective Processes (Total)
V100. Positive Emotions (Total)
V101. Negative Emotions (Total)
V102. Cognitive Mechanisms (Total)
V103. Insight
V104. Cause
V105. Discrepancy
V106. Tentative Words
V107. Certainty
V108. Agreement (Assent)
V109. Inhibition
V110. Inclusive Words
V111. Exclusive Words
V112. Perception (Total)
V113. See
V114. Hear
V115. Feel
V116. Relativity (Total)
V117. Motion
V118. Space
V119. Time
V120. Articles
V121. 1st person singular pronouns
V122. 1st person plural pronouns
V123. 2nd person pronouns
V124. 3rd person singular pronouns
V125. 3rd person plural pronouns
Additional Text Variables
V126. Idioms
V127. Phrasal Verbs
V128. Question Marks
V129. Function Words
V130. Past Tenses
V131. Present Tenses
V132. Future Tenses
V133. Auxiliary Verbs
V134. Prepositions
V135. Numbers

Efficient Handwritten Numeral Recognition System Using Leaders of Separated Digit and RBF Network

Thangairulappan Kathirvalavakumar*, M. Karthigai Selvi, and R. Palaniappan

Research Center in Computer Science
V.H.N.S.N College (autonomous), Virudhunagar-626 001, India
{kathirvalavakumar,svrpalani}@yahoo.com, karthigavishaal@gmail.com

Abstract. In this paper an efficient method has been proposed to classify handwritten numerals using leader algorithm and Radial Basis Function network. Handwritten numerals are represented in matrix form and clusters with leaders are formed for each row of each digit separately. Every leader is with single target digit. Duplication patterns are avoided from the cluster leaders by combining those in a single pattern with target vectors having corresponding bits in on mode. Now resultant target vectors are with 10 bits corresponding to the number of digits considered for classification. Constructed leaders are trained using Radial Basis Function network. Experimental results show that the minimum number of patterns are enough for training compared to total patterns and it has been observed that convergency is fast during training. Also the number of resultant leaders after avoiding duplication patterns are less and the number of bits in each resultant pattern is 12.

Keywords: Handwritten numerals, Radial basis function, Clustering, Leader algorithm, Classification.

1 Introduction

Now-a-days more attention is needed in processing large data as it needs more memory and computations. More methods have been designed to minimize data size, computation and recognition[3,4,11,16,18]. classification of handwritten numerals need more memory and computation as Handwritten numerals data size is large.

Ravindra Babu et al. [16] have used run length encoded binary data for classification.This method focus on minimizing storage and computation on data. This method representing data in compressed form and classifying the compressed data without decompression. This method performs clustering and classification just using distance method. They have shown that this method works well for handwritten numerals. Ananthanarayana et al. [3] have proposed a clustering scheme for prototype selection based on PC-Tree data structure. The results of

---

* Corresponding Author. The work of T. Kathirvalavakumar is supported by University Grants Commission for Major Research Project, Government of India.

R. Prasath et al. (Eds.): MIKE 2014, LNAI 8891, pp. 135–144, 2014.

the work show that handwritten data are recognized efficiently and accurately. Synthetic pattern recognition is used to overcome the dimensionality problem but it involves some drawback. Monu Agrawal et al. [11] have proposed a new strategy to reduce time and memory requirements and also to overcome the drawback involved in synthetic pattern by applying prototype as an intermediate step in the synthetic pattern generation technique. An efficient hierarchical clustering algorithm is proposed by Vijaya et al. [18] for effective clustering and prototype selection for pattern classification. This method uses incremental clustering principles to generate a hierarchical structure for finding the subgroups/subclusters with each cluster. They have have presented Leader-Sub-leader an extension of the leader algorithm. Yi-Ching Liaw [20] have presented an effective fast method for clustering. This algorithm uses the property of cluster distance increases as the cluster merge process proceeds and it adopts fast search algorithm to reject impossible candidate clusters. This method reduces the number of distance calculations and computation time.

Jaemo Sung et al. [9] have presented a method for recognizing handwritten numerals using hierarchical Gabor feature and Byesian network classifier. Optimal Gabor features are extracted by maximizing Fisher Linear Discriminant and are used in Bayesian network classifier. Ehtesham Hassan et al. [8] have presented a novel frame work for the application of multiple features for recognizing handwritten based identity. New multi kernel learning has been formulated using genetic algorithm. Bin Chen et al. [4] have presented a method for recognizing on-line handwritten Japanese character based on Markov Random Field model by artificially generating large amount of training patterns. Artificial patterns are generated by combining six type of linear distortion model among themselves to identify handwritten with real distortion. Ahmad-Montaser Awal et al. [1] have presented an online handwritten mathematical expression recognition system as a simultaneous optimization of expression segmentation, symbol recognition, and 2D structure recognition under the restriction of a mathematical expression grammar by combining syntactic and structural information. Al-Omari and Al-Jarrah [2] have presented a system to recognize handwritten numerals using Probabilistic neural network. It involves feature vector based on centre of gravity and a set of vectors to the boundary points of the digit object. Nibaran et al. [12] have proposed a system by combining Principal component analysis, modular principal component analysis and Quad-tree based hierarchically derived longest run features to recognize handwritten numerals. Pal et al. [13] have proposed a method for recognizing the numerals of Devanagari, Bangala, Telugu, Oriya, Kanada, and Tamil by the directional features computed from the blocks of a bounding box of numerals. Shivanand Rumma et al. [17] have proposed a method for recognizing handwritten Kannada numerals by Radial basis function. Kumar et al. [10] have proposed unconstrained offline handwritten numeral recognition system using local and global features of profile of numeral image, majority voting scheme and neural network. Vijayakumar et al. [19] have proposed a novel algorithm for recognizing handwritten digits. They have grouped the digits into blobs with stem and without stem. The blobs are identified using morphological region filling methods with connected components. Rajashekararadhya and Vanaja Ranjan [14] have proposed a zone based feature extraction method for recognizing handwritten numerals. This method involves character centroid and average distance of pixel present in the zone and the centroid of the character. Nearest neighbor, feedforward neural network and support vectors are also used in this work. Rajashekararadhya and Vanaja Ranjan [15] have proposed an efficient zone based feature extraction method for recognizing handwritten numerals. In this they have proposed zone centroid and image centroid based distance metric feature extraction system. Dhandra et al. [5] have proposed an approach for recognizing Kannada, Telugu and Devanagari handwritten numerals using Probabilistic neural network. Dhandra et al. [6] have proposed script independent automatic numeral recognition system using Probabilistic neural network. This system uses directional density estimation, water reservoirs, maximum profile distances and fill hole density features. Desai [7] have proposed a system for recognizing Gujarathi handwritten optical character numeral using feedforward neural network by extracting features from four different profiles of digits.

2   Clustering with Leader

Clustering technique with leader concept [18] is used to group meaningful patterns so as to improve classification accuracy with minimum input-output operations. In this method [18], first pattern is treated as a cluster leader. Remaining patterns are compared with the leaders of existing clusters and is assigned to member of a cluster whose leader is with minimum distance. If the distance between pattern and the leader is greater than predefined distance then the pattern is a leader of a new cluster. Distance between pattern is computed by the Manhattan formula as follows:

$$Manhattan distance = |X - Y| \tag{1}$$

3   Radial Basis Function

Radial Basis Function(RBF), a feedforward neural network, as shown in figure 1, is used for classification. In this figure n neurons are used in the input layer and m neurons in the output layer. Gaussian basis functions are used in the hidden layer. It consists of three layers namely input, hidden and output layer. Number of nodes in the input layer depends on the number of bits in the input pattern. Hidden layer contains Q basis functions. The Number of neurons in the output layer equals number classification. Basis functions are calculated by

$$\phi_i(X) = exp\left(\frac{||X - X_i||^2}{2\sigma^2}\right) \tag{2}$$

The output neuron values are calculated by

$$f_j(X) = \sum_{i=1}^{Q} w_{ij}\phi(||X - X_i||), \tag{3}$$

where w_{ij} is the weight between i^{th} basis function and j^{th} output neuron and X_i is the center for the ith basis function. Here the number of basis function is equivalent to the number of centers. Weights between hidden and output layer are adjusted using

$$W(t+1) = W(t) + \lambda * (d_j - y_j)\phi_j(X), \qquad (4)$$

where t represents iteration and λ is the positive learning parameter, d_j and y_j represent desired and obtained value respectively for the jth neuron. Network error is calculated by

$$SSE = \sum_{i=1}^{N} \frac{1}{2} \sum_{j=1}^{m} (d_j - y_j)^2, \qquad (5)$$

where m, N represents the number of classes and pattern respectively.

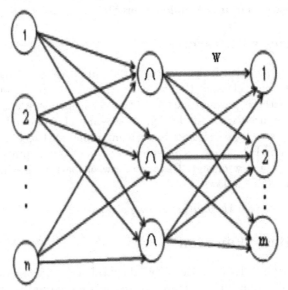

Fig. 1. Radial Basis Function network

3.1 RBF Training Algorithm

step 1. Initialize weight of the network randomly.

step 2. For an input pattern calculate basis function value for hidden neurons using (2).

step 3. Compute the output for each neuron in the output layer using (3).

step 4. Find the error for each output neuron by finding the difference between desired and obtained value for the given input pattern.

step 5. Update weights of the network, that is between hidden and output layer using (4).

step 6. Find the error of the network using (5).

step 7. Repeat steps 2 to 6 until error reaches the defined tolerance.

```
0 0 1 1 1 1 1 1 1 1 0 0
0 0 1 1 1 1 1 1 1 1 0 0
0 0 1 1 1 0 0 0 1 1 0 0
0 1 1 1 0 0 0 1 1 1 0 0
1 1 1 0 0 0 0 1 1 0 0 0
1 1 1 0 0 1 1 1 0 0 0 0
1 1 1 0 0 1 1 1 0 0 0 0
1 1 1 1 1 1 1 1 1 1 1 0
0 1 1 1 1 1 1 1 0 1 1 0
0 0 1 1 1 1 0 1 0 1 1 0
0 0 1 1 1 0 0 0 0 0 1 0
0 0 1 1 1 1 1 1 1 1 1 0
0 0 1 1 0 0 0 0 1 1 1 0
0 0 1 1 1 1 1 1 1 1 1 0
0 0 1 1 1 1 1 1 1 1 0 0
0 0 0 0 1 1 1 1 0 0 0 0
```

Fig. 2. Handwritten digit 8 in matrix form

4   Proposed Procedure

Every pattern of handwritten digit can be viewed in a matrix form as in figure 2. Here every pattern of handwritten digit is represented by 16×12 matrix form and its target class is represented by a single digit. Every pattern to be divided into more patterns equivalent to number of rows in a digit. So, every row of a matrix of a digit becomes a pattern. Clusters to be formed among every row of individual digits independently using cluster with Leader algorithm [18]. Leader of every cluster to be collected. Leaders of a particular digit may be matched with leaders of other digits. If every pattern is represented with the target of single digit, and a pattern of a digit occurs as a patterns of different digits then neural network may be get confused during training. To avoid confusion, every pattern to be represented with the target value of 10 bits showing its membership in other digits. So instead of having single target digit for each pattern, 10 bits are to be used to represent 10 different digits of decimal number system. First position of the target vector is meant for digit zero, second bit for digit one, third bit for digit two and similarly tenth bit of the target vector is meant for the digit 9. The value 1 in the first bit of the target vector represent corresponding pattern is available in the digit zero. Similarly the value 1 in the kth position of the target vector represent pattern is also available in the digit

$k + 1$. If the target vector of any one of the pattern is in the form 0011010111 then it represents this pattern is available in the digits $2, 3, 5, 7, 8, 9$. Because of this procedure same pattern available in different digits become a single pattern which avoid duplication and hence the number of patterns are minimized. After all leaders patterns are collected, every leader to be matched with other leaders to check equivalency. Matched leaders with equivalency are to be converted into single leader with target vector of 10 bits based on its target digit. The leaders not matched with others are also to be converted into target vector with 10 bits having value 1 in one bit and zero in other bits based on its target digit. Because of this procedure lesser number of patterns with lesser number of bits are obtained for further processing. Now these resultant patterns are to be used in the RBF neural network for classification.

4.1   Training Algorithm

step 1. Convert first 192 bits of 193 bit patterns as 16×12 patterns and treat 193rd bit as target value for the 16 patterns.

Step 2. Repeat step 1 for all training patterns of the problem.

step 3. For i= 0 to 9 {this loop is for accessing 10 different digits}

step 4. For j = 1 to 16

step 5. Form clusters for the patterns of jth pattern of ith digit using Manhattan distance formula and leader algorithm. end end

step 6. Collect leaders of each cluster with its digit value as target value.

step 7. Now make target of each pattern as 10 bits with a value 1 or 0. Value 1 in the ith bit represents that it is the leader of a digit $i - 1$ and value 0 otherwise.

    step 7.1. Find redundant leaders among the collection by matching bit by bit.

    step 7.2. Convert redundant leaders into a single leader with 10 target bits each of value 1 or 0 based on the redundant leader belonging to the digit.

    step 7.3. Remaining unique leaders with target digit i are to be attached with 10 target bits with a value 1 in the $(i + 1)^{th}$ bit and 0 in other bits.

step 8. Apply this resultant patterns in RBF feedforward network.

step 9. Train the RBF network using RBF Training algorithm.

5   Experimental Result

The proposed method is applied on Handwritten digit data[11] having 667 patterns per class. Totally 6670 patterns each with 193 bits are used for processing. The last bit of the pattern represent target class of the pattern. The experiment is carried out using MathLab13 software in the Intel Quad core system.

Every pattern is converted into 16 number of patterns with single digit as target value. Now 6670 patterns become $1,06,720$ patterns. clusters are formed among every row of individual digit separately with different values. Number of leaders generated are tabulated in second column of Table 2. After avoiding duplication of the patterns and forming patterns with 10 target bits, the patterns are used in the RBF network for training with different thresholds. Number of leaders obtained after avoiding duplications for different thresholds are shown in third column of Table 2. In the RBF network, 12 input and 10 output neurons are used for training. Number of basis functions are equivalent to number patterns used for training. The patterns used for training are considered as centers to calculate the value of the basis functions. The experiment is executed with different learning parameters and is identified that $\lambda = 0.01$ gives fast convergence during training. The time required for generating clusters for different thresh-

Table 1. Time needed to generate Leaders for different Thresholds

Threshold	Number of Clusters	Time (s)
2	5690	0.00695495
3	2221	0.00540601
4	1251	0.00486841
5	653	0.00525100
6	444	0.00469269

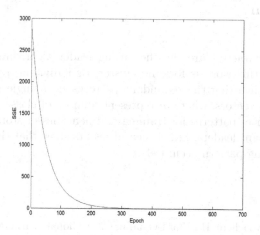

Fig. 3. Learning Curve

Table 2. Leaders Generated, Training Time for different Thresholds

Threshold	Number of Leaders	Number of Leaders after Removing Redundancy	SSE	Epoch	Training Time (s)
2	5690	624	0.0099	660	5.184235
3	2221	479	0.0099	630	2.993268
4	1251	364	0.0100	623	1.630014
5	653	228	0.0098	592	0.456554
6	444	161	0.0099	569	0.216737

Table 3. Leaders Generated for Digits when Threshold = 3

Digit	Number of Leaders
0	261
1	47
2	322
3	226
4	235
5	256
6	243
7	167
8	265
9	199

olds are shown in the Table 1. Table 3 shows the number of Leaders generated for different digits when threshold = 3. Figure 3 shows Learning curve of the RBF training algorithm when threshold = 4 and it shows the convergency of the training. The RBF training is carried out for 25 different trials. The average of accuracy, epoch and training time for different thresholds are listed in Table 2.

6 Conclusion

Handwritten characters are classified using leader algorithm and RBF network. Novelty of this work is forming cluster for individual row of each digit separately and converting the redundant patterns into single pattern with m component target vectors, where m represent number of classes. It leads to obtain lesser number of patterns for training and hence lesser storage capacity for keeping the resultant leaders. Experimental results show that the RBF network classify the training pattern accurately.

References

1. Awal, A.-M., Mouchere, H., Viard-Gaudin, C.: A Global Learning Approach for an Online Handwritten Mathematical Expression Recognition System. Pattern Recognition System 35, 68–77 (2014)

2. Al-Omari, F.A., Al-Jarrah, O.: Handwritten Indian Numerals Recognition System Using Probabilistic Neural Networks. Advanced Engineering Informatics 18, 9–16 (2004)

3. Ananthanarayana, V.S., Narasimha Murty, M., Subramanian, D.K.: Efficient Clustering of Large Data Sets. Pattern Recognition 34, 2561–2563 (2001)

4. Chen, B., Zhu, B., Nakagawa, M.: Training of an on-line Handwritten Japanese Character Recognizer by Artificial Patterns. Pattern Recognition Letters 35, 178–185 (2014)

5. Dhandra, B.V., Benne, R.G., Hangarge, M.: Kannada, Telugu, and Devanagari Handwritten Numeral Recognition with Probabilistic Neural Network: A Novel Approach. IJCA Special Issue on Recent Trends in Image Processing and Patten Recognition, 83–88 (2010)

6. Dhandra, B.V., Benne, R.G., Hangarge, M.: Kannada, Telugu, and Devanagari Handwritten Numeral Recognition with Probabilistic Neural Network: A Script Independent Approach. International Journal of Computer Applications 26, 11–16 (2011)

7. Desai Apurva, A.: Gujarati Handwritten Numeral Optical Character Recognition through Neural Network. Pattern Recognition 43, 2582–2589 (2010)

8. Hassan, E., Chaudhury, S., Yadav, N., Karla, P., Gopal, M.: Off-line handwritten Input Based Identity Determination Using Multi Kernel Feature Combination. Pattern Recognition Letters 35, 113–119 (2014)

9. Sung, J., Bang, S.-Y., Cho, S.: A Bayesian Network Classifier and Hierarchical Gabor Features for Handwritten Numeral Recognition. Pattern Recognition Letters 27, 66–75 (2007)

10. Kumar, R., Goyal, M.K., Ahmed, P., Kumar, A.: Unconstrained Handwritten Numeral Recognition using Majority Voting Classifier. In: IEEE Int. Conf. on Parallel Distributed and Grid Computing (PDGC), pp. 284–289 (2012)

11. Agrawal, M., Gupta, N., Shreelekshmi, R., Narasimha Murty, M.: Efficient Pattern Synthesis for Nearest Neighbour Classifier. Pattern Recognition 38, 2200–2203 (2005)

12. Das, N., Reddy, J.M., Sarkar, R., Basu, S., Kundu, M., Nasipuri, M., Basu, D.K.: A statistical Topological Feature Combination for Recognition for Handwritten Numerals. Applied Soft Computing 12, 2486–2495 (2012)

13. Pal, U., Sharma, N., Wakabayashi, T., Kimura, F.: Handwritten Numeral recognition of Six Popular Indian Scripts. In: Ninth Int. Conf. on Document Analysis and Recognition, ICDAR 2007, vol. 2, pp. 749–753 (2007)

14. Rajashekararadhya, S.V., Vanaja Ranjan, P.: Handwritten Numeral/Mixed Numerals Recognition of South-Indian Scripts: The Zone Based Feature Extraction Method. Journal of Theoretical and Applied Information Technology 7, 63–79 (2009)

15. Rajashekararadhya, S.V., Vanaja Ranjan, P.: Efficient Zone Based Feature Extraction Algorithm for Handwritten Numeral Recognition of Four Popular South Indian Scripts. Journal of Theoretical and Applied Information Technology 6, 1171–1181 (2008)

16. Ravindra Babu, T., Narasimha Murty, M., Agrawal, V.K.: Clasification of Run Length Encoded Binary Data. Pattern Recognition 40, 321–323 (2007)

17. Rumma, S., Vishweshwarayya, C.H., Bhuvaneshwari, B.D.: Handwritten Kannada Numeral Recognition using Radial Basis Function. International Journal of Computer Applications 98, 18–20 (2014)

18. Vijaya, P.A., Narasimha Murty, M., Subramanian, D.K.: Leaders-Subleaders: An Efficient Hierarchical Clustering Algorithm for Large Data Sets. Pattern Recognition Letters 25, 505–513 (2004)
19. Vijaya Kumar, V., Sri Krishna, A., Raveendra Babu, B., Radhika Mani, M.: Classification and Recognition of Handwritten Digits by using Mathematical Morphology. Sadhana 35, 419–426 (2010)
20. Liaw, Y.-C.: Improvement of the Fast Exact Pairwise-Nearest-Neighbor Algorithm. Pattern Recognition Letters 42, 867–870 (2009)

Iterative Clustering Method for Metagenomic Sequences

Isis Bonet[1,*], Widerman Montoya[1], Andrea Mesa-Múnera[1],
and Juan Fernando Alzate[2]

[1] Escuela de Ingeniería de Antioquia, Envigado, Antioquia, Colombia
{ibonetc,widerman.montoya,amesamu}@gmail.com
[2] Centro Nacional de Secuenciación Genómica-CNSG,
Facultad de Medicina, Universidad de Antioquia
jfernando.alzate@udea.edu.co

Abstract. Metagenomics studies microbial DNA of environmental samples. The sequencing tools produce a set of genome fragments providing a challenge for metagenomics to associate them with the corresponding phylogenetic group. To solve this problem there are binning methods, which are classified into two sequencing categories: similarity and composition. This paper proposes an iterative clustering method, which aim at achieving a low sensitivity of clusters. The approach consists of iteratively run k-means reducing the training data in each step. Selection of data for next iteration depends on the result obtained in the previous, which is based on the compactness measure. The final performance clustering is evaluated according with the sensitivity of clusters. The results demonstrate that proposed model is better than the simple k-means for metagenome databases.

Keywords: Metagenomics, clustering, sequences binning, k-means.

1 Introduction

Metagenomics is a new science that combines different research field as genomics, bioinformatics and system biology. The objective of this field is to study genomes of many microbial organisms from a specific environment, which cannot be cultivated in laboratory. Understanding microbial communities' structure is a challenge in different areas such as biomedical, agriculture, environmental and life sciences [1].

The fast development of DNA sequencing techniques using different technologies generations, such as GS-FLX (454) /Roche, Solexa /Illumina, ABI SOLID /Applied Biosystems of second generation, and Helicos TSMS / Helicos BioSciences, Pacific BioSciences /Pacific BioSciences, of third generation; has led to new challenges in metagenomic studies [2]. Such studies are looking for identify the microorganisms in a sample to determine its metabolic functions [2]. Sequencing tools produce a puzzle of sequence fragments, which are known in this field with the name of reads (genome fragments). Following studies of the reads are performed, with the purpose of assembling them by a process of overlapping using large sequences named contigs [3].

---

* Corresponding author.

R. Prasath et al. (Eds.): MIKE 2014, LNAI 8891, pp. 145–154, 2014.
© Springer International Publishing Switzerland 2014

There is an even bigger problem resulting from the contigs that is related with the ongoing assembly process to obtain the complete genome, because of the analysis of an environmental sample contained diverse individuals.

Metagenomics also requires a binning process that allows contigs assignment of different species to their corresponding phylogenetic group. There are some methods of binning, which are classified into two sequencing categories: similarity and composition. Within this category it's found different software such as BLAST [4] and Phylopythia [5]. MEGAN [6] is one of the most widely used binning methods based on similarity sequence, which assigns reads to taxa, based on BLAST results. These kinds of methods are supported by a database of known species genome and they use similarity techniques as alignment to find similar sequences. For this reason, binning algorithms based on similarity are very time consuming. On the other hand, binning methods based on composition sequence made analyzes of genomes features, such as GC content, codon usage or oligonucleotide frequencies to describe the sequences and find clusters that represent the different taxonomic groups. Other features commonly used are called k-mers and represent the characteristic of oligonucleotide frequency of fragments sequence with size k, so that to compare them with a reference set of complete genomes [7]. The k-mer feature, with $k=4$, is widely used, also known as tetranucleotide frequencies. For example, TETRA method yields a statistical analysis using tetranucleotide patterns based on the characteristic of the GC content [8].

Binning algorithms based on similarity need this kind of features to compare the sequences between each other and grouped them based on their similarity into different clusters in order to seek the taxonomic groups in the data sample. Leading with this problem some author had been used different strategies; one of them is used clustering methods. In [9] a Self-Organizing Maps (SOM) method was used for efficiently cluster complex data using the oligonucleotide frequencies calculation. MetaCAA also is a clustering method based on tetranucleotides frequencies [3]. In [10] a comparison of some clustering methods is done.

Binning methods using a complete genomes knowledge-based classifier are referred to supervised learning methods, while methods that do not depend on training data are referred to unsupervised learning methods. Unsupervised learning methods are focused on major classes of collected data and do not perform well with data samples that don't have a significant population. On the other hand, supervised learning methods have a better performance in classifying the data of small populations [11].

In metagenomics, supervised learning methods are more precise, but they are time consuming because of the amount of different organisms present in the sample. Reducing organisms in the sample can improve their performance. That means, if binning method using knowledge-based classifier gets a set of subsequences of the same organism as input, the process to find the specify organism is easier and faster. A previous clustering process can be a way to provide different groups as inputs for supervised learning methods of binning. However, the aim is to find a clustering method which builds pure clusters. That is, members of each cluster belong to the same organism. This doesn't mean all subsequences of one organism are in the same cluster.

This paper is focused on an unsupervised method for assignment of genomic fragments into pure clusters based on composition sequence. Some of the widely-used sequence-based measures, such as GC content nucleotides usage and k-mers frequencies, have been used to represent the genomic fragments. Further, for clustering fragments to cluster that represent the different genomes in the sample, a clustering iterative process based on k-means is proposed. The method has several iterations in the subset of data with more "error", that is the instances that belong to less compact clusters. For each iteration of the method, the improvement of the compactness of clusters is shown.

2   Methods and Data

2.1   Data

Assembled genomic sequences at contig level of different organisms including viruses, bacteria and eukaryotes were downloaded from the FTP site of the Sanger institute as is shown in table 1.

Selected viral sequences include Influenza and Dengue virus genomes. Sixty four dengue genomes ranging from 10,785 to 10,392 bp and an average GC content of 45.95%. Eight influenza genomic sequences that ranged between 2309 and 853 bp and an average GC content of 43.06%. No ambiguous "N" nucleotides were present in these contigs.

Bacterial sequences come from Bacteroides dorei and Bifidobacterium longum. For B. dorei, a total of 1948 contigs that summed 6,771,958 bases was analyzed. The contig N50 calculated value was 11,054   bases and only 8 "N" ambiguous bases were present. The largest contigs have 83484 bases. For B. longum, a total of 2,377,370 bases contained in 33 contigs that ranged between 580,034 and 540 bases were analyzed. The calculated contig N50 value was 154,900 and no ambiguous "N" bases were detected. The GC content was 42.3% for B. dorei   and   59.93% for B. longum.

Table 1. Sequences and data source

Organism	Data source
Aspergillus fumigatus	ftp://ftp.sanger.ac.uk/pub/project/pathogens//A_fumigatus/AF.contigs.031704
Ascaris suum	ftp://ftp.sanger.ac.uk/pub/project/pathogens//Ascaris/suum/genome/assembly/contigs.fasta
Dengue	ftp://ftp.sanger.ac.uk/pub/project/pathogens//Dengue/Dengue.fasta
Glossina	ftp://ftp.sanger.ac.uk/pub/project/pathogens///Glossina/morsitans/Assemblies/tsetseGenome-v1.tar.gz
Bacteroides dorei	ftp://ftp.sanger.ac.uk/pub/project/pathogens//Bacteroides/dorei/D8/454LargeContigs.fna
Bifidobacterium longum	ftp://ftp.sanger.ac.uk/pub/project/pathogens//Bifidobacterium/longum/454LargeContigs.fna
Candida parasilopsis	ftp://ftp.sanger.ac.uk/pub/project/pathogens//Candida/parapsilosis/contigs/CPARA.contigs.fasta
Influenza	ftp://ftp.sanger.ac.uk/pub/project/pathogens//Influenza/Santiago_7981_06.fasta

The selected eukaryotes included 2 fungi, 1 nematode and 1 insect. The analyzed fungi were the mold Aspergillus fumigatus and the yeast candida parasilopsis. A total of 29,416,758 bases of A. fumigatus were analyzed. Theses sequences were contained in 344 contigs that ranged between 2,962,289 and 1,001 bases. The calculated N50 value was 1,120,772 bases and 3995 ambiguous "N" bases were detected. In the case of C. parasilopsis, a total of 13,265,923 bases contained in 1592 contigs were used. The calculated contig N50 value was 14,196 and the sizes ranged between 66,655 and 1,003 bases. 2919 ambiguous "N" bases were counted. The GC content was 49.55% for A. fumigatus and 38.86% for C. parasilopsis.

The analyzed nematode was Ascaris suum. A total of 527,713,826 bases contained in 138,557 contigs were analyzed. The contig N50 value was 8,524 and the count of ambiguous "N" was 7,668. The GC content was 37.89%.

The insect genomic sequences belong to Glossina morsitans fly. A total of 363,109,041 bases contained in 24,072 contigs that ranged between 538224 and 101 bases. The calculated contig N50 value was 49,769 and no ambiguous "N" bases were detected. The GC content was 34.12%.

2.2   Features

For the experiment some features were selected:

- GC: G + C content

$$GC = \frac{G + C}{A + T + G + C}$$

where A, T, G and C are the count of different nucleotides in the sequence.

- Nucleotide frequencies: Number of occurrences of A, T, G and C in the sequence. It was normalized by the size of the sequence.
- Codon frequencies: Number of each possible codon in the sequence. It was normalized by the total of codons (64 codons)
- k-mer (k=4): are represented for the 256 possible tetranucleotides. It was compute as the number of each tetranucleotide and normalized with the total of tetranucleotides in the sequence.

Features were used in all combinations, producing 15 databases.

2.3   *K*-means

K-means is one of the most popular clustering methods, despite the problem to estimate the parameter k (number of cluster). This algorithm finds a set of k centroids, and associates each instance in the data to the nearest centroid, based on a distance function [12]. Here we proposed a clustering method based on k-means.

Euclidean (Equation 1) and Cosine (Equation 2) distance were used to compare the sequences.

$$Euclidean(X,Y) = \sqrt{\sum_{i=1}^{N}(x_i - y_i)^2} \tag{1}$$

$$Cosine(X,Y) = 1 - \frac{\sum_{i=1}^{n}(x_i \times y_i)}{\sqrt{\sum_{i=1}^{n}(x_i)^2} \times \sqrt{\sum_{i=1}^{n}(y_i)^2}} \tag{2}$$

Where X and Y are the instance to compare, with dimension N (features number), and x_i and y_i denote the ith feature of X and Y respectively.

For the implementation of the clustering method, we used Weka [13], which is a free machine learning package that has implemented k-means. Furthermore, it has the advantage that it is easy to add a new clustering method.

3   Iterative Clustering Method

The process of clustering is based on the following steps:

Step 1: Select a tentative k, preferably a higher value than expected. Run k-means with the data.

Step 2: After getting the first set of clusters, they are evaluated based on measures of compactness and separation of clusters. Clusters with low separation between theirs centroids are merged into one. By other hand the compactness is used to divide the database, this means that clusters with low compactness are used to build the new database to repeat the clustering process returning to step 1.

Step 3:   Once the process is stable, that means the compactness and separation are lower than a threshold, the last step is to minimize, if possible, the number of clusters. Clusters evaluation is repeated, for all clusters resulted of each iteration of k-means.

At the beginning of the process if necessary select the appropriated features and distance measure. For this problem we use the sensitivity of clusters to evaluate them.

In short, the general idea of this clustering method is seek clusters with a high sensitivity. In metagenomics the aim to assign the sequences to a phylum is associated with the sensitivity taking into account the phylum that best represents each cluster. That means the sensitivity is measured focus on the percentage that represents each organism in each cluster.

3.1   Performance Measures

There are some measures in the literature to assess performance of clusters. Here we use measures based on the pairwise difference of between and within-cluster distances. These measures are used to evaluate the cluster in each step of the proposed method and join similar clusters.

Furthermore, to assess the performance of clustering we focused on the final composition of the cluster, that is the number of different groups, and the purity of clusters, understanding by "pure cluster" a cluster with genomic fragments that belong to only one organism. Considering that clustering in metagenomics is a way to reduce the time-consumer of methods based on similarity of sequences it is more important getting clusters with a predominance of a phylogenetic group. Keeping this in mind, we use binomial estimator (equation 3) as a measure of sensitivity to evaluate the results. Although, this measure is used for binary problem, here we suppose the predominant sequences in each cluster as the positive, and the other as negative. The sensitivity is computed for each cluster meaning a range of pureness. The general sensitivity is computed by the average of all cluster sensitivity.

$$Sensitivity = \frac{number\ of\ positive}{total} \tag{3}$$

Each cluster is labeled with the organism which has the greatest number of sequences inside. The organism belonging to one cluster with different label is considered wrong. The sensitivity can be computed by cluster and average the results of other clusters.

4   Results and Discussion

In this paper a clustering method based on repeating a classical clustering algorithm (k-means algorithm) consecutively by a set of data composed of the "bad" clusters is proposed. A cluster is considered "bad" when its compactness is low.

A metagenome database built from 8 different organisms is used to evaluate the method.

Some different attributes are used to describe the sequences: GC content, nucleotides frequencies, codon frequencies and tetranucleotides.

Euclidean and Cosine distances were used for the k-means algorithms.

The first step was to select the best features to describe the data. This selection was focused on the result of a k-means with k between 5 and 15. The estimation was only based on the sensitivity of clusters. As explained before, our aim is to obtain pure clusters despite some organism can be divided in different clusters. Later, these clusters of genomic fragments can be classified using a supervised algorithm easier and faster. It is more important to have clusters with only one organism than to group genomic fragments of the same organisms together. For this reason the sensitivity, which here represents the percentage of the predominant organisms in each cluster, is a good measure for clustering.

The best result was obtained with k=15, tetranucleotides as features and Cosine distance. Figure 1 shows this result. The left part of the figure represents the number of clusters, the organisms assigned and the number of fragments associated with each organism. It can be seen most of clusters have a percentage relative to the predominant organism superior of 90%. The sensitivity was 92.85%, nevertheless the organism are very scattered.

Once selected the representation of data and the parameters of the k-means we test the proposed method. Starting from the result obtained before we go to the step 2 in order to evaluate the different clusters to merge closer clusters and separate clusters less compact for the next step. The process was repeated five times until to achieve the stability of the model.

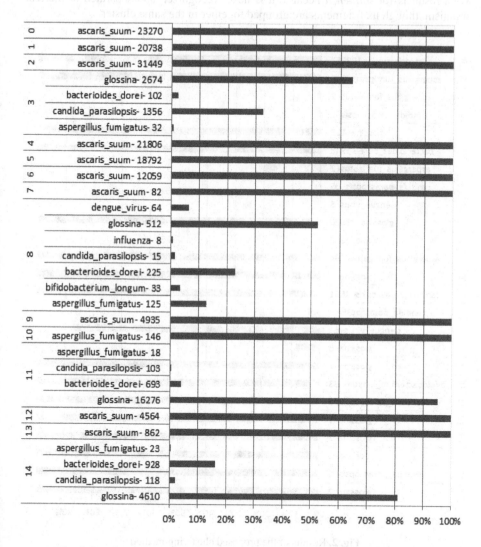

Fig. 1. Results of classical k-means with tetranucleotides frequencies as features

Final result is better compared with the first one (simple k-means). Figure 2 shows the results of the last step of the model yielding a 99.1% of sensitivity of the clusters, which results are in the range of 87.14 and 100%. The error of misassigned sequences is 5.516%.

With a more deep analysis of the results, we can see that *ascaris suum, bacterioides_dorei, bifidobacterium_longum* are completely grouped in clusters 0, 3 and 10 respectively. *Aspergillus_fumigatus* is grouped into 4 different clusters, but has two clusters complete for it. *Dengue* is divided into two groups. *Glossina* and *Candida* are more partitioned, although, *Glossina* leads all the clusters to which it belongs. The worst result is for *Influenza* because it is never recognized and separated of the rest organism, though its fragments are grouped together in the same cluster.

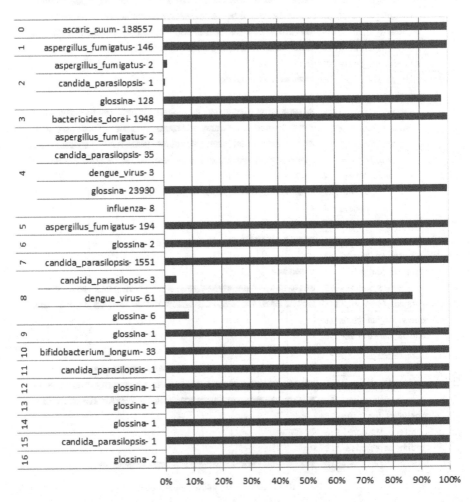

Fig. 2. Results of the proposed clustering method

In short, the results presented by the iterative use of k-means are superior of the only one running of k-means. By the application of metagenomics means an advantage this kind of group, although the patterns are divided in different group. Taking into account this, the error of the model based on the count of sequence misassigned is 0.045.

This paper is not intended to show the best clustering method for metagenomics, but rather to show a promising method to bear in mind for this area.

This iterative algorithm can be used with other base clustering method such as SOM or Expectation Maximization. In future work we expect compare the proposed method with other base methods and other metagenome databases.

5   Conclusions

In this paper we present an approach based on the iterative application of k-means to pattern that belongs to "bad" cluster. The classification of cluster is focused on validation measures of the compactness and separation cluster. The proposed method is applied to a metagenome dataset composed of 8 different organisms. The result achieved by the proposed method, in line with the objective of obtaining clusters with high sensitivity, outperforms result obtained with a simple k-means. Taking into account the error, the proposed method improves the purity of clusters by 5.471%. The results presented here do not mean that the method described here is better than other clustering methods for any metagenomic problem, but it is a promising method to bear in mind.

Other clustering methods can be used as the base for the proposed algorithm. This proposed method can also be applied to other metagenome databases.

References

1. Council, N.R.: The New Science of Metagenomics: Revealing the Secrets of Our Microbial Planet. The National Academies Press (2007)
2. Wu, Y.-W., Ye, Y.: A Novel Abundance-Based Algorithm for Binning Metagenomic Sequences Using l-Tuples. In: Berger, B. (ed.) RECOMB 2010. LNCS, vol. 6044, pp. 535–549. Springer, Heidelberg (2010)
3. Reddy, R.M., Mohammed, M.H., Mande, S.S.: MetaCAA: A clustering-aided methodology for efficient assembly of metagenomic datasets. Genomics 103, 161–168 (2014)
4. Camacho, C., Coulouris, G., Avagyan, V., Ma, N., Papadopoulos, J., Bealer, K., Madden, T.: BLAST+: architecture and applications. BMC Bioinformatics 10, 421 (2009)
5. McHardy, A.C., Martin, H.G., Tsirigos, A., Hugenholtz, P., Rigoutsos, I.: Accurate phylogenetic classification of variable-length DNA fragments. Nat. Meth. 4, 63–72 (2007)
6. Huson, D.H., Auch, A.F., Qi, J., Schuster, S.C.: MEGAN analysis of metagenomic data. Genome Research 17, 377–386 (2007)
7. Chan, C.-K., Hsu, A., Halgamuge, S., Tang, S.-L.: Binning sequences using very sparse labels within a metagenome. BMC Bioinformatics 9, 215 (2008)
8. Teeling, H., Waldmann, J., Lombardot, T., Bauer, M., Glockner, F.: TETRA: a webservice and a stand-alone program for the analysis and comparison of tetranucleotide usage patterns in DNA sequences. BMC Bioinformatics 5, 163 (2004)
9. Abe, T., Kanaya, S., Kinouchi, M., Ichiba, Y., Kozuki, T., Ikemura, T.: Informatics for Unveiling Hidden Genome Signatures. Genome Research 13, 693–702 (2003)
10. Li, W., Fu, L., Niu, B., Wu, S., Wooley, J.: Ultrafast clustering algorithms for metagenomic sequence analysis. Briefings in Bioinformatics 13, 656–668 (2012)

11. Kunin, V., Copeland, A., Lapidus, A., Mavromatis, K., Hugenholtz, P.: A Bioinformatician's Guide to Metagenomics. Microbiology and Molecular Biology Reviews 72, 557–578 (2008)
12. MacQueen, J.: Some methods for classification and analysis of multivariate observations. In: Proceedings of the Fifth Berkeley Symposium on Mathematical Statistics and Probability. Statistics, vol. 1, pp. 281–297. University of California Press, Berkeley (1967)
13. Witten, I., Frank, E.: Data Mining: Practical Machine Learning Tools and Techniques. Morgan Kaufmann, San Francisco (2005)

Is There a Crowd? Experiences
in Using Density-Based Clustering and Outlier Detection

Mohamed Ben Kalifa, Rebeca P. Díaz Redondo,
Ana Fernández Vilas, Rafael López Serrano, and Sandra Servia Rodríguez

Information & Computing Lab. AtlantTIC Research Center,
School of Telecommunications Engineering. University of Vigo, 36310 Spain
{mbk,rebeca,avilas,rafael,sandra}@det.uvigo.es

Abstract. The massive growth of GPS equipped smartphones coupled with the increasing importance of Social Media has led to the emergence of new location-based services over LBSNs (Location-based Social Networks) which allow citizens to act as social sensors reporting about their locations. This proactive social reporting might be beneficial for researchers in a wide number of scenarios like the one addressed in this paper: monitoring crowds in the city involving an assembly of individuals in term of size, duration, motivation, cohesion and proximity. We introduce a methodology for crowd-detection that combines social data mining, density-based clustering and outlier detection into a solution that can operate on-the-fly to *predict public crowds*, i.e. to foresee, in short term, the formation of potential multitudes based on the prior analysis of the region. Twitter is mined to analyze geo-tagged data in New York at New Year's Eve, so that those predictable public crowds are discovered.

Keywords: data mining, Location-based social network, crowd detection, citizen-as-a-sensor, density-based clustering, Twitter.

1   Introduction

The large amount of data generated through LBSNs allows researchers to investigate human behavior in this spatial, temporal and social context by combining geographical trajectories with the online social friendship [1–3]. Furthermore, in smart communities and cities, citizens really act as social sensors, reporting on their physical locations and social interactions with others. One usage scenario for this data may be one of monitoring multitudes by considering smartphones and social media as a comprehensive source of all kinds of information. In the field of crowd research there is no consensus on the definition of a crowd, to give some examples - *"a crowd is a temporary gathering of individuals who share a common focus of interest"* [4]. For Reicher on the other hand, a crowd is only a crowd when *"individuals share a social identity"* [5]. Regardless of the differences in the core of these definitions, in this paper, a "crowd" is defined as a group of individuals at the same physical location at the same time. However, there is no compromise about the minimum amount of people to be considered a crowd, the membership to a crowd and even the different

R. Prasath et al. (Eds.): MIKE 2014, LNAI 8891, pp. 155–163, 2014.

types of crowds and their classification. Le Bon [6] held that crowds existed in two kinds: (i) heterogeneous crowds that can be anonymous (street crowds, etc.) or not anonymous (juries, parliamentary assemblies, etc.) and (ii) homogeneous crowds that include sects (political sects, religious sects, etc.), castes (the military caste, the priestly caste, the working caste, etc.) and classes (the middle classes, the peasant classes, etc.). In our opinion, these traditional classifications should have some reflect in Social Media that deserves to be researched.

Crowding phenomena is an interesting topic studied in the context of human sciences [5], models and tools to satisfactory describe behaviors and interactions between individuals into a crowd. Prior research in field Social media have shown that it is possible to detect some kind of disasters, such as earthquakes [7] and forest fires [8] by analyzing the information post by users (text, audio, messages, etc.). For example, TweetTracker [9] a Twitter-based analytic and visualization application that helps humanitarian assistance and disaster relief (HADR) organizations to acquire situational awareness during disasters and emergencies to aid disaster relief effort. Besides using Twitter, [10] aims to perform a nation-wide geo-social event detection by analyzing the content of a great number of messages. Our proposal, although using Twitter information, it is not analyzing the messages content, which is part of our further work, we have analyzed only the density of messages in specific areas by using the GPS location of the available tweets.

Following the classification proposed by Le Bon [4], our work focuses on the discovery of heterogeneous-anonymous crowds from geo-located Twitter data which representing an assembly of people together traced at the same period of time in order to present a solution to detect predictable public crowds, i.e. public crowds that may be foreseen from some vague knowledge of the time and the area in where they will happen, based on the combination of the density-based algorithm DBSCAN [11] together with outlier detection [12]. We tested the suitability of using these data mining techniques to discover crowds in New Year's Eve at the center of New York City by using a normal day for training the system.

The remainder of this paper is organized as follows. After overviewing technologies related with our approach (Section 2), Section 3 shows an overview of the strategy used to identify crowds and explains the technical processing of the dataset. Section 4 details the results of the experiment, finally, we interpret and describe the significance of our findings in Section 5.

2   Background on Clustering and Outlier Detection

LBSNs have emerged as a valuable source for data mining and analytics in order to infer the spatio-temporal patterns of people's daily routine and provide real-life applications, for instance emergency management, urban computing or disease propagation [13]. In the field of data analytics, clustering, as an important technique of exploratory data mining, is definitely appropriate to the aim of this paper: detecting crowds, i.e. group of people which is geographically near to each other. The goal of clustering is to group the data into meaningful classes according to some distance measure (geographical in our case). From the data mined on LBSNs, it is desirable that the clustering algorithm is able: (i) to discover clusters of arbitrary shapes; (ii) to handle noise; and (iii) to cluster without knowing the number of clusters in advance.

With all these considerations in mind, we consider *Density-based clustering* a good alternative; where clusters are formed as dense areas that are separated from sparse regions. Even inside the category of density-based clustering, there are different approaches: (i) DBSCAN [11] (*Density-Based Spatial Clustering of Applications with Noise*) makes clusters grow according to a density-based connectivity analysis; (ii) OPTICS [14] extends DBSCAN to produce a cluster ordering obtained from a wide range of parameter settings; and (iii) DENCLUE [15] clusters data samples based on a set of density distribution functions. Since neither a cluster ordering is needed nor distribution functions are available for the individuals, we consider the simpler DBSCAN to group individual in the smart city. One additional advantage of Density-based clustering, so including DBSCAN, is its ability to handle outliers to discard and/or detect failures in the dataset. In our proposal the application is clear: we need to detect when the size of a group of people located in a specific area at a given range of time is unexpected, i.e. it is clearly higher than the usual size.

3   Methodology and Experiment

Our analysis is mainly based on the density-based clustering algorithm DBSCAN, which is a data clustering algorithm that finds clusters through density-based expansion of seed points. DBSCAN's definition of cluster is based on the concept of density reachability: a point q is said to be directly density reachable by another point p if the distance between them is below a specified threshold ϵ and p is surrounded by sufficiently many points. Then, q is considered to be density able by p if there exists a sequence $p_1, p_2, ..., p_n$ such that $p_1 = p$ and p_{i+1} is directly density reachable from p_i. A cluster, which is a subset of the given set of points, satisfies two properties:

- All points within the cluster are mutually density-connected, meaning that for any two distinct points p and q in a cluster, there exists a point o such that both p and q are density reachable from o.
- If a point is density connected to any point of a cluster, it is also part of the cluster, otherwise, it is considered a noise point.

The application of DBSCAN requires fixing ϵ and *MinPts* as input parameters. Zhou et al. proposed a method to determine both input parameters by using a maximum likelihood estimation technique [16]. However, their algorithm is strongly conditioned by the application context and needs to be reformulated to work properly for crowd detection. In our application context, we propose ϵ to be fixed by specifying the geographical region under study as a circle with center p_0 and radius d (Fig. 1). Then, we define ϵ as the arithmetic average of the minimum distances $dist_i$ between any data point p_i in this region and all the data points in the dataset D.

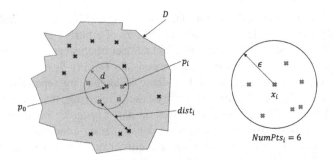

Fig. 1. Obtaining parameters ϵ and *MinPts*

It is needed to provide an indicator to decide when a group of people is a crowd. This measure has to detect high concentrations of people in a cluster on a specific location and its environment with respect to number of people in other clusters. It is not possible to assign a fixed number of persons as general indicator. The idea is to determine the number of tweets in a normal situation and consider it as a reference to detect multitudes in a special situation. This notion is similar to the calculation of outliers [12], considering the one-dimensional case of this technique as suitable for our circumstance. To implement this method, where the interquartile range (*IR*) is defined as the difference between the third quartile (Q_3) and the first quartile (Q_1), the outlier is calculated in a normal situation using the following formulation:

$$outlier := Q_3 + \delta \cdot IR$$

Considering (i) moderate outlier if $\delta = 1.5$ and (ii) extreme outlier if $\delta = 3$. Using the technique of outliers, we can say that points considered extremes are low likely to appear, and if it presents may will be considered strange.

Once we have introduced a crowd detection methodology for the smart city, which combines social data mining, density-based clustering and outlier detection, we report a particularized experiment for detecting predictable public crowds: public crowds whose formation is foreseen in short-term based on a prior analysis of citizens' habits according to both time and geographical area. Once fixed the area, we propose obtaining a set of values (number of clusters and average cluster size) for each hour and location in two different contexts. First, what we named "a normal day": a day that can be used as a reference of the typical distribution of the citizens all around the city. Second, what we named "a special day": a day when a special and massive event is going to take place, which is going to be used to compare the changes on the normal behavior of the citizens.

For this experiment, we have selected New York City as region of study and a very especial event "New Year's Eve" celebration, where people go to *Times Square*. We have studied how citizens behave for 24 hours in both circumstances (normal day and special day) in order to infer both: a normal pattern, conditioned by the time of the day and the geographical area, and an extraordinary pattern, conditioned by the same parameters, which could be considered an upper threshold. For that, we collected a sample of the geo-tagged tweets posted in New York City during a period of 24 hours starting on December 31st, 2013 at 3:00pm in GMT, ending up with a sample of 180.111 geo-located tweets. In order to compare the crowds obtained in New Year's Eve with the ones in an ordinary day, we collected a new sample, again for a period of 24 hours, starting on February 24th, 2014 at 3:00pm in GMT, obtaining 124.378 geo-located tweets in this new sample. As the most important crowd during New Year's Eve is expected to take place around Times Square, we only keep those tweets posted in a radius of 500 meters from the center of Times Square. We finally group tweets by hour of publication and run our experiment to detect the crowds every hour.

For EXTRACTING TWITTER DATA, this work uses the Twitter Streaming API. In New York City, we collect approximately 191.000 tweets for a period of 24 hours starting on the 3pm of December 31st of 2013. Being that our predictable public crowd (special day), to compare New Year's Eve to a normal day (no predictable public events), we collected 137.874 tweets, between 15:00 of February 24th and 15:00 of February 25th 2014 (normal day). After that, some PRE-PROCESSING is needed, mainly (1) fixing the DBSCAN input parameters by using a distance of 500 meters around the center of Times Square and (2) splitting the one day dataset to obtain 24 hours divided by hour period of time. Finally, for DETECTING CROWDS, we (3) apply DBSCAN with the fixed input parameters for every hour; and (4) apply the method of outlier to detect multitudes using normal day as reference of crowd measurement.

4   Findings

As mentioned, we have applied our methodology to crowds for New Year's Eve at New York City. After estimating ϵ and *MinPts* (a total of 24 parameters), we applied DBSCAN algorithm in an hourly basis, both in the special day and in the normal day. We have represented in Fig. 2 number of clusters in New York City at midnight, in a special day and in a normal day. This figure represents all the clusters around the city. From the graphic, in the New Year's Eve day, the number of clusters is observably higher than in a normal day, expecting the working period from 06:00 a.m. to 10:00 p.m.

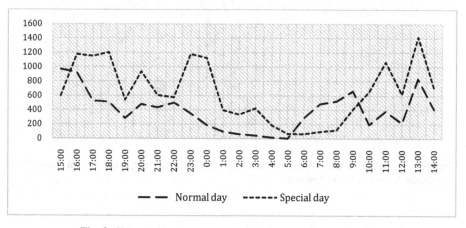

Fig. 2. Clustering results - Number of clusters in New York City

In special day, clusters number has also decreased between 00:00 a.m. and 08:00 a.m. Fig. 3 illustrates the variation in the number of clusters for 24 hours in New York around 500 meters from the center of the city. It shows obviously a dissimilar behaviour in both days. Otherwise, in a normal day, number of clusters has reached its maximum in the morning of the next day comparing to a special day. This gives an impression about the scheduling of crowd activity and the place where people tend to move around frequently.

Fig. 4 provides a descriptive view of all the clusters at Midnight in the city. Apart from estimating parameters and applying DBSCAN to the data, outlier detection is considered as primary source for detecting crowds

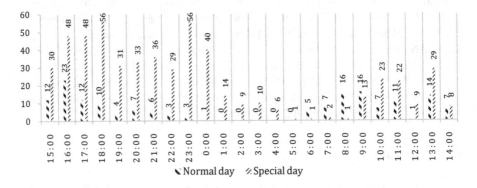

Fig. 3. Number of clusters in 500 meters from Times Square in normal day and special day

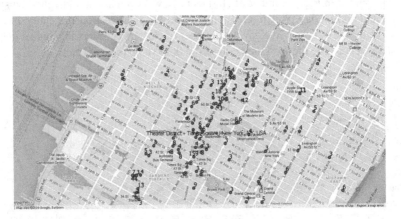

Fig. 4. Geographic representation of clusters at midnight in special day in New York City(*Times Square* is the yellow star)

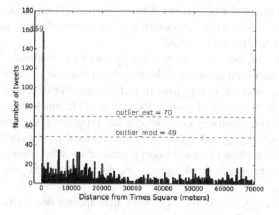

Fig. 5. Number of tweets per distance from Times Square – New year's Eve – 00:00 a.m.

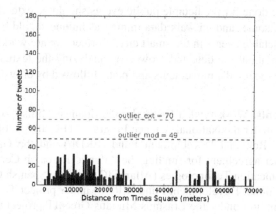

Fig. 6. Number of tweets per distance from Times Square – Normal day – 00:00 a.m.

Figures 5 and 6 show the variation of the number of tweets at Midnight of New Year's Eve and of the normal day from Times Square to 70 Km. Superior horizontal line represents an outlier extreme and inferior line an outlier moderate. At Midnight in New Year's Eve (Fig. 5), there is one cluster exceeding extreme outlier with 159 tweets. In a normal day (Fig. 6), no cluster reaches neither the moderate nor the extreme condition for outliers.

5   Discussion and Further Work

In this paper, we have proposed a crowd detection methodology based on Twitter Data Mining, density-based clustering and outlier detection to detect crowds in the smart city. Although infrastructure sensors in the smart city may detect crowds, their coverage can hardly reach the one obtained for the growing number of Internet-connected devices and the increasing participation of citizens at Social Media. Thus, citizens turn into sensing citizens (citizen-as-a-sensor), which constitute a more cost-effective solution by providing real-time and geo-located readings.

In our experiment, information provided by Twitter is randomly selected from the Total Twitter data, which means that the possibility that the same user's tweet is duplicated, on hourly interval, is very low. In fact, the number of people posting tweets is less than the actual number of people located in NYC but still the number of geo-located tweets is significatively considerable and represents the complete set of Twitter data sample when geographic boundary box are used for data collection as authors of [17] claim. We assume that each tweet represent a person. However, we are aware that there is more people there (those who are not tweeting). So, our approach is conservative.

To our knowledge, this is the first work that applies density-based clustering to crowd detection from Twitter. Furthermore, on this basis, we report a summary of the result of the experiment conducted in New York City. As the mechanisms for estimating DBSCAN parameters and outlier condition highly depend on the kind of event, the work we have done for predictable public events should be extended to other kind of crowds. Also, some random walk data mining technique should be established for detecting unpredictable events in the smart city. Further, we are working on obtaining more information about the detected crowds by analyzing the textual information in the tweets and analyzing the movements and paths followed by crowds in the cities.

Acknowledgements. Work funded by the by the Spanish Ministry of Economy and Competitiveness under the National Science Program (TEC2013-47665-C4-3-R); and by the European Regional Development Fund (ERDF) and the Galician Regional Government under agreement for funding the Atlantic Research Center for Information and Communication Technologies (AtlantTIC); and the Spanish Government and the European Regional Development Fund (ERDF) under project TACTICA; and the European Commission under the Erasmus Mundus GreenIT project (GreenIT for the benefit of civil society. 3772227-1-2012-ES-ERA MUNDUS-EMA21; Grant Agreement n° 2012-2625/001-001-EMA2). The authors also thank GRADIANT for its computing support.

References

1. Cheng, Z., Caverlee, J., Lee, K., Sui, D.: Exploring Millions of Footprints in Location Sharing Services. In: ICWSM 2011, pp. 81–88 (2011)
2. Zheng, Y., Xie, X.: Ma GeoLife: A Collaborative Social Networking Service among User, Location and Trajectory. IEEE Data Eng. Bull. 33, 32–39 (2010)
3. Gao, H., Liu, H.: Data Analysis on Location-Based Social Networks. In: Mob. Soc. Netw., pp. 165–194. Springer, New York (2014)
4. Forsyth, D.: Group dynamics, 5th edn., vol. 40, p. 9823 (2009)
5. Reicher, S.: The Psychology of Crowd Dynamics. Psychol. Soc. 44, 113–128 (2012)
6. Le Bon, G.: The Crowd. Transaction Publishers (1994)

7. Sakaki, T., Okazaki, M., Matsuo, Y.: Earthquake shakes Twitter users: real-time event detection by social sensors. In: Proc. 19th Int. Conf. World Wide Web, pp. 851–860 (2010)
8. De Longueville, B., Smith, R., Luraschi, G.: OMG, from here, I can see the flames!: a use case of mining location based social networks to acquire spatio-temporal data on forest fires. In: Proc. 2009 Int. Work. Locat. Based Soc. Networks, pp. 73–80 (2009)
9. Kumar, S., Zafarani, R., Liu, H.: Understanding User Migration Patterns in Social Media. In: AAAI 2011 (2011)
10. Lee, R., Sumiya, K.: Measuring geographical regularities of crowd behaviors for Twitter-based geo-social event detection. In: Proc. 2nd ACM SIGSPATIAL Int. Work. LBSNs, pp. 1–10 (2010)
11. Ester, M., Kriegel, H., Sander, J., Xu, X.: A density-based algorithm for discovering clusters in large spatial databases with noise. In: KDD 1996, pp. 226–231 (1996)
12. Beniger, J.R., Barnett, V., Lewis, T.: Outliers in Statistical Data. Contemp. Sociol. 9, 560 (1980)
13. White, D.J., Chang, H.G., Benach, J.L., et al.: The geographic spread and temporal increase of the Lyme disease epidemic. JAMA 266, 1230–1236 (1991)
14. Ankerst, M., Breunig, M.M.M., Kriegel, H.H., Sander, J.: Optics: Ordering points to identify the clustering structure. ACM SIGMOD Rec., 49–60 (1999)
15. Hinneburg, A., Gabriel, H.H.: Denclue 2.0: Fast clustering based on kernel density estimation. In: Berthold, M., Shawe-Taylor, J., Lavrač, N. (eds.) IDA 2007. LNCS, vol. 4723, pp. 70–80. Springer, Heidelberg (2007)
16. Zhou, H., Wang, P., Li, H.: Research on Adaptive Parameters Determination in DBSCAN Algorithm. J. Inf. Comput. Sci. 9, 1967–1973 (2012)
17. Morstatter, F., Pfeffer, J., Liu, H., Carley, K.: Is the Sample Good Enough? Comparing Data from Twitter's Streaming API with Twitter's Firehose. In: ICWSM (2013)

Detecting Background Line as Preprocessing for Offline Signature Verification

K. Rakesh[1,2] and Rajarshi Pal[1]

[1] Institute for Development and Research
in Banking Technology (IDRBT), Hyderabad, India
[2] School of Computer and Information Sciences,
University of Hyderabad, Hyderabad, India
rakesh.k0710@gmail.com, iamrajarshi@yahoo.co.in

Abstract. Hand-written signature is commonly used for authenticating a person. Extraction of desired features from the captured signature image is crucial for automated signature verification. Presence of a background line (which may be a part of a paper form) is common in an offline signature. Removal of this kind of background line is necessary for correct extraction of features. But sometimes, a signature contains a line as part of it. This paper shows how intensity distributions can distinguish a background line from a line which is part of a signature.

1 Introduction

With increased usage of computers in every sphere of life, the electronic verification of an individual's identity has become essential and this has inspired the development of various automatic identification systems. Different types of identification systems are: possession based identification systems (key or badge), knowledge based identification system (password), and biometric based identification systems [6]. Biometric based identification again can be classified as physical biometric based (e.g., fingerprint, iris, palm, etc) and behavioral biometric based (e.g., voice, signature) [16]. Signature verification is classified as a behavioral identification because it is based on behavioral traits of individuals [12].

Signature is legally and socially accepted means of identifying a person [10]. Therefore, a good amount of research efforts have been directed to come up with methods of automated signature verification [1,5,6,8,11,15,17]. Automated signature verification methods can again be of two types: online and off-line verification [15]. Only static information such as shape related information is available for off-line signature verification system. Off-line signature verification systems are of interest in scenarios where only hard copies of signatures are available [12]. It is commonly used in verifying cheques in banks, passport verification system, document authentication, authenticating the candidates in public examinations, etc. [14]. The work reported in this paper is restricted to off-line signature verification system.

R. Prasath et al. (Eds.): MIKE 2014, LNAI 8891, pp. 164–171, 2014.
© Springer International Publishing Switzerland 2014

Success of signature verification depends on accurate extraction of discriminating features. Various preprocessing techniques like separation from background, background line removal, smoothing, segmentation are often applied in the acquired signature image. Figure 1 shows a few signatures having line at the background. Overlapping of signature with background line leads to serious recognition problems. So in an automatic signature verification system, the line is to be removed as a pre-processing step. But there may be instances of the line being part of the signature as shown in figure 2.

Fig. 1. Signatures with line as a background pattern

Fig. 2. Signatures with line as part of it

The line present in the signature image can be detected using techniques such as morphological filters, horizontal projection profile, an Hough transform [13]. Works in [3,4] show how morphological filters can be applied to detect lines from hand-written text/signature. The horizontal projection profile can be used to detect the horizontal line in the signature image as in [2]. Horizontal projection profile (HPP) is nothing but the number of black pixels in each row of a signature image. Hough transform can also be used for detecting lines present in the signature image. Whereas horizontal projection profile only detects horizontal lines, Hough transform is general enough to detect lines at any orientation.

But these techniques remove the line irrespective of whether it is part of the signature or the background. The work, reported in this paper, attempts to detect whether a line is a part of the signature or the background. The line should not be removed if it is a part of the signature. Therefore, the contribution of this paper is to determine whether a line is part of the signature or the background by analysing the difference of intensity distributions of the pixels belonging to the line and rest of the signature.

An outline of the rest of the paper is given here: A method to identify presence of a line is described in Section 2. Section 3 explains the theory behind distinguishing the line as part of the background or the signature. Section 4 presents and analyzes the experimental results. Finally, conclusive remarks are drawn in Section 5.

2   Identifying the Presence of a Line

At First, the signature image is binarized and thinned. Then, the line present in the thinned signature image is identified using edge point detection followed by Hough transformation. These steps are explained in this section.

Binarization is used to locate the signature strokes from the signature image. Binarization is the process of converting gray scale image into binary image. This has been done by classifying the pixels based on an appropriate threshold. This threshold has been calculated using Otsu's method [9]. Figure 3 is a gray level signature image that have a line as background pattern. Figure 4 is a binary version of the signature image in figure 3 after applying Otsu's thresholding.

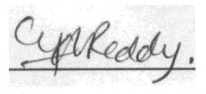

Fig. 3. A gray level signature image

Fig. 4. The binary image obtained from the image in figure 3

To make signature verification invariant to the thickness of the pen, thinning is applied on the signature image [7]. The black pixel's positions (from the binary image) and corresponding gray values (from the input signature image) are stored in gray level thinned image as shown in Figure 5.

Then Canny edge detector is used for identifying a set of edge points. Hough transform uses these edge points to detect the line that is present in the thinned signature image. In figure 6, the line segment is highlighted using a different color.

3   Categorizing the Line: Theoretical Fundamentals

The steps described in previous section determines whether a line is present in a signature image or not. If the line is not present in the signature image,

Fig. 5. The thinned gray level image

Fig. 6. Signature image with highlighted line segment

then the signature image can directly be used for feature extraction for off-line signature verification system. If a line is present in the signature image, then it is needed to find out whether the line is part of the signature or the background. A background line has to be removed for accurate extraction of signature features, whereas a line can also be an integral part of a signature as shown in figure 2. This section provides the fundamental theory of distinguishing a line as part of the background from a line as part of the signature.

The signature pixels are extracted from the thinned gray level image (figure 5) by excluding line pixels. The gray level distribution of the line pixels is obtained by finding the probability of each gray level that is present in the thinned line. The gray level distribution of the signature pixels is obtained by finding the probability of each gray level that is present in thinned signature strokes except pixels belonging to the line. To distinguish whether both the distributions of line and signature pixels are same or not, Behrens-Fisher statistics t is estimated using the following expression:

$$t = \frac{|\bar{x}_1 - \bar{x}_2|}{\sqrt{\frac{\sigma_1^2}{n_1} + \frac{\sigma_2^2}{n_2}}} \qquad (1)$$

where $\bar{x}_1$ and $\bar{x}_2$ are gray level means of signature pixels and line pixels, respectively. σ_1^2, σ_2^2 are variances of signature pixels and line pixels, respectively, and n_1, n_2 are number of pixels in signature and line, respectively. If Behrens-Fisher score is low, then both distributions are same, i.e., the line is part of the signature. If Behrens-Fisher scores are high, then both distributions are different, i.e., line is not part of the signature, rather it is part of the background.

4 Experimental Analysis

In figure 7, distributions of obtained Behrens-Fisher scores are presented for 175 signature samples. The line is part of the background in 100 out of 175 signature samples and the line is part of the signature in 75 signature samples. X-axis represents Behrens-Fisher score and Y-axis represents the fraction of cases (with respect to 175 signature samples) getting that score. The distribution in red color is the Behrens-Fisher score distribution of signatures where the line is part of the signature. The other distribution in blue color represents the Behrens-Fisher score distribution of the signatures where the line is part of the background. This figure suggests the following:

- Low Behrens-Fisher score is achieved where the line is part of the signature.
- High Behrens-Fisher score is obtained if the line is not part of the signature, rather part of the background.
- It can be observed in figure 7 that there is a slight overlapping region in the middle. If a Behrens-Fisher score falls in this region, it can not be surely claimed whether the line is part of the signature or part of the background. This leads to a fuzzy theoretic approach of explaining the result.

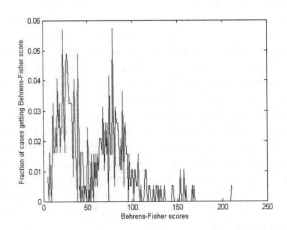

Fig. 7. Distributions of Behrens-Fisher statistics for line as part of the signature (in red) and line as part of the background (in blue)

Let the obtained Behrens-Fisher score be t. Moreover, let $m_1(t)$ and $m_2(t)$ be the number of cases with Behrens-Fisher statistic t where the line is a part of the signature and the line is not part of the signature, respectively.

The fact that a line is part of the signature can be claimed with a fuzzy membership value $\mu_1(t)$

$$\mu_1(t) = \frac{m_1(t)}{m_1(t) + m_2(t)} \tag{2}$$

Similarly, a line is part of the background can be claimed with a fuzzy membership value $\mu_2(t)$

$$\mu_2(t) = \frac{m_2(t)}{m_1(t) + m_2(t)} \tag{3}$$

Table 1. Results on a test set of signature samples

	Line is part of the background	Line is part of the signature	Total
Number of samples	76	46	122
Correctly identified	67	43	110
Wrongly identified	9	3	12
Accuracy (%)	88.16	93.48	90.16
Error rate (%)	11.84	6.52	9.84

The following observations can be drawn from Figure 7:

- If the Behrens-Fisher score is very low, then, $m_2(t)$ is 0. Then $\mu_1(t) = 1$. It can be claimed with surety that the line is part of the signature.
- If the Behrens-Fisher score is very high, then, $m_1(t)$ is 0. Then $\mu_2(t) = 1$. It can be claimed with surety that the line is not part of the signature.
- If the Behrens-Fisher statistic t lies on the overlapping portion then ambiguity arises. For example, if $m_1(t)$ is 0.005 and $m_2(t)$ is 0.01, then $\mu_1(t) = 0.33$ and $\mu_2(t) = 0.67$. It indicates the belongingness of the line within the signature as a fuzzy membership value 0.33. Moreover, belongingness of the line as not part of the signature can be represented as a fuzzy membership value 0.67. If a binary decision is required for practical applications, then a comparison between values of $\mu_1(t)$ and $\mu_2(t)$ decides in favor of the higher value.

The proposed method has been tested using another set of 122 signature samples. The line is a part of the background in 76 out of these 122 signature samples. The line is an integral part of the signature in rest of the 46 signatures. The experimental results are summarized in table 1. When the line is part of the background, 9 cases out of 76 signature samples were wrongly identified and 88.16% of accuracy is obtained using the proposed method. When the line is part of the signature, 3 cases out of 46 signature samples were wrongly identified and 93.48% of accuracy is obtained using the proposed method. Overall, 12 cases out of 122 signature samples are wrongly identified and an accuracy of 90.16% is obtained using this method.

Figure 8 represents a few signature samples on which the proposed method could not classify the line correctly. For top two signatures in figure 8, the line is originally part of the background but the proposed method has identified it as part of the signature. On the other hand, for bottom two signatures in figure 8, the line is originally part of the signature, but the proposed method has suggested it to be part of the background.

Figure 9 presents the intensity distributions of pixels belonging to the lines (in blue) and rest of the signatures (in red) from the signature samples in figure 8. It can be seen that these intensities are distributed over similar gray values. Thus, difference in intensity distributions can not provide correct conclusion in these 9.84% of cases (please see table 1).

Fig. 8. A few failure cases

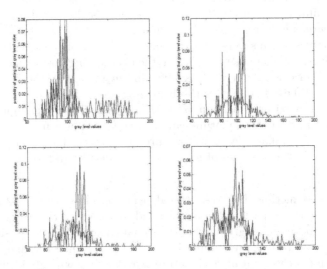

Fig. 9. Intensity distributions of pixels belonging to the lines (in blue) and rest of the signatures (in red) for the failure cases in figure 8

5   Conclusion

Main contribution of this paper is to show how difference in intensity distributions of pixels belonging to a line and rest of the signature can detect whether the line is part of the signature or not. Experimental results reveal a good accuracy in support of the proposed method though a few failure cases have also been analyzed. If binary classification is not required, then a fuzzy theoretic explanation can also be provided for the ambiguous cases as shown in Section 3.

Introducing a method to conclude a line (which may be present in a captured image of a signature) as part of the signature or the background, this work enriches the set of preprocessing techniques prior to feature extraction of offline signature verification.

References

1. Armand, S., Blumenstein, M., Muthukkumarasamy, V.: Off-line signature verification based on the modified direction feature. In: Proc. of IEEE International Conference on Pattern Recognition, pp. 509–512 (2006)
2. Arvind, K.R., Kumar, J., Ramakrishnan, A.G.: Line removal and restoration of handwritten strokes. In: Proc. of IEEE International Conference on Computational Intelligence and Multimedia Applications, pp. 208–214 (2007)
3. Bansal, A., Garg, D., Gupta, A.: A pattern matching classifier for offline signature verification. In: Proc. of First International Conference on Emerging Trends in Engineering and Technology, pp. 1160–1163 (2008)
4. Dimauro, G., Impedovo, S., Pirlo, G., Salzo, A.: Removing underlines from handwritten text: An experimental investigation. In: Downton, A.C., Impedovo, S. (eds.) Progress in Handwriting Recognition, pp. 497–501 (1997)
5. Ferrer, M.A., Vargas, J.F., Morales, A., Ordóñez, A.: Robustness of offline signature verification based on gray level features. IEEE Transactions on Information Forensics and Security 7(3), 966–977 (2012)
6. Jain, U.A., Patil, N.N.: A comparative study of various methods for offline signature verification. In: Proc. of International Conference on Issues and Challenges in Intelligent Computing Techniques, pp. 760–764 (2014)
7. Lam, L., Lee, S., Suen, C.Y.: Thinning methodologies - a comprehensive survey. IEEE Transactions on Pattern Analysis and Machine Intelligence 14(9), 869–885 (1992)
8. Munich, M.E., Perona, P.: Visual identification by signature tracking. IEEE Transactions on Pattern Analysis and Machine Intelligence 25(2), 200–217 (2003)
9. Otsu, N.: A threshold selection method from gray-level histograms. IEEE Transactions on Systems, Man, and Cybernetics 9(1), 62–66 (1979)
10. Pal, S., Alireza, A., Pal, U., Blumenstein, M.: Off-line signature identification using background and foreground information. In: Proc. of International Conference on Digital Image Computing: Techniques and Applications, pp. 672–677 (2011)
11. Pirlo, G., Impedovo, D.: Verification of static signatures by optical flow analysis. IEEE Transactions on Human-Machine Systems 43(5), 499–505 (2013)
12. Plamondon, R., Shihari, S.N.: On-line and off-line handwriting recognition: A comprehensive survey. IEEE Transactions on Pattern Analysis and Machine Intelligence 22(1), 63–84 (2000)
13. Prokopiou, K., Kavallieratou, E., Stamatatos, E.: An image processing self-training system for ruling line removal algorithms. In: Proc. of International Conference on Digital Signal Processing, pp. 1–6 (2013)
14. Randhawa, M.K., Sharma, A.K., Sharma, R.K.: Offline signature verification with concentric squares and slope based features using support vector machines. In: Proc. of IEEE International Advance Computing Conference, pp. 600–604 (2013)
15. Sanmorino, A., Yazid, S.: A survey for handwritten signature verification. In: Proc. of 2nd International Conference on Uncertainty Reasoning and Knowledge Engineering, pp. 54–57 (2012)
16. Wajid, R., Mansoor, A.B.: Classifier performance evaluation for offline signature verification using local binary patterns. In: Proc. of 4th European Workshop on Visual Information Processing, pp. 250–254 (2013)
17. Zhu, G., Zheng, Y., Doermann, D., Jaeger, S.: Signature detection and matching for document image retrieval. IEEE Transactions on Pattern Analysis and Machine Intelligence 31(11), 2015–2031 (2009)

Application of Zero-Frequency Filtering
for Vowel Onset Point Detection

Anil Kumar Vuppala

Language Technologies Research Centre
International Institute of Information Technology Hyderabad, A.P., India
anil.vuppala@iiit.ac.in

Abstract. Vowel onset points in speech signals, are the instances where the voicing of the vowels begin. These points serve as important landmarks for the analysis as well as synthesis of speech signals. These landmarks help to identify the information about the behaviour of transition of several different sounds into and out of the vowel regions. In this paper, we propose a new method to identify vowel onset points for a speech signal using the zero frequency filtered (ZFF) speech signal and its frequency spectrum. The ZFF signal is obtained by passing the speech signal through a resonator with central frequency as 0 Hz. Therefore, ZFF signal essentially contains the low pass components of a given speech signal. Vowels are mostly characterized by the significant energy content in the relatively low frequency bands. Significant improvement in VOP detection performance is observed using proposed method compared to existing methods.

Keywords: Vowel onset point (VOP), Vowels, zero frequency filtering, frequency spectrum.

1 Introduction

Vowel onset point (VOP) is defined as the instant at which the onset of vowel takes place in the speech signal. The importance of VOP for speech processing is observed in [1–9]: (i) Recognition of consonant-vowel (CV) units in Indian languages, (ii) Spotting CV segments in continuous speech, (iii) Determining the durations of the vowels for recognizing Indian languages and (iv) Speech rate manipulation. Performance of the above applications depend on accurate detection of VOP. VOP detection methods available in literature are based on raising trend of resonance peaks in the amplitude spectrum [10]; zero-crossing rate, energy and pitch information [11]; wavelet transform [12]; neural network [5]; dynamic time warping [13]; and excitation source information [14]. Recently, a method has been proposed by combining the evidence from excitation source, spectral peaks, and modulation spectrum [2] for the detection of VOP. Each of these evidence carries complementary information with respect to VOP.

Existing methods determine most of the VOPs within 40 ms deviation. Applications such as speech rate manipulation and CV unit recognition, require the

R. Prasath et al. (Eds.): MIKE 2014, LNAI 8891, pp. 172–177, 2014.

VOP detection with minimal deviation for their better performance. In particular, for speech rate manipulation (i.e., for slow down or speed up the speech rate) precise location of vowel onset points are very useful in providing the naturalness for the desired speech rate. Here, naturalness in output speech is retained due to preserving the consonant regions from the manipulation either for speeding up or slowing down the speech rate. In the same way, accurate VOP locations are important for CV unit recognition as well. In this paper, we propose a method for improving the VOP detection accuracy using zero frequency filtering (ZFF). ZFF signal is essentially the low pass component of a given speech signal and may be useful for detection of VOP onset points. This paper is organized as follows: Section 2 describes the baseline methods for the detection of VOP. Proposed method for detection of of VOPs is presented in section 3. The performance of the proposed method is analyzed using TIMIT database, and the results are presented in section 4. Section 5 provides the summary of the paper and further issues that need to be addressed.

2 Proposed VOP Detection Method

The zero frequency filter [15] has been proposed to extract the epochs or glottal closure instants for a given speech signal. The method involved passing the speech signal twice through a ideal digital resonator centred at 0 Hz. The resulting signal observes a polynomial type growth/decay imparted by the filter. A local mean subtraction for a repeating window of length approximately equal to the pitch period of the signal. The zero frequency signal is a low pass signal where the positive zero crossings indicate the location of the epochs. Based on this analysis we propose a new method to identify the vowel onset points in a given speech signal. Vowel onset points are generally characterized by the onset of high energy in low frequency bands. Because of the of the onset of periodicity resulting in low frequency bands incurring harmonic energies. Because of these reasons, we find a sudden onset of energy in the lower frequency bands. Application of a resonator at zero frequency might therefore be helpful towards identifying such an onset of energy near the low frequency bands. For the proposed method to detect VOPs, the speech signal is passed once through a resonator centred at 0 Hz. Resonator at 0 Hz is achieved by using the equation shown in 1.

$$h[n] = \frac{1}{(1 - z^{-1})^2} = \frac{1}{1 - 2z^{-1} + z^{-2}} \tag{1}$$

The output of this resonator is essentially a low pass signal observing a polynomial type growth/decay. The ZFF signal is then processed in blocks of 20 ms with a shift of 10 ms. For each block, a 256-point DFT is computed, and ten largest peaks are selected. The sum of these spectral peaks is plotted as a function of time. The change at the VOP available in the spectral peaks energy is further enhanced by computing its slope using FOD. These enhanced values are convolved with FOGD operator. The convolved output is the VOP evidence using spectral peaks energy. The advantages of this method lies with the use of zero frequency filtered signal instead of speech signal to calculate the summed energy of ten prominent peaks. As the ZFF signal is essentially low pass signal, most of the prominent peaks lie in low frequency regions compared to other segments of speech. Therefore, there is a sharp rise of energy in the sum value of first ten peaks in DFT spectrum at onset of vowel. Sequence of steps in the proposed VOP detection method are shown in Fig. 1.

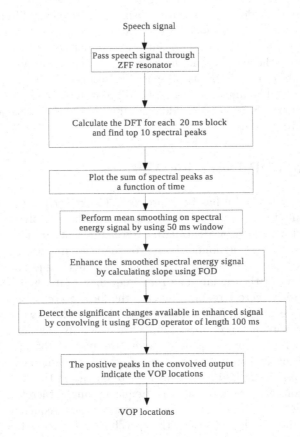

Fig. 1. Sequence of steps in the proposed VOP detection method

Proposed VOP detection method is demonstrated in Fig 2 using the continuous speech utterance *"Seeds of soybean cotton corn sesame and rape yield semidrying oils"*. Figs 2(a) shows the speech signal with manually marked VOPs. The final VOPs detected by the proposed method are observed to be very close to manually marked VOPs. Sum of spectral peaks in ZFF signal and its smoothed signal are shown in Figs. 2(b) and (c) respectively. 2(d) shows the enhanced signal correspond to the signal present in Fig. 2(c). Fig. 2(e) shows the VOP evidence signal obtained by convolving the enhanced spectral energy signal with FOGD. We can observe that manual VOPs marked in Fig. 2(a) and detected VOPs marked in 2(e) are close to each other.

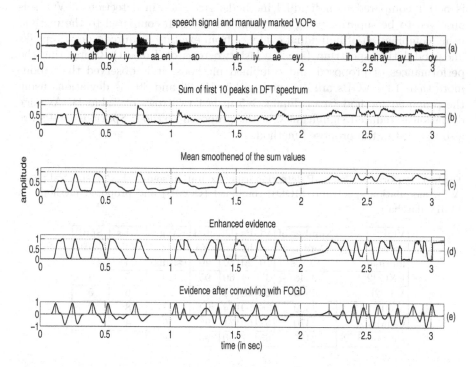

Fig. 2. VOP detection using proposed method for the utterance *"Seeds of soybean cotton corn sesame and rape yield semidrying oils"*. (a) Speech signal with manually marked VOPs. (b) Sum of 10 spectral peaks in ZFF speech signal. (c) Mean smoothed spectral energy. (d) Enhanced spectral energy signal. (e) Proposed VOP evidence signal.

3   Results and Discussion

Experiments are conducted on TIMIT database [16] for analyzing the performance of the proposed VOP detection method. About 220 sentences (120 sentences are spoken by female speakers and 100 sentences are spoken by male speakers) having 2407 manually marked VOPs are considered for analyzing the performance of the proposed VOP detection method. Among 2407 VOPs, 1013 VOPs correspond to the utterances spoken by male speakers, and the rest 1394 VOPs correspond to the utterances spoken by female speakers. Table 1 shows the accuracy in detection of VOPs using different methods. Column-1 indicates different methods considered in the analysis for detecting the VOPs. Column-2 indicates the total number of VOPs detected using various methods. Columns 3–6 indicate the percentage of VOPs detected within the specified deviations.

Column-7 indicates the average deviation (in ms) with respect to the manual marked VOPs. Columns 8 and 9 indicate the % of missed and spurious VOPs respectively.

From the results, it is evident that the performance of the combined method is better compared to individual methods. Accuracy in detection of VOPs is observed to be superior using the proposed method, compared to the method using combination of individual evidence from excitation source, spectral peaks and modulation spectrum (see the last two rows of Table 1). Comparing the performances of proposed and combined methods, it is observed that about more than 13% VOPs are detected within 10, 20 and 30 ms deviations using the proposed method (see columns 3–5 in last two rows of Table 1). Average deviation is reduced significantly in the proposed method. Spurious VOPs are also reduced in the proposed method.

Table 1. Performance of VOP detection using excitation source (EXC), spectral peaks (VT), modulation spectrum (MOD), combined (COMB) and proposed methods on TIMIT database

VOP detection method	Number of Detected VOPs	VOPs detected within ms (%)				Avg. dev. ($\approx$ ms)	Miss VOPs (%)	Spu. VOPs (%)
		10	20	30	40			
EXC [2]	2448	35	54	60	95	20	5	4
VT [2]	2500	26	46	70	94	21	6	5
MOD [2]	2330	36	50	74	92	18	8	2
COMB[2]	2470	52	60	71	96	16	4	3
Proposed [2]	2435	65	77	90	96	9	4	2

4   Summary of the Work

In this paper, accuracy VOP detection is improved using the low frequency characteristics of the zero frequency filtered speech signal. Performance of prosed method is compared using excitation source, spectral peaks energy, modulation spectrum and combined methods. Performance of the proposed method is evaluated using TIMIT database. More than 13 % improvement in VOP detection performance is observed using proposed method compared to existing methods. Significant reduction is average VOP detection performance is also observed. In future work, performance of proposed method needs to be analyzed for degraded speech.

References

1. Rao, K.S., Vuppala, A.K.: Non-uniform time scale modification using instants of significant excitation and vowel onset points. Elsevier Speech Communication 55(6), 745–756 (2013)

2. Prasanna, S.R.M., Reddy, B.V.S., Krishnamoorthy, P.: Vowel onset point detection using source, spectral peaks, and modulation spectrum energies. IEEE Trans. on Audio, Speech, and Language Processing 17(4), 556–565 (2009)
3. Prasanna, S.R.M., Gangashetty, S.V., Yegnanarayana, B.: Significance of vowel onset point for speech analysis. In: Proc. of Int. Conf. Signal Processing and Communications, Bangalore, India, pp. 81–88 (2001)
4. Vuppala, A.K., Rao, K.S., Chakrabarti, S.: Improved consonant-vowel recognition for low bit-rate coded speech. Wiley International Journal of Adaptive control and Signal processing 26(4), 333–349 (2012)
5. Gangashetty, S.V., Sekhar, C.C., Yegnanarayana, B.: Detection of vowel onset points in continuous speech using autoassociative neural network models. In: Proc. Int. Conf. Spoken Language Processing, Jeju Island, Korea, pp. 401–410 (2004)
6. Vuppala, A.K., Rao, K.S., Chakrabarti, S.: Spotting and recognition of consonant-vowel units from continuous speech using accurate vowel onset points. Springer Circuits, Systems and Signal Processing 31(4), 1459–1474 (2012)
7. Rao, K.S., Yegnanarayana, B.: Duration modification using glottal closure instants and vowel onset points. Speech Communication 51, 1263–1269 (2009)
8. Vuppala, A.K., Rao, K.S.: Speaker identification under background noise using features extracted from steady vowel regions. Wiley International Journal of Adaptive control and Signal processing 29(9), 781–792 (2013)
9. Vuppala, A.K., Yadav, J., Rao, K.S., Chakrabarti, S.: Vowel onset point detection for low bit rate coded speech. IEEE Transactions on Audio, Speech and Language Processing 20(6), 1894–1903 (2012)
10. Hermes, D.J.: Vowel onset detection. J. Acoust. Soc. Amer. 87, 866–873 (1990)
11. Wang, J.-F., Wu, C.H., Chang, S.H., Lee, J.Y.: A hierarchical neural network based C/V segmentation algorithm for Mandarin speech recognition. IEEE Trans. on Signal Processing 39(9), 2141–2146 (1991)
12. Wang, J.-H., Chen, S.-H.: A C/V segmentation algorithm for Mandarin speech using wavelet transforms. In: Proc. IEEE Int. Conf. Acoust., Speech, Signal Processing, Phoenix, Arizona, pp. 1261–1264 (1999)
13. Gangashetty, S.V., Sekhar, C.C., Yegnanarayana, B.: Extraction of fixed dimension patterns from varying duration segments of consonant-vowel utterances. In: Proc. of IEEE ICISIP, pp. 159–164 (2004)
14. Prasanna, S.R.M., Yegnanarayana, B.: Detection of vowel onset point events using excitation source information. In: Proc. of Interspeech, Lisbon, Portugal, pp. 1133–1136 (2005)
15. Murty, K.S.R., Yegnanarayana, B.: Epoch extraction from speech signals. IEEE Trans. on Audio, Speech, and Language Processing 16(8), 1602–1613 (2008)
16. Garofolo, J.S., Lamel, L.F., Fisher, W.M., Fiscus, J.G., Pallett, D.S., Dahlgren, N.L., Zue, V.: TIMIT acoustic-phonetic continuous speech corpus linguistic data consortium. In: Proc. of IEEE ICISIP, Philadelphia, PA (1993)

A Hybrid PSO Model for Solving Continuous
p-median Problem

Silpi Borah[1] and Hrishikesh Dewan[2]

[1] Department of Management Sciences,
Oxford College of Engineering, Bangalore
silpi.borah@ieee.org
[2] Department of Computer Science & Automation
Indian Institute of Science, Bangalore
hrishikesh.dewan@csa.iisc.ernet.in

Abstract. p-Median problem is one of the most applicable problem in the areas of supply chain management and operation research. There are various versions of these problems. Continuous p-median is one of them where the facility points and the demand points lie in an 'n' dimensional hyperspace. It has been proved that this problem is NP-complete and most of the algorithms that have been defined are mere approximations. In this paper, we present a meta-heuristic based approach that calculates the median points given a set of demand points with arbitrary demands. The algorithm is a combination of genetic algorithms, particle swarm optimization and a number of novel techniques that aims to further improve the result. The algorithm is tested on known data sets as and we show's its performance in comparison to other known algorithms applied on the same problem.

1 Introduction

Location based problems [1] are one of the most applied problems in the areas of supply chain management (SCM) and operation research (OR). The problem itself is an optimization problem, where we seek to find the most optimal location in a space where the distance between the demand points and the facilities are minimized. The simplest location problem is the weber problem [2] where a single facility or a point in a space needs to be determined that minimizes the distance of the points from the facility point. The weber problem is escalated to a multi-weber problem [3] where instead of a single facility multiple such facilities are required to be found. Although, a specific area of management science, location based problems are also an interesting area of research in data mining where cluster analysis is involved. Traditionally for a multi facility problem, there are three variants - continuous, discrete and network. In the continuous problem the facility locations are required to be identified in a hyperspace of d dimensions. The identifier locations may or may not be the demand points but could be any points within a defined search space. The discrete variant is the identification of facilities from the given set of demand points and in the network

R. Prasath et al. (Eds.): MIKE 2014, LNAI 8891, pp. 178–188, 2014.

problem the set of facilities is composed of both linear and continuous sets. In this paper, we attempt to solve the continuous multi facility location problem. Ideally, in any multi-facility location problem when applied for a real use case, there are constraints. Constraints could be areas in space which are unsuitable for use in the space due to competition or health hazard. We relax constraints in this work and assume that all areas are equally selectable. In several literature, this problem of identify facilities given a set of demand points is also known as p-median problem.

It has been proved that the p-median problem is NP hard [4]. As a result, a number of meta heuristic algorithms [5] were being applied to solve this problem. Simulated Annealing (SA) [6], Genetic Algorithm (GA) [7] , Tabu Search (TS) [8] etc. some of the algorithms that have been applied to solve this problem. In our paper, we have used a hybrid model compromising of both GA and a swarm intelligence technique called particle swarm optimization (PSO) [9] to find the best facility locations. The PSO algorithm is being modified to support better local search and we include Weiszfeld's [10] iterative method to further fine tune the results. We have modified the PSO algorithm by introducing mother particles that gives birth to new ones and perish or cease to exist when they no longer be able to produce better values for a period of time. The new born particles are restricted to a specific search space thereby providing us with more local search facilities to converge better and faster when the function space is convex. We demonstrate the usefulness of the algorithm by using different data sets that are available in public OR domain.

The rest of the paper is organized as follows. Section 2 introduces the problem more formally and our algorithm, Section 3 demonstrates the result and compares with the prior work in this regard. Finally in section 4, we conclude with an emphasis on the problems and future work in this regard.

2   Proposed Solution

Our solution to solve large un-capacitated problem is designed by using a swarm intelligence technique called Particle Swarm Optimization (PSO). PSO has been used for solving a wide variety of problems in areas such as machine learning, data mining, structural engineering etc. [11] provides a comprehensive overview of the different types of applications wherein PSO has been successfully applied so far. The traditional algorithm for PSO however suffers from the curse of dimensionality and doesn't lend itself for a better local search [12]. Therefore, instead of using the traditional PSO method, we have modified the algorithm by mixing some of the ideas from GA and several new techniques. PSO along with GA provides an elegant mix of both exploration and exploitation. For solving large scale problems using meta-heuristics, a combination of both exploration of the search space and exploitation of the best space is of utmost importance. To enhance further the results, we have introduced the notion of mother particles that has the ability to give birth to new particles and also delete them from the search space. In the sub-sections that follow we describe our solution in detail.

However, before we describe our solution, we first introduce the most essential aspect of the problem statement.

2.1 The p-median Problem

The p-median problem is a facility location problem where in there are N demand points with each demand point having a demand W and we are required to find K facility locations. The K facility locations are either demand points itself or they lie in the same hyperspace as in the demand points. In the former case, it is a discrete combinatorial optimization problem where as in the later we have a continuous problem. Mathematically, the continuous p-median problem can be expressed as in 1.

$$F(X) = \sum_{i=1}^{n} w_i \times d(x_i, z_k)$$
$$X = \{x_1, x_2, \cdots x_n\}$$
$$Z = \{z_1, z_2, \cdots z_k\}$$
$$x_{(im)} \in \left[l_m, u_m\right]$$
$$z_{(km)} \in \left[l_m, u_m\right] \tag{1}$$

$$m = 1, 2 \cdots M$$
$$l_m \in R$$
$$u_m \in R$$

As shown in equation 1, there are a total of N demand points. Each demand point x_i has a weight w_i. There are K facility locations and each facility location z_k is defined in the same space as the demand points. In other words, the dimensions (in total M dimensions) in which the demand points are described are exactly the same as that of the facility points. The dimension vector of each demand and facility location is defined and bounded by a distinct lower (l_m) and upper bound (u_m). The term $d(x_i, z_k)$ is a distance function and is defined as in equation 2. That is, we seek for the facility location z_k from Z which has the minimum distance from the demand point x_i. Thus, the function $F(X)$ is a minimization function.

$$d(x_i, z_k) = \min d(x, z) \tag{2}$$

In the most usual scenarios, the continuous p-median problem is restricted to a two dimensional space thus befitting the models of a location space. However, the number of dimensions could be increased to represent other choices apart from just the distance. The choice of a distance function also varies and is problem dependent choice. For example, the p-median problem solutions are also used as a basis for clustering large data sets. Clustering large data sets where in each data has a finite dimension of attributes (≥ 2) and each attribute can have both nominal, ordinal and numerical values are very common. Problems of such types are generally solved using manhattan (equation 4), cosine (equation 5) etc. In our problem, we have used '2' dimension and the Euclidean distance as shown in equation 3. In all the equations, x and y are two points in an n dimension space.

$$d = \sqrt{\sum_{i=1}^{n} (x_i - y_i)^2} \tag{3}$$

$$d = \sum_{i=1}^{n} |x_i - y_i| \tag{4}$$

$$d(x, y) = \frac{x.y}{||x|| * ||y||} \tag{5}$$

There are a number of solutions that have been found to solve this problem by using deterministic techniques. The Weiszfeld's solution, Lagrangian Relaxation etc. are some of the example solutions that have been tried so far. In the Weiszfeld's method initially selects arbitrary median locations in the space and iteratively modifies to reach the most optimal solution. The Weiszfeld's update equations is shown in equation 6.

$$T(z) = \frac{\sum_{i=1}^{n} \frac{w_i \cdot x_i}{d(z, x_i)}}{\sum_{i=1}^{n} \frac{w_i}{d(z, x_i)}}$$
$$z_{i+1} = T(z) \tag{6}$$

In the equation 6, the arbitrary median points are updated first using the demand points and corresponding weights. The iteration is continued till there is no change or a definite number of iterations are exhausted. We use Weiszfeld's solution in every iteration to further fine the result.

2.2 Genetic Algorithm

Genetic Algorithm (GA) is a meta-heuristic that is being used in a wide range of problems that are NP-hard. The idea behind the genetic algorithm is to encode solutions as chromosomes and successively modify it to yield the most optimal solution. The modification of the chromosome is done by using three operators : selection, cross over and mutation. Initially, we choose a fixed number of distinct points in the search space. Each point, which is a solution to the objective function is then encoded as a string. Each such representation is then evaluated with the objective function. The different values of the functions provides us a quantitative value of the solution. The selection operator selects a subset from this solution set. A simplest select operator is to select randomly a subset of solutions from the solution set. We use an elitist method, wherein, the best solution is retained and rest of the solution is randomly chosen to move to the next phase. In the next phase, the cross over operator is used. Two selected chromosomes from the last operation then exchange their encoded solution string. There are many varieties of cross over operators: single link, multi link etc. are few of the most used ones. We use double link crossover operator Finally, mutation operator is used to change the solution string to a certain degree. The three basic operators modifies the solution set obtained at each iteration to reach the local minima.

We use genetic algorithms as because they have a very good exploration capacity. When, the best chromosomes are selected and cross over is applied with the weak solution sets, the weak solution sets receives a part of the best solution set. As a result, they tend to move towards the better solution space. However, we do not apply genetic algorithms with all the solution sets. We select GA only with the neighborhoods where the best solution is not found. The GA algorithm is described in Algorithm 1.

Algorithm : GA Algorithm
Description : Algorithm computes the best solution in a sequence of iterations
Input : Number of Solutions 'NS', Total Number of Iterations 'TI'
Result: Best Solution Vector 'BS', Best Vector Value 'BV'
Begin

Initialize current iteration 'CI' ← 0.
Initialize vector of solutions
while *(CI ≤ NS)* **do**
> Evaluate all solutions
> select the best solution in the neighborhood
> elilist solution← best solution
> Add elilist solution to the next solution set
> **for** *(All other solutions)* **do**
> > select any two solutions
> > Do Double linkage crossover
> > Do mutation on both solution
> > Add it to the solution list
> **end**
> Increment CI
end

Algorithm 1: GA Algorithm

2.3   PSO

PSO is a meta-heuristic that has been inspired by the behavior of a flock of birds or boids of fish as they move along searching for food. In PSO, each solution in the search space is encoded as a particle. Each particle is evaluated using a fitness function and they interact with one another to propagate it's value. Unlike GA, every particle in PSO has a memory. Each particle stores the best value seen so far by itself and best value seen in the neighborhood. In the case where the particle itself is producing the best value in the neighborhood, both the memories are same. Formally, the PSO method can be described as in equation 7. Equation 7 is the most basic PSO update equations. p_{best} and $p_{current}$ are the particle's best position seen so far and current position respectively. They are vectors of dimension 'd' and each dimension update is independent of the other dimensions. g_{best} is the best solution seen so far. p_i and g_i are personal and social factors. c1 and c2 are constant values and they are generally in the range of [1.5-2.5]. v_{new} and v_{old} are the new and old velocity of the particles and they are of the same dimension as position vector. The algorithm moves in iterations and in each iteration it tries to move towards the best solution seen so far in the neighborhood. A particle uses both the best solution seen so far and the neighborhood best to update it's new position. The velocity parameters are initialized to a random value in [0-1] and at every iteration it is re-calculated to obtain a new value. The new velocity and old position is added to yield a new position. With the new position in place, the fitness function is used to calculate the value. The process continues till a certain desired value is reached or for a fixed number of iterations.

$$
\begin{aligned}
p_i &= 2 \times c_1 \times *(p_{best} - p_{current}) \\
g_i &= 2 \times c_2 \times (g_{best} - p_{current}) \\
v_{new} &= v_{old} + p_i + g_i \\
x_{new} &= x_{old} + v_{new}
\end{aligned}
\tag{7}
$$

In a PSO model, each particle is connected to other particles. Depending on the connection, a number of network topologies are formed. Mesh, Ring, Random and Dynamic are popular models of network topology that are used in evaluation of search space. The topology of the network has a lot of bearing in the trajectory of the particle and hence the solutions that are yielded. For example, in a mesh topology where each particle is connected to every other particle, the final solution so obtained is a local minima. This is due to the reason that the best solution has a direct impact on every other particle. As a result, in most experiments, it is the random or ring topology that is being used. Generally, the topology of the network remains static through out the network, but it has been observed that dynamic network topologies, where the topology of the network changes randomly or by iterations also yields better values.

Consequent to this traditional method of PSO, there are large number of variations [13] that have been designed. Improvement in velocity calculation are the most and the algorithm has been adapted to solve discrete combinatorial problems [14] as well. In this work, we use such a modified velocity calculation procedure where in the velocity is clamped to a certain value without letting the particle to explode from the search space. The equation for velocity update is shown in 8 and a more detail exposition of the same can be found in [15].

$$
\begin{aligned}
p_i &= 2 \times c_1 \times *(p_{best} - p_{current}) \\
g_i &= 2 \times c_2 \times (g_{best} - p_{current}) \\
v_{new} &= w * v_{old} + p_i + g_i \\
x_{new} &= x_{old} + v_{new}
\end{aligned}
\tag{8}
$$

2.4   A Hybrid PSO

In order to explore the search space more effectively and also have a distinct local search capability, we have modified the traditional PSO in a number of ways. First, is the inclusion of the GA. GA's are used in those spaces where the best neighborhood position doesn't have the global best location. In case of the neighborhood which has the global best, we take the best particle and allow it to have a number of new born particles. The new born particles are created within a distinct radius. It is within 5% of the best position. Thus, if the function is convex at this point, one of the children will have better fitness value. The new born children will be the new mother and there would be a re-birth of new children corresponding to the position of the new mother. The older children will die and they would be removed from the search space. Thus, if we start a PSO with N particles, then at next iteration, $N + M$ number of particles will be generated. Here M is the number of children produced. If on the other hand during the progress of the iteration, a new neighborhood has the best position, then the children of the old mother die and the process is continued for the new mother. The new algorithm that is executed in the PSO neighborhood is shown in Algorithm 2.

Algorithm : Hybrid PSO
Description : Computes the best solution and value using a modified PSO method
Input : Maximum Number of Iteratons, Number of Particles, Number of New Born Particles, Radius
Result: Best Solution, Best Value
Begin

Initialize Particles with random values in the search space
Initialize current iteration to 0
while *(current iteration¡ number of iterations)* **do**
  Evaluate all the particles
  Select the best solution in the neighborhood
  if *!(current best particle = preceding best particle)* **then**
    if *current best particle is not one among the children* **then**
      Delete the old children set
      Term the best particle as mother
    end
    Else **if** *current best particle is one among the children* **then**
      Term the best particle as mother
      Delete other children
    end
  end
  Create a circle with radius R around the mother's position
  Initialize N children within the created circle
  Evaluate the children
  Add them to the particles list
end

Algorithm 2: Hybrid Pso

2.5   Proposed Solution

Our algorithm is developed based on the Hybrid PSO as described in the preceding section and uses Weiszfeld's equation as shown in equation 6. The Weiszfeld's equation is used in every iteration to further enhance the solution. As shown in Algorithm 3, we first initialize a total of 'P' number of particles in the space. Each particle initialized 'K' number of points in the space. K is the number of medians that are sought. Thus each particle is responsible for computing the p-medians of the problem. The first iteration of the algorithm is always an PSO based evaluation. Each particle belongs to a neighborhood. A total of 'N' number of neighborhoods are defined. After the initial run, all particles are evaluated. The neighborhood with the best solution uses the hybrid PSO algorithm as shown in 2. Rest of the neighborhood uses a GA model as shown in 1 At every iteration, the Weiszfeld's method is used to further update the new vectors. If during the execution of the algorithm, a new best particle is found in a different neighborhood, the old best neighborhood is converted from the PSO to the GA mode, all its children die and the new neighborhood is converted to a PSO model. The execution continues till we reached the maximum number of iterations and the final result is produced.

Algorithm : *p*-median
Description : Computes the p-median given a set of demand points
Input : Demans Points, Number of Medians, Number of Neighborhood, Maximum Number of Iterations
Result: Median Location Vector, Median Value
Begin

Initialize current iteration to 0
Initialize particles
Create N Neighborhood
Assign Particles to distinct Neighborhood
while *(current iteration¡ number of iterations)* **do**
│ Evaluate each particle
│ Find the best particle
│ Find the best neighborhood
│ **if** *(best neighborhood method=GA)* **then**
│ │ Change to the hybrid PSO method
│ │ Apply Hybrid PSO to the best neighborhood
│ **end**
│ Else do nothing
│ **for** *(All other neighborhood)* **do**
│ │ **if** *(neighborhood method= PSO)* **then**
│ │ │ Change to GA method
│ │ **end**
│ │ Else do nothing
│ **end**
│ Store Best Value and Position
end

Algorithm 3: Algorithm for p-median problem

3   Experiment and Results

To evaluate the effectiveness of the algorithm, we used two data sets that are popularly used in testing continuous uncapacited *p*-median problems. The test set includes a set of 287 and 654 points. The demand of all these points are drawn from [16] and the points itself are used from [17]. These test set is

Table 1. Comaprison of various solutions of 287 points

p	StPSO	NPSO	StPSO+LS	NPSO+LS	MS+LS	Optima	HPSO1	HPSO2
2	14488.70	14448.88	14434.48	14431.92	14434.41	14427.59	14431.26	144228.91
3	13687.60	12563.79	12143.47	12161.42	12137.85	12095.44	12133.86	12101.36
4	12784.38	11734.25	10895.05	10735.74	10969.41	10661.48	10663.41	10662.81
5	12205.65	10588.56	10129.17	9836.13	10128.65	9715.63	9938.53	9921.73
6	11790.50	10243.52	9581.41	9068.72	9800.56	8787.56	9156.11	8991.57
7	11245.15	9732.47	8972.99	8354.21	9403.97	8160.32	9301.10	8821.32
8	11435.88	9428.73	8943.19	8047.47	9055.68	7564.29	8665.31	7975.19
9	10948.41	8877.79	8717.08	7844.99	8791.66	7088.13	8234.76	7093.13
10	10901.40	8869.72	8574.84	7492.21	8456.48	6705.04	7917.56	6911.32
11	10177.23	7938.26	8608.71	7016.39	8313.94	6351.59	7685.14	6961.80
12	10195.75	7535.52	8344.82	6609.31	7957.18	6033.05	7129.83	6632.81
13	9970.47	7509.22	8128.71	6705.37	7844.16	5725.19	6983.21	6372.17
14	9741.95	7345.54	7766.41	6279.56	7567.33	5469.65	6743.29	5961.82
15	9824.27	7540.35	7829.33	6195.82	7767.35	5224.70	5331.78	59918.97
16	9760.08	6984.96	7691.48	6472.31	7683.83	4981.96	7715.89	5893.78
17	9574.54	7270.75	7393.86	6132.09	7450.52	4755.19	6689.93	5319.89
18	9549.47	6934.07	7130.35	5677.44	7232.64	4547.37	5321.73	5961.63
19	8742.14	6642.70	7064.38	5683.48	6888.95	4342.06	5432.82	5121.73
20	9148.22	6634.67	7132.89	5436.79	7244.67	4148.84	4283.44	6135.12

Table 2. Comparison of various solutions of 654 points

p	StPSO	NPSO	StPSO+LS	NPSO+LS	MS+LS	Optima	HPSO1	HPSO2
2	816039.01	815637.89	815313.30	815313.30	815313.30	815313.30	825316.61	815313.20
3	571972.87	557574.70	551062.91	551062.91	551062.91	551062.88	551062.97	551062.10
4	380162.03	346377.24	288203.57	288200.68	288195.18	288190.99	288194.32	288191.71
5	314705.56	285146.13	209080.62	209076.34	209079.72	209068.79	209071.82	209069.33
6	305368.74	284262.86	180610.41	180551.82	180547.79	180488.21	180533.86	180499.33
7	269924.58	234270.74	164105.13	163951.16	163991.10	163704.1	163902.34	163708.21
8	239756.84	218274.21	148306.09	147634.04	148132.56	147050.79	147961.75	147851.70
9	235506.89	237490.09	131724.85	132028.37	132655.15	130936.12	132351.25	131012.06
10	235394.36	229802.04	117747.76	117041.09	116775.22	115339.03	116385.33	115899.09

Table 3. Sensivity Analysis of the Neighborhood Size

Neighborhood Size	Mean	Std Deviation
5	6903.07	±0.0032
8	6807.07	±0.0081
10	6801.03	±0.0321
15	6786.03	±0.0318
20	6803.04	±0.0081

popular and has been used to compare algorithms in other works such as [18], [19], [20]. We evaluate the test set using two algorithms. First, by using the hybrid pso without Weiszfeld' equations and second by using the Weiszfeld' equations. They are referred as HPSO1 and HPSO2 in the tables. The result sets and their comparison are shown in Tables 1 and 2 for 287 and 654 demand points respectively. There are nine columns in the tables. The first column (p) is the number of medias sought, second is the standard PSO implementation that uses constriction based velocity update rule. The third is the solution offered by [20]. Fourth, fifth are modifications to the PSO's using the Weiszfeld' equations. Sixth is using the multi-start method and seventh is the actual optimal solution reported so far for the test problems. Compared to the algorithm as defined in [21] and [18], [19] and [20] , we have better results. The results are better due to the fact that we position new particles in the search space corresponding to the best location. As a result, the local search get's further intensified and hence new best values are readily extracted. However, not all neighborhood types produce better results. In our experiments we found out that the neighborhoods created randomly produces better values when compared to the star topology. The star topology is prone to initialization set of the particles and they converge quickly in the local minima in case of star topology. However, with the random neighborhood and with application of GA such early convergence is not at all seen as each neighborhood acts as an independent swarm.

We have set the number of particles as 100 and total number of iterations as 50,000 . The experimental settings are same as that of the [21]. Each neighborhood is set to random fixed size. To check for the neighborhood size, we do a sensivity test by varying the neighborhood size from 5- 20. The number of particles in each experiment remains the same and each experiment is done for 25 runs. The mean, std. deviation of the experiments are shown in Table 3.

4 Conclusion

In this paper, we tried providing a solution to the unconstrained continuous p-median problem. However in real life, constraints are always a part of the problem. The solution to solve the constraint continuous p-median problem cannot be solved alone by the algorithm that we have suggested here. For solving constraint objective functions, there are a number of solutions that are available. For example, one could use penalty functions and modify the same objective function to solve it. Defining the penalty function however is an art and require sometimes deep domain expertise. Solving the constraint p-median problem is left as a future work. Further, it has been observed that with large number of points, solving the p-median would take an in-ordinate amount of time. We choose to build a parallel version of the same so that large problems of such kind can be tackled effectively.

References

1. Gudehus, T., Kotzab, H.: Comprehensive logistics. Springer (2009)
2. Wesolowsky, G.O.: The weber problem: History and perspectives. Computers & Operations Research (1993)
3. Rosing, K.: Towards the solution of the (generalised) multi-Weber problem. Economisch Geografisch Instituut, Erasmus Universiteit (1990)
4. Kariv, O., Hakimi, S.L.: An algorithmic approach to network location problems. ii: The p-medians. SIAM Journal on Applied Mathematics 37(3), 539–560 (1979)
5. Talbi, E.G.: Metaheuristics: from design to implementation, vol. 74. John Wiley & Sons (2009)
6. Van Laarhoven, P.J., Aarts, E.H.: Simulated annealing. Springer (1987)
7. Goldberg, D.E.: Genetic algorithms. Pearson Education India (2006)
8. Glover, F., Laguna, M.: Tabu search. Springer (1999)
9. Kennedy, J.: Particle swarm optimization. In: Encyclopedia of Machine Learning, pp. 760–766. Springer (2010)
10. Weiszfeld, E.: Sur le point pour lequel la somme des distances de n points donnés est minimum. Tohoku Math. J. 43(355-386), 2 (1937)
11. Eberhart, R.C., Shi, Y.: Particle swarm optimization: developments, applications and resources. In: Proceedings of the 2001 Congress on Evolutionary Computation, vol. 1, pp. 81–86. IEEE (2001)
12. Parsopoulos, K.E., Vrahatis, M.N.: Recent approaches to global optimization problems through particle swarm optimization. Natural Computing 1(2-3), 235–306 (2002)
13. Imran, M., Hashim, R., Khalid, N.E.A.: An overview of particle swarm optimization variants. Procedia Engineering 53, 491–496 (2013)
14. Banks, A., Vincent, J., Anyakoha, C.: A review of particle swarm optimization. part ii: hybridisation, combinatorial, multicriteria and constrained optimization, and indicative applications. Natural Computing 7(1), 109–124 (2008)
15. Shi, Y., Eberhart, R.: A modified particle swarm optimizer. In: The 1998 IEEE International Conference on Evolutionary Computation Proceedings. IEEE World Congress on Computational Intelligence, pp. 69–73. IEEE (1998)
16. Bongartz, I., Calamai, P.H., Conn, A.R.: A projection method for l p norm location-allocation problems. Mathematical Programming 66(1-3), 283–312 (1994)
17. Beasley, J.E.: Or-library: distributing test problems by electronic mail. Journal of the Operational Research Society, 1069–1072 (1990)
18. Aras, N., Özkısacık, K., Altınel, İ.K.: Solving the uncapacitated multi-facility weber problem by vector quantization and self-organizing maps. Journal of the Operational Research Society 57(1), 82–93 (2006)
19. Gamal, M., Salhi, S.: A cellular heuristic for the multisource weber problem. Computers & Operations Research 30(11), 1609–1624 (2003)
20. Salhi, S., Gamal, M.: A genetic algorithm based approach for the uncapacitated continuous location–allocation problem. Annals of Operations Research 123(1-4), 203–222 (2003)
21. Brito, J., Martínez, F.J., Moreno, J.A.: Particle swarm optimization for the continuous p-median problem. In: 6th WSEAS International Conference on Computational Intelligence, Man-Machine Systems and Cybernetics, CIMMACS, pp. 14–16. Citeseer (2007)

Bees Swarm Optimization
for Web Information Foraging

Yassine Drias and Samir Kechid

LRIA, USTHB,
USTHB, BP 32 El Alia, Bab Ezzouar Algiers, Algeria
YassineDrias@Gmail.com, skechid@usthb.dz

Abstract. The present work is related to Web intelligence and more precisely to Wisdom Web foraging. The idea is to learn the localization of the most relevant Web surfing path that might interest the user. We propose a novel approach based on bees behaviour for information foraging. We implemented the system using a colony of cooperative reactive agents. In order to validate our proposal, experiments were conducted on MedlinePlus, a benchmark dedicated for research in the domain of Health. The results are promising either for those related to some Web regularities and for the response time, which is very short and hence complies with the real time constraint.

Keywords: Information Foraging, Web intelligence, Wisdom Web, Bee Swarm Optimization (BSO), MedlinePlus.

1 Introduction

Nowadays, Web intelligence tends to evolve in permanence. A major concern is the development of the Information Foraging (IF) paradigm. It consists in surfing on the Web to get useful information under a time constraint. This issue may appear at a first glance simple and with no major interest. However, its importance is stimulating nowadays the Web users as the latter is in an incessant growing and the human ability to explore the astronomical amount of data on the web is relatively very limited. Besides, tackling such issue is very welcomed in domains like business, finance, health and science. The potential users not only will spend less time getting the localization of the needed information but they can even get it in real time.

1.1 The Web Structure

Web mining is the analysis and discovery of data, documents and multimedia information existing on the Web. It includes studies of features such as the structure of hyperlinks, Web usage statistics and search of the Web sites contents. Studies on the structure including hyperlinks allow the detection of the pages that have authority on a certain topic. Web Usage statistics used techniques to discover patterns in the log files of users. Meanwhile searching the Web contents

R. Prasath et al. (Eds.): MIKE 2014, LNAI 8891, pp. 189–198, 2014.

aims at getting the closest possible information needs for users by presenting the most appropriate Web pages.

1.2 Information Foraging

The concept of information foraging (IF) shares the same goal as Information Retrieval (IR), which is information search. While IR uses a complex process (indexing and matching), IF consists in navigating from one page to another for the same purpose.

The present study deals with simulating IF by taking into consideration the complex structure of the Web. We designed a system for IF and a tool for helping users to undertake IF with efficiency.

The rest of the paper is organized as follows. In section 2, we report the studies that are the most related to our concern. In section 3, we expose the model of Web surfing that we used in our study and that is inspired from our own practices of real-world Web surfing. In section 4, we present the contribution we developed on authorities ranking learnt using bees behaviour. Section 5 summarizes the experiments we conducted on a real Web source, which is MedlinePlus and the results we obtained.

2 Context and Related Literature

Liu in his talk at IJCAI03 [6], suggests new directions for research in the new field of Web Intelligence (WI) that has emerged a decade ago from artificial intelligence and information technology. The main goal of WI is to develop theories and technologies towards using optimally the connectivity of the Web.

The speaker then predicts that the next biggest paradigm that will occupy a large part in the research domain for the next era is the Wisdom Web. He describes the objectives of Wisdom Web and its most important concerns. He enchains with two recent challenges, which are respectively IF and ubiquitous communities. The speaker then exposes the work he did with Zhang on a model for IF agents on the Web and which is described next.

In [7] and [8], the authors proposed an agent-based model for IF and validated it using empirical Web log datasets. They consider Web topology, information distribution and interest profile in building a Wisdom IF agent. They found out that the unique distribution of agent interest leads for regularities in Web surfing and that Web regularities are interrelated. They also undertook an interesting study on three categories of users according to their interest and familiarity with the Web: A random user with no intention in surfing, a rational one with an objective but with no familiarity on surfing and a recurrent user who is familiar with the Web and who has a goal for surfing. The result is that independently from the kinds of users, the regularities of user surfing are the same, which means that the user ability for predicting the surfing chain is predominant.

Strong regularities in Web surfing were also studied by Huberman et al. [3] from the theoretical point of view. They proposed a model for studying surfing behaviours and the experiments they held showed common surfing behaviours. The study conducted by Huberman et al. in [4] shows that the Web pages are distributed over the sites according to a universal power law, which is an example among the other strong regularities.

In [5], Ibekwe-SanJuan describes in a clear and nice manner the recent concepts, methods and applications of text mining. The chapter on Web mining contains a rich documentation on notions that concern the new developments of the Web technologies such as Web topology, sites popularity, sites ranking and propagation of metadata to co-links.

Chi and Pirolli in [1] introduce the concept of social information foraging and its understanding. They explored models for social IF and focused on the importance of the benefits of cooperative foraging.

From the above literature review, we remark that the rare existing papers related to IF offer new ideas on how to develop theories and technologies about IF, which means that the area is still in an early stage. The theory developed by Pirolli et al. is an important advancement and can stimulate future works in the field. On the other hand, the agent-based model proposed by Liu et al. can be considered as the commencement of future IF technologies.

The original contribution of the present paper consists in developing an approach based on Multi-Agent Systems (MAS) for IF. To our knowledge, there is no existing work that uses MAS for IF. In this paper, we propose a multi-agent architecture that fits the real surfing process, and that aims at keeping good performance compared to a centralized solution. For this purpose, we investigate bees swarm optimization for constructing a tool for IF. Before detailing the approach, the model of Web surfing used is presented in the next section.

3   Web Navigation Model

The navigation model we propose, attempts to simulate the real-world process of Web surfing. We first give a mathematical representation of the real Web site we experiment, which is MedlinePlus. Then we formulate the statement we are addressing. At the end, we describe the mathematical model we adopt.

3.1   Web Site Representation and Notations

MedlinePlus as well as many other real-word Web sites can be represented by an oriented hyper-graph H = (P,L,T), which is a generalization of a graph. P is a set of pages or nodes, L a set of links and T a set of hyper-edges or topics. Unlike in a graph, a hyper-edge in a hyper-graph is a set of nodes. The structure of a hyper-graph is suitable for a Web server because:

- each node represents a Web page,
- each arrow represents a link between two pages and
- each hyper-edge represents a topic, which is a subset of P.

MedlinePlus is described with more details in the section dedicated to the experiments.

3.2   Problem Statement

Navigating on the Web with a user interest profile corresponds to visiting a branch of the hyper-graph, one node at a time starting at the homepage of the Web site and ending at the authority, which is a page containing the best relevant information concerning the user interest. When moving from one page p_i to another p_j, the following features are considered:

- $Uscent_i(t)$: the user information scent at page p_i at click t, which is related to the user interest V, which is a vector of keywords.
- $Sim(p_i, p_j)$: the similarity between page p_i and page p_j using their respective title keywords and p_i synonyms if available.
- N_i: the number of outgoing links of page p_i.

The aim is to determine the authority for a given user interest, that is, at each click corresponding to a page p_i, the question is to find the best move to another page p_j in order to reach the authority.

3.3   The Model

We consider three types of surfing strategies, corresponding to the three kinds of users reported in [7] and [8]: pseudo random, recurrent and rational.

At each click t, we propose a surfing rule for each strategy to compute the probability (P_{ij}) of surfing from page p_i to page p_j using the user interest profile and a heuristic. When surfing, the user is guided by his/her scent that depends on his/her interest profile but also by a heuristic that measures the similarity between the current page and the next one. We introduce then a factor called $UScent_i$ to measure the user scent at page i and a second one called $heur_{ij}$ that calculates the resemblance of $page_i$ with $page_j$. Power law distributions are adopted for these factors. Indeed, the user scent is important for only a very few pages and is weak for many of them. In the same way, a few pages are very similar to the current page and the others are almost dissimilar.

Pseudo Random Agent Strategy. Rule (1) exhibits the computation of the probability P_{ij}. The surfing page is drawn at random among the outgoing pages with probability (P_{ij}).

$$P_{ij}(t) = \frac{(Uscent_j)^\alpha (heur_{ij})^\beta}{\Sigma_{l \in N_i}[(Uscent_l)^\alpha (heur_{il})^\beta]} \tag{1}$$

Recurrent Agent Strategy. In this case, Rule (2) is considered. The system selects all times the surfing page that presents the best score.

$$P_{ij}(t) = \{ \begin{array}{l} 1 \text{ if } j = arg\ max\{(Uscent_j)^\alpha (heur_{ij})^\beta\} \\ \\ 0 \text{ otherwise} \end{array} \tag{2}$$

Rational Agent Strategy. We introduce a noise called q_0 to disturb the surfing and a variable q. During the process, a value from the interval $[0,1]$, is drawn at random for q. If it is lower than or equal to q_0, then the page with the best score is selected, otherwise another is chosen at random. Rule (3) models this situation.

if
$$q < q_0 \ then \ P_{ij}(t) = \begin{cases} 1 \ if \ j=argmax\{(Uscent_j)^\alpha (heur_{ij})^\beta\} \\ 0 \ otherwise \end{cases}$$
else
$$P_{ij}(t) = \frac{(Uscent_j)^\alpha (heur_{ij})^\beta}{\Sigma_{l\in N_i}[(Uscent_l)^\alpha (heur_{il})^\beta]} \tag{3}$$

The score for a page is measured by the similarity between the interest vector and the title keywords and synonyms of the page.

Description of the Used Variables, Parameters and Functions

* q : a random variable uniformly distributed over the interval $[0,1]$.
* $q_0 \in [0,1]$: a tunable parameter. Arg max calculates the page j having the maximum value $(Uscent_j)^\alpha (heur_{ij})^\beta$.
* $heur_{ij}$: the heuristic value of the link (p_i, p_j).
* α : a parameter that controls the relative weight of the scent information.
* β : a parameter that controls the relative weight of the heuristic.

User Information Scent Modelling. The user information scent at page p_i can be measured by a quantity that increases as the user gets close to the authorities. Starting with a weak amount of scent, each time the user surfs from one page to another, a quantity relative to the quality of the page in consideration to the user interest is added. $UScent_j(t + 1)$ is then computed using Formula (4):

$$UScent_j(t+1) = UScent_i(t) + sim(V,p_j) \tag{4}$$

Where

$V = (v_1, v_2, ..., v_u)$ is the vector of keywords of the user interest profile.
$p_i = (t_1, t_2, ..., t_s)$ is the vector of title keywords and eventually synonyms of p_i.
$sim(V,p_j) = \Sigma_{(V_i=t_l)}1 \quad V_i \in V \ \& \ t_l \in p_j$ (the number of times $V_i = t_l$)

Heuristic. The heuristic we use consists of the similarity that may exist between page p_i and page p_j. It is then computed using formula (5).

$$heur_{ij} = sim(p_i, p_j) = \Sigma_{(t_k=t_k)}1 \ _{k\in p_i \ \& \ l\in p_j} \ (\ the \ number \ of \ times \ t_k = t_l) \tag{5}$$

4   Authorities Mining Using Bee Swarm Optimization

A colony of artificial bees is launched to seek authorities according to their behaviour that guarantees finding the richest places of the target. The authorities mining takes into account the Web topology and the user interest profile. The group of bees work for one user, which is different from simulating a group of users behaviour. The latter issue was studied in [1]. Whereas in our case, the cooperative feature is handled in the implementation by the bees.

The search starts at the site homepage, then guided by their senses, the bees are directed to the goal and after several generations, they find the right outcomes. The approach is detailed in the next section after presenting the general BSO algorithm.

4.1   Bee Swarm Optimization

Bee Swarm Optimization (BSO) was designed as a nature and bio-inspired approach to solve complex problems and was published the first time in 2005. Afterwards, it showed its efficiency through its adaptation to several problems such as information retrieval [2].

BSO translates as accurately as possible the behaviour of bees when searching for food. Its framework is based on the experience held in 1991 by Seely, and Sneyd Camazine, who showed that when two food sources are equidistant from the hive, the bees exploit those whose concentration is highest. In order to attract its congeners, the bee exploiting the source of greater concentration makes a dance, which is proportional in strength to the source concentration.

A bee called BeeInit, represents a potential solution to the problem and initializes a solution reference called Sref. A search space, namely SearchArea is then determined from Sref using a well-defined diversification generator. Each bee considers one solution from SearchArea as the starting point from which it performs a local search. After the bees have accomplished their search, they communicate their best results through a table called *Dance*. The best solution stored in this table is taken as the reference solution for the next iteration. In addition, a tabu list for backup solutions references *Sref* at each iteration is used to avoid stagnation and local optimum. The framework of BSO is outlined in Algorithm 1.

```
Algorithm 1. BSO
Input: a search space
Output: the best solution
begin
    1. Sref := BeeInit, drawn randomly;
    2. while stop condition is not reached do
        2.1 insert Sref in TL the Tabu List;
        2.2. determine SearchArea from Sref;
        2.3. assign for each bee a solution of SearchArea;
        2.4. for each bee do
```

```
                   2.4.1. search in the corresponding area;
                   2.4.2 insert result in Dance;
              2.5 Choose the best solution from Dance for Sref;
end.
```

4.2   BSO for Authorities Mining

The search space for the colony of bees will be the MedlinePlus database. In BSO, bees encapsulate solutions and a solution for our application is a surfing path. So bees will seek for Web surfing paths that end with authorities. The adaptation of BSO to IF is called BSO-IF and is outlined in Algorithm 2.

```
Algorithm 2. BSO-IF
Input: MedlinePlus and the user interest profile
Output: surfing path ending with an authority
begin
    1. Queue := empty;
       Sref := BeeInit, a surfing path starting at the homepage
       and drawn randomly;
    2. while stop condition is not reached do
       2.1. insert Sref in TL the Tabu List;
       2.2. searchArea = a subset of k surfing paths starting at
       the homepage and such that the first node is distant from
       the first node of Sref;
       2.3. assign for each bee a surfing path of SearchArea;
       2.4. for each bee do
           2.4.1. search locally for a better surfing path using
           formula (2); (or (1) for random or (3) for rational)
           2.4.2. insert result in Dance;
       2.5. Choose the path with the highest score from Dance for
       next Sref;
       2.6. Insert the path in Queue in a decreasing order of the
       score
end.
```

The first statement consists in initializing the process, which create randomly a surfing path starting at the homepage corresponding to topic p_0 of the hypergraph H, p_0 has links to all existing topics.

Instruction 2 loops until a maximum number of iterations is reached or when a surfing path is found with a satisfactory score. Instruction 2.1 consists in inserting Sref in the tabu list in order to avoid revisiting the same surfing paths. Instruction 2.2 computes the set of surfing paths that will be assigned to bees (instruction 2.3). It is drawn randomly from the set of outgoing links. Instruction 2.4.1 consists in determining a better surfing path using the transition rule (1), (2) or (3) depending on whether a random, recurrent or rational surfing strategy is adopted. Rule (2) is considered in our experiments. 2.4.2 allows the artificial bee to store its result in the table Dance. The surfing path with the highest score among those stored in Dance is assigned to Sref in order to reiterate the process. This path is then inserted in a ranked structure called Queue in a decreasing order of the score such as the best path appears at the head of Queue.

5   Experimental Results

5.1   Description of the Real-World Benchmark

Extensive experiments were performed on an on-line medical database provided by the U.S. National library of Medicine called *MedlinePlus*. It provides information on over 900 diseases, health conditions and wellness issues. Our experiments deal only with health topics described in an XML database that includes pages describing medical topics. The data is available at `http://www.nlm.nih.gov/medlineplus/xml.html`. We worked on the version of the 23th July 2014 where the number of nodes was equal to 1903 with a total volume of 27 MB. Each topic is specified by a title and contains the following elements: an URL, an identifier, the language of the topic (English or Spanish), the date of its creation, eventually tags specifying among others, topic synonyms, translation to other languages, a full summary, related topics, which are internal links to similar topics and external sites.

Only links to related topics are exploited because they belong to the database. External sites are ignored as they direct to pages outside the database.

5.2   Experimentations and Results

BSO-IF was implemented in Java Eclipse help System Base, version 2.0.2 on a PC with an Intel core I5-3317U Processor (1.70 GH) with 4 GB RAM. Figure 1, Figure 2 and Table 1 exhibit the results of the experiments of selecting the the optimal parameters, which are shown in Table 2.

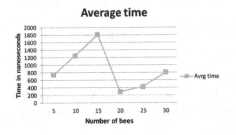
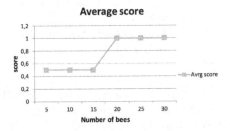
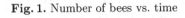

Fig. 1. Number of bees vs. time   Fig. 2. Number of bees vs. score

Then different user interests were experimented for each type of surfing strategies. The results we focused on are: the authority page (the surfing page with the highest score), its URL, its score, the surfing depth and the surfing time in nanoseconds. They are shown in Table 3 for the recurrent strategy.

Table 1. Choosing α and β parameters

α	β	P_{ij}
1	1	1.41
1	2	0.5
2	1	4.94
2	2	$3.43 * 10^{19}$

Table 2. The parameters that yields the best results

Number of bees	20
Maximum number of generations	30
TL	30
α	2
β	1

Table 3. Experimental Results for different user interest profiles for the recurrent agent strategy

User interest	Relevant Page		Score	Surfing depth	Surfing time
	Title	URL			
Pain, Abdominal	Abdominal Pain	*/abdominalpain.html	1.0	2	122
H5N1	Bird Flu	*/birdflu.html	1.0	2	140
Heart, Diseases	Heart Diseases	*/heartdiseases.html	1.0	5	155
Hypersensitivity	Allergy	*/allergy.html	1.0	1	98
Cancer	Cancer	*/cancer.html	1.0	2	138
Poor, Blood, Iron	Anemia	*/anemia.html	1.0	2	152
High, Blood Pressure, Medicines	High Blood Pressure	*/highbloodpressure.html	0.75	3	1291
Skin, Allergies	Skin Conditions	*/skinconditions.html	0.5	2	844
MCI	Mild Cognitive Impairment	*/mildcognitiveimpairment.html	1.0	4	313
Anorexia	Body Weight	*/bodyweight.html	0.15	20	5641
Recovery, surgery	After Surgery	*/aftersurgery.html	1.0	2	199
Pimples	Acne	*/acne.html	1.0	1	301

* : http://www.nlm.nih.gov/medlineplus

5.3   Validation of the Model

In order to validate our model, we experimented real-word surfing performed by the help of students. The adopted strategy was the recurrent one since the students are familiar with the Web environment. Each proposed user interest profile was assigned to 20 students with the aim to use it in the surfing test. The results are the means of feedbacks returned from students. They are impressive as they are close to those of the proposed model. More precisely the authorities formed were the same (when they exist) and 90% of the students' surfing depth outcomes were identical within one unit to the results provided by our program.

6   Conclusion

In this work, we proposed an approach for Web information foraging using multi-agent system. Unlike previous works that use a single agent surfing model, our approach which is inspired from nature and biological psychology, adopts an analogy with animals groups hunting, which simulates real-world surfing. The second originality of the present study is in the use of a real-word site server MedlinePlus for the experiments instead of an artificial one.

We implemented the system using Bee Swarm Optimization (BSO) on MedlinePlus. At our best knowledge, there is no studies on information foraging using a multi-agent system. This idea is inspired from nature where animals hunt together in a group and rarely alone. The results are promising.

The perspectives for this study are numerous. For the short term, we are thinking about integrating the user preferences in the user interest profile in the surfing model.

References

1. Chi, E.H., Pirolli, P.: Social Information Foraging and Collaborative Search. In: HCIC Workshop, Fraser CO (2006)
2. Drias, H., Mosteghanemi, H.: Bees Swarm Optimization based Approach for Web Information Retrieval. In: IEEE/WIC/ACM International Conference on Web Intelligence and Intelligent Agent Technology, pp. 6–13 (2010)
3. Huberman, B.A., Pirolli, P.L.T., Pitkow, J.E., Lukose, R.M.: Strong regularities in World Wide Web surfing. Science 280, 96–97 (1997)
4. Huberman, B.A., Adamic, L.A.: Growth dynamics of the World-Wide Web. Nature 410, 131 (1999)
5. Iberkwe-SanJuan, F.: Fouille de textes méthods, outils et apllications, Hermès, Lavoisier (2007)
6. Liu, J.: Web Intelligence: What Makes Wisdom Web? Invited Talk. In: IJCAI 2003 (2003)
7. Liu, J., Zhong, N., Yao, Y.Y., Ras, Z.W.: The Wisdom Web: New challenges for Web Intelligence (WI). Journal of Intelligent Information Systems 20(1), 5–9 (2003)
8. Liu, J., Zhang, S.W.: Characterizing Web usage regularities with information foraging agents. IEEE Transactions on Knowledge and Data Engineering 16(5), 566–584 (2004)
9. Zhong, N., Hua Ma, J., He Huang, R., Ming Liu, J., Yu Yao, Y., Xue Zhang, Y., Hui Chen, J.: Research challenges and perspectives on Wisdom Web of Things (W2T). J. Supercomputing (2010)

Modeling Cardinal Direction Relations in 3D for Qualitative Spatial Reasoning

Chaman L. Sabharwal and Jennifer L. Leopold

Missouri University of Science and Technology
Rolla, Missouri, USA – 65409
{chaman,leopoldj}@mst.edu

Abstract. Many fundamental geoscience concepts and tasks require advanced spatial knowledge about the topology, orientation, shape, and size of spatial objects. Besides topological and distance relations, cardinal directions also can play a prominent role in the determination of qualitative spatial relations; one of the facets of spatial objects is the determination of relative positioning of objects. In this paper, we present an efficient approach to representing and determining cardinal directions between free form regions. The development is mathematically sound and can be implemented more efficiently than the existing models. Our approach preserves converseness of direction relations between pairs of objects, while determining directional relations between gridded parts of the complex regions. All the essential details are in 2D. Yet the extension to 3D is seamless; it needs no additional formulation for transition from 2D to 3D. Furthermore, the extension to 3D and construction of a composition table has no adverse impact on the computational efficiency, as the technique is akin to 2D.

Keywords: spatial reasoning, 3D objects, cardinal directions, minimum bounding rectangle.

1   Introduction

Spatio-temporal reasoning is useful in several areas: geophysical systems, cognitive sciences, robotics, and networking. There are two forms of spatial knowledge, usually referred to as quantitative and qualitative knowledge. While qualitative reasoning methods encompass approximate, imprecise, incomplete information, quantitative reasoning methods rely on numerical metric-based knowledge. For example, determining the intersection between two objects of different dimensions (0D, 1D, 2D or 3D space) is computationally intensive, and an application, where there may be thousands of pair-wise intersections, leads to a huge volume of data to be processed. Regardless of whether qualitative or quantitative reasoning is the focus in an application, the task of processing spatial information is computationally intensive.

Most of the existing models for cardinal directions (which are discussed in more detail in Section 2) generate only partial solutions, hence they yield incomplete and sometimes inaccurate solutions. For example, in point-based models the point approximation

R. Prasath et al. (Eds.): MIKE 2014, LNAI 8891, pp. 199–214, 2014.

of objects has inherent error in the representation of the objects. In this paper, we consider objects of 0D, 1D, 2D, and 3D heterogeneously for determining cardinal directions between any two objects, see Fig. 1. When we refer to a spatial object, it may encompass any of those four dimensions. Cardinal directions are one aspect of spatial information, and typically are defined with respect to relative position of a pair of spatial objects. Our analysis is applicable to different dimensions and spatial environments.

The qualitative directional relation may depend on the user's intrinsic frame of reference relative to which directions are specified. Also it is accepted that for purposes of a 3D object coordinate frame, the north is in positive y direction, east is in the positive x-direction, and above is in the positive z- direction. The cardinal directions are universally accepted, and they may be related to the reference frame independent of its orientation. In this paper, we use cardinal directions relative to the coordinate frames for determining direction between a pair of objects.

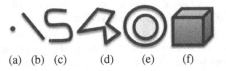

(a) (b) (c)   (d)   (e)   (f)

Fig. 1. The objects under consideration for directional relation may be: (a) a point 0D, (b) a line 1D, (c) a 1D curve, (d) a 2D non-convex polygon, (e) a 2D disk with a hole (shaded), and (f) a 3D cube. If we use all of the object types (a) to (f), we have a complex object with disjoint parts.

Spatial models that consider cardinal directions also must carefully consider the converseness property. Casually, this notion seems trivial: the converse of the "direction from A to B" is the "direction from B to A", the converse of the direction east is the direction west, of north is south, of up is down, and the converse of above is below. Algorithmically we can have one method to derive the direction relation from A to B, and use the same algorithm by interchanging the parameters for converse relation (e.g. direction from B to A). In natural language communications, it is desirable to have the converseness property for pairwise distinct spatial relations. Except for point objects, most of the previous work lacks a precise computation of the direction converseness property.

The paper is organized as follows. Section 2 describes the background on space partitioning, grid creation, representation, and interpretation. Section 3 gives the mathematical foundations for grid calculation, grid representation semantics, atomic and complex directional relations, and a general computation technique. Section 4 discusses extension of the techniques to 3D. Section 5 specifies future work. Section 6 presents our conclusions followed by references in Section 7.

2   Background

There are several models for cardinal directions. We compare our work with other strategies, particularly with the recent advanced work of [1] and [2]. Evolution of cardinal direction stems by considering regions as points, representing them using minimum bounding rectangles, and tiling that handles reference and target objects equally. Simplest cardinal relations are in the form of a cross or a cone, see Fig. 2.

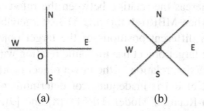

Fig. 2. (a) Cross- (b) cone-shaped direction relations

Along each axis there are only three mutually exclusive possibilities: along x-axis: origin, east or west; along y-axis: origin, north or south; along z-axis: origin, above or below; along y=x: northeast, southwest; along y=-x: northwest or southeast, see Fig. 2(a, b).

More complex directions are represented in the Oriented Point Relation Algebra (OPRA$_m$) [4], and [5]. However, point approximation models do not account for the region extent or shape. In these approaches, the reference object and target object are approximated with single points. This coarse approximation leads to direction computation errors.

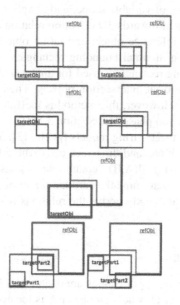

Fig. 3. The relation between the MBRs is unchanged, but the directional relation between the objects changes

The Minimum Bounding Rectangle (MBR) approach uses four basic directions (N, S, E, W); four compound directions (NW, NE, SW, SE) are a result of composition. This approach leads to erroneous results because of the inherent error in representing an object with an MBR, because several different objects may have the same MBR [7], [8], see Fig. 3; whereas the relation between the reference object and the target object can vary even when MBR(refObj) and MBR(targetObj) are unchanged [6]. In Fig. 3, there are seven different positions of the target object, but it has the same MBR. The MBR of the target object has the same relation with the MBR of the reference object, yet the direction relation of the target object is different for each configuration. Thus this MBR model is inadequate for determining the accurate directional relation. The Direction-Relation Model (DRM), [8] and [9], improves this by using nine cardinal directions: eight compass directions CD ={N, S, E, W, NW, NE, SW, SE}, and center, denoted by O, for a minimum bounding rectangle for the reference object, see Fig. 4. The target object has no role in creating this grid. The target object may intersect several cells, but the reference object is located in cell O. As a result, the cardinal directions are not symmetric and do not preserve the converseness property of cardinal direction relations.

Tiling-based models include projections and MBRs. The projection-based model [10] uses tiling to partition the space based on projections on the axes of the reference frame. This is a hybrid of Allen's 13 relation algebra and projections of regions on the coordinate axes [6]. It translates into 169 comprehensive relations, but may yield erroneous results in some cases [1]. New emerging applications including multimedia, and geospatial technologies require the handling of complex application objects that are highly structured, large, of variable shape, and simply or multiply connected with holes. A major improvement towards 3D direction relation has been made [1] whereby the grids for both the reference object and target object are computed, and integrated to create a common minimal bounding rectangle. The composite grid, OIG, for this composite bounding rectangle is used for analysis. An object intersection matrix (OIM) is used to record the intersection objects. Then an interpretation phase is used to interpret relations. However, this method is inefficient and also is not without ambiguity in the interpretation phase (see Section 3.5(3)).

Our approach is closer to 2D-string models ([12], [13], and [2]) using negative indexes. We create strings, dir(A) and dir(B), of cells where the objects A and B intersect the composite minimal grid(A,B) containing objects A and B. An intelligent indexing scheme is used to determine the cardinal direction relations and composition tables. It is also possible to quickly derive the relations between any two cells in the grid(A,B).

2.1   Space Partitioning

In 2D space, cardinal or compass directions are denoted by NW, N, NE, W, E, SW, S, and SE relative to the center O, where an object A is located, see Fig. 4. A 2D region divides the space into nine parts, one bounded region and eight unbounded sections. In the partition of the 2D plane in nine parts, the center section O is a bounded rectangle, whereas sections NW, N, NE, W, E, SW, S, SE have unbounded extent, see Fig. 4.

The symbols will be represented by indexes that play a dual role: (1) they represent the orientation direction vectors for the location relative to O, and (2) they represent the location area for analysis, see Fig. 4. Thus these directional symbols may be used to represent location and direction. With no loss of generality, we refer to all parts in Fig. 4 as regions or sections or locations or directions relative to O interchangeably in this discussion as they are represented by indexes.

2.2   Grid Creation and Representation

2.2.1   Creation of Grid

Each region partitions the 2D space into nine parts, forming a 3x3 grid associated with its minimum bounding rectangle. In Fig. 4, the solid blue lines form the minimum bounding rectangle enclosing an object A in red, MBR(A); blue dashed lines pertain to the semi-infinite rectangular sections adjoining to MBR(A).

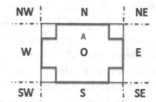

Fig. 4. 3x3 grid for Object A

In Fig. 5(a), 6(a), (b), the solid-line rectangles represent the minimum bounding rectangles for objects A and B individually. The dotted lines in conjunction with minimum bounding rectangles form the 3x3 grids associated with A and B. That is, the compound MBR(A,B) is partitioned into a grid so that each of the grid components is identified with a position as well as a direction attribute simultaneously relative to the center of the grid.

When the spatial grids of two objects are integrated to form a common grid containing both the objects, the two 3x3 grids are merged, resulting in an at most 5x5 grid. This means there are at most 16 unbounded sections and at most 9 bounded sections. These nine bounded sections form the composite grid(A,B) for both objects A and B, see Fig. 5(a). The red minimum bounding rectangle is the extent of A, and the green minimum bounding rectangle is the extent of the object B. The red and green dotted lines refer to the semi-infinite rectangular sections associated with grid(A) and grid(B), respectively. Now both the objects A and B have a common MBR(A,B) and associated unbounded regions. The unbounded parts do not contribute to direction determination, so the grid(A,B) will refer to regions contained in MBR(A,B) only. The composite MBR(A,B) is a grid(A,B) composed of bounded segments of grid lines from both A and B (i.e., gridlines enclosed by the black dotted rectangle). For spatial reasoning, the optimal worst case composite grid(A,B) for two regions is 3x3 (generated by four horizontal and four vertical lines) for A and B in Fig. 5(a). The best case optimal grid(A,B) is 1x1, where the minimum bounding rectangles for both A and B coincide with the composite MBR(A,B) and grid(A,B), see Fig. 5(b). For another example of a 2x3 grid for A and B see Fig. 6. In all, there are nine possible grids for MBR(A,B) depending on the location of objects.

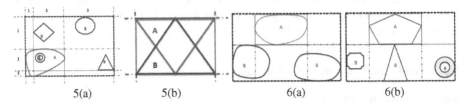

5(a)   5(b)   6(a)   6(b)

Fig. 5. (a) 3x3 gird enclosing A and B   (b) 1x1 grid enclosing A and B.   **Fig. 6.** Example of 2x3 grid enclosing objects A and B.   In (a) object B is simply connected, whereas in (b) object B has disjoint, multiply connected parts.

2.2.2   Grid Interpretation

In order to determine the directional relation between two regions, a significant amount of computational effort may be spent. Previous approaches (e.g., [1] and [6]) have required the creation of specialized data structures and functionality such as: an object intersection matrix, an object intersection grid space matrix, an intersection location function, an intersection location table, a direction interpretation signature function, and a hand-crafted 9x9 table to determine the direction between any two locations in the grid. In this previous work, no insight into the general composition of relations is given. The grid representation and interpretation is not without ambiguities. For example, in Fig. 6(a), 6(b), A is completely on one side of B. The interpretation in [1] would be: A is partly to the north of B, partly to northeast of B, partly northwest of B. An accurate interpretation would be that the whole A is to the north of a part of B, northwest of a part of B, and northeast of a part of B. B is partly to the southwest, partly to the south, and partly to the southeast of whole of A. Thus the converse relation interpretation does not hold accurately in the original statement. For nine orientation directions between two simple objects, the 9x9 table is generated with an interpretation function. To determine each entry in the table, eight tests are made by the signature function [1].

We present a rigorous method for accommodating any configurations for a pair of objects A and B. Thus we show that our approach is unique, complete, analytically sound, and extensible to higher dimensions. Our approach is not an extension of previous approaches such as [1], [12]; it is novel and is more efficient.   In Section 3, we provide a very simple, clean, cognitively consistent, mathematically sound, provably correct, one-step formulation to store and determine cardinal directions between objects. We also show that our formulation extends naturally to 3D with no additional conceptual development of formulation. Even the slightest gain in computational efficiency has a significant impact everywhere when a computation is done at many places and when the computation is repeated thousands of times.

3   Mathematical Analysis

3.1   Composite Grid Calculation

Each object can be represented with an MBR coupled with a 3x3 grid. We use the notation, x^A_m for the greatest lower bound of x-coordinates of all points in A; x^A_M represents the least upper bound of all the x-coordinates in A; similarly we use y^A_m and y^A_M for the y-coordinates of all points in A. All the points in the region A lie in the rectangle $\{(x,y): x^A_m \leq x \leq x^A_M, y^A_m \leq y \leq y^A_M\}$. The other eight unbounded parts are to the N, S, E, W, NW, SW, NE, SE as shown in Fig. 5. The sections corresponding to cells O, NE, SE, NW, SW are closed sets and those for the cells N, E, S, and W are open sets forming a partition of the whole plane. For the object A, the minimum bounding rectangle, MBR(A) becomes O(A) or simply O for the center of grid(A)

$$O(A) = \{(x,y): x^A_m \leq x \leq x^A_M, y^A_m \leq y \leq y^A_M\}$$

The other eight sections become

$$N(A) = \{(x,y): x^A_m < x < x^A_M, y > y^A_M\}$$
$$S(A) = \{(x,y): x^A_m < x < x^A_M, y < y^A_m\}$$
$$E(A) = \{(x,y): x^A_M < x, y^A_m < y < y^A_M\}$$
$$W(A) = \{(x,y): x < x^A_m, y^A_m < y < y^A_M\}$$
$$NE(A) = \{(x,y): x \geq x^A_M, y \geq y^A_M\}$$
$$NW(A) = \{(x,y): x \leq x^A_m, y \geq y^A_M\}$$
$$SE(A) = \{(x,y): x^A_M \leq x, y \leq y^A_m\}$$
$$SW(A) = \{(x,y): x \leq x^A_m, y \leq y^A_m\}$$

These definitions are slightly different from those in [1]; in that work they are defined in ad hoc manner devoid of symmetry where a part of the boundary belongs to O(A) and a part of it does not. Here the plane is partitioned into nine sections of which five sections are closed sets and four sections are open sets as depicted by the above definitions. Now MBR(A) = $\{(x,y): x^A_m \leq x \leq x^A_M, y^A_m \leq y \leq y^A_M\}$ and it is enclosed by four grid lines $x=x^A_m$, $x=x^A_M$; $y=y^A_m$, $y=y^A_M$ as well. The grid lines for MBR(B) are: $x=x^B_m$, $x=x^B_M$; $y=y^B_m$, $y=y^B_M$. These grid lines are used to create a grid for the composite minimum bounding rectangle for both A and B denoted by MBR(A,B)=$\{(x,y): min(x^A_m, x^B_m) \leq x \leq max(x^A_M, x^B_M), min(y^A_m, y^B_m) \leq y \leq max(y^A_M, y^B_M)\}$. In general, the composite grid(A,B) is pxq for $1 \leq p, q \leq 3$ where p and q are defined by

$$p = 3 - |\{y^A_m, y^A_M\} \cap \{y^B_m, y^B_M\}|$$
$$q = 3 - |\{x^A_m, x^A_M\} \cap \{x^B_m, x^B_M\}|$$

From now on, we will refer to grid(A,B) as only that part which is within MBR(A,B), and the grid(A,B) is at most 3x3. Let us start with 3x3 grid(A,B), then we will generalize our findings to any possible pxq grid(A,B), $1 \leq p, q \leq 3$.

3.2   Grid Representation Semantics

For two objects A and B, MBR(A,B) is

$$\{(x,y): \min(x^A_m, x^B_m) \le x \le \max(x^A_M, x^B_M), \min(y^A_m, y^B_m) \le y \le \max(y^A_M, y^B_M)\}$$

The grid(A,B) is at most 3x3. For 3x3 grid, the cells are conventionally labeled as O, N, S, E, W, NW, NE, SW, SE, consistent with Fig. 5 and Table 1.

Table 1. Table of Directions Relative to the Center O

NW	N	NE
W	O	E
SW	S	SE

The grid(A,B) with direction values NW, N, NE, W, O, E, SW, S, SE, can be indexed in several ways. Some authors [1] use the standard matrix indexing scheme shown in Table 2. A signature function is used to map the direction from one cell to another cell; such a signature function uses up to 9 tests to determine a direction relation between a pair of cells [1].

Table 2. Grid(A,B) Indexing Scheme

(1,1)	(1,2)	(1,3)
(2,1)	(2,2)	(2,3)
(3,1)	(3,3)	(3,3)

We present a different indexing scheme using negative indexes, see Table 3. The NW cell is indexed with (-1,1), and converse(NW) = - (-1,1)=(1,-1), which is the representation for the SW cell. The value (-1,1) also represents the position vector for direction of NW with respect to O. Similarly, other indexes are position vectors for directions of grid cells with respect to O. In general, these cells are represented with ordered pairs (u,v) where u = -1, 0, 1 denotes west, center, and east, respectively, and v = -1, 0, 1 denotes south, center, and north, respectively.

Later we will generalize this indexing technique in two ways: (1) how to index pxq grids when p≠3 or q≠3, and (2) how to label cells in 3D. Since the NW section is unbounded, any point can be represented by (-u,v) for positive real u and v. To accommodate all such representations consistent with indexing, our technique really identifies the whole NW region with (-1,1) by NW ≡ (sign(u), sign(v)) where (u,v) is the index representing a point in section NW. Thus Table 3 identifies the orientation vector of every cell in Table 1 with an ordered pair. Thus Table 1 and Table 3 are synonymous element by element in our analysis.

Table 3. Indexing the Cells in Grid(A,B)

(-1,1)	(0,1)	(1,1)
(-1,0)	(0,0)	(1,0)
(-1,-1)	(0,-1)	(1,-1)

We can generalize this indexing scheme to any grid(A,B) which is pxq; p≠3 or q≠3, that is, a grid that is not 3x3. Indexing tables of different dimensions is as easy as the 3x3 case. Unlike [2] which uses ad hoc methods for indexing tables of size pxq; p≠3 or q≠3, here we use the same technique as for 3x3. That is, lowest left most corner is labeled (-1,-1) and the rest of the cells are labeled sequentially.

3.3   Defining Grid Index Arithmetic

The pxq grid, $1 \leq p,q \leq 3$ is indexed with elements from the set CD = {(-1, 1), (0, 1), (1, 1), (-1, 0), (0, 0), (1, 0), (-1, -1), (0, -1), (1, -1)} and is identified with compound cardinal directions set

$$CD = \{NW, N, NE, W, O, E, SW, S, SE\}.$$

In our calculations, we will need the algebraic operations of negation, addition and subtraction on elements of an index set of ordered pairs. With the following definitions, the set CD of indexes is closed with respect to negation, addition, and subtraction:

(1) From the symmetry of CD, it follows that if $(u,v) \in CD$, then $(-u,-v) \in CD$
     $(-u, -v)$ can simply be written as $- (u, v)$ and conversely.

(2) If $(u,v) \in CD$ and $(a,b) \in CD$, then $(u,v)+(a,b)$ is defined as $(sign(u+a), sign(v+b))$
     which trivially belongs to CD, where $sign(x)= 1,0,-1$ according as $x>0$ or $x=0$ or $x<0$.

(3) Similarly, if $(u,v) \in CD$ and $(a,b) \in CD$, then $(u,v)-(a,b)$ is defined as $(u,v)+(-a,-b)$ or
     $(sign(u-a), sign(v-b))$ trivially which again belongs to CD.

The following example confirms that these operations are valid.
Example 1. If A is in the west cell (-1,0) of the grid, and B is in the southeast cell (1,-1) of the grid, then A is to the northwest of B. It may be confirmed from

$$W - SE = (-1,0) - (1,-1) = (sign(-2),sign(1))= (-1,1) = NW$$

which represents northwest.

3.4   Cardinal Direction Relations

3.4.1   Atomic Relations
The cell location, object intersection, and orientation can be treated uniformly here, all in one step. For example, in Table 3 symbol (-1,1) refers to the northwest cell of the grid(A,B); also this cell is to the northwest direction relative to the center cell, O. The direction of A relative to O is denoted by dir(O,A) or simply dir(A) and is represented by dir(A)={(-1,1)}. Other directions are defined similarly consistent with Table 3 when an object A intersects a corresponding cell.

Definition. The relation between two objects A and B is atomic if they belong to single cells of grid(A,B).

This is simple for an object intersecting only one cell. The only thing needed is the location where the object intersects in the grid(A,B). This eliminates all the nine tests done by the signature function in [1]. Let dir(A) and dir(B) be sets of indexes where the qualitative intersection of A and B is detected in the grid(A,B).

Definition. When dir(A) and dir(B) are singleton, the objects intersect the grid in single cells a and b where $a=(a_1,a_2)\in$ dir(A) and $b=(b_1,b_2)\in$ dir(B), and the direction of B relative to A is defined as
    dir(A,B)=dir(a,b)={(sign(b_1-a_1),sign(b_2-a_2)) : $a=(a_1,a_2)\in$ dir(A), $b=(b_1,b_2)\in$ dir(B) }

Our definition and indexing scheme can be easily applied to complex objects as well. We will first identify the cells where we know that a complex object intersection occurs, although not what that intersection is. Then we will give a more comprehensive definition in Section 3.4.2 for dir(A,B), the direction between compound objects.

3.4.2   Complex Relations

Here we extend the definition of cardinal relations to include objects intersecting more than one cell, see Fig. 6. When two complex objects A and B intersect the grid in multiple cells, the composite direction of object B relative to A is the set of pairwise directions of gridded components of B relative to the gridded components of A. It is defined as

$$dir(A,B) = \cup_{a\in dir(A),\ b\in dir(B)}\ dir(a,b)$$

This definition is independent of the size, pxq, of the grid(A,B). No additional testing is required for the correctness of the result. It encompasses the definition for *atomic* relation also. It is possible that grid(A,B) is not 3x3, see Section 3.5 for examples. Finally we show how to extend this indexing scheme to construct a composition table. The advantage of our indexing scheme over existing approaches is the ease with which many queries can be resolved.

3.5   Cardinal Direction Relation Examples

Here we use several illustrative examples to resolve queries of the form:
    (1) What do we do when an object intersects more than one grid cell?
    (2) How should relations be interpreted if a part versus the whole object intersects a grid cell?
    (3) What do we do when the grid(A,B) is not 3x3?
    (4) How do we determine the composition of dir(A,O) and dir(O,B)?

The answers to these questions are very simple and straightforward; they do not necessitate the significant computational effort required by methods presented in [1] and [6]. We resolve these questions with examples for ease in understanding followed by formal analytical descriptions.

Definition. If dir(A) is known, then dir(A,O) is defined as the set   of negative indexes in dir(A),
    $$dir(A,O) \equiv \{-u: u\in dir(A)\} \equiv - dir(A).$$

This definition applies even if the object is complex and intersects more than one cell.

(1) What do we do when the object intersects more than one grid cell?

We have defined the *complex* directional relation of an object relative to O in Section 3.3.2; it encompasses the definition for atomic relation also.

Example. In Fig. 7, the object B has parts, B_1, B_2, B_3 such that B_1, B_2, B_3 intersect cells N, NE, and E, respectively. Then we readily know that

$$dir(O,B_1)=\{(0,1)\}, dir(O,B_2)=\{(1,1)\}, dir(O,B_3)=\{(1,0)\},$$

Recall, the direction of B relative to O is the union of these sets

$$dir(B)=\{ (1,0), (0,1), (1,1)\}$$

and is interpreted as B is *partially* to the north, *partially* to the northeast of O, and *partially* to the east of O. This is consistent with [1], and there is no overhead in determining the compound object direction determination.

(2) How should relations be interpreted if a part versus the whole object intersects a grid cell?

Example
Let $dir(A) = \{(0,1)\}$ and $dir(B) = \{(-1,0),(0,0),(1,0)\}$, as shown in Fig. 6. Then from resolution to query (4)

$$dir(A,B) = dir(B) - dir(A)$$
$$=\{(-1,0) - (0,1), \quad (0,0) - (0,1),(1,0) - (0,1) \}$$
$$= \{(-1,-1),(0,-1),(1,-1)\}$$

and similarly

$$dir (B,A) =\{(1,1),(0,1),(-1,1)\}.$$

Since $dir(A,B)$ has more than one element, then B is partially to the southwest, partially to the south, and partially to the southeast of A. According to previous work [1], [11], since $dir(B,A)$ has more than one element, A is partly to the north of B, partly to northeast of B, and partly northwest of B, which is not accurate. An accurate interpretation in this case would be that the whole of A is to the north of part of B, northwest of part of B, and northeast of part of B. To correct this shortcoming, we provide the following simple algorithm. *Algorithmically*, the semantics would be:
if $|dir(A)| =1$
    if $|dir(B)| =1$,
whole of A is on one CD of whole of B determined by dir(B,A).
      else
        whole of A is on multiple CDs of B determined by dir(B,A).
else
      A is partly on multiple CDs of B determined by dir(B,A).

Similarly we test for dir(A,B) for accurate interpretation for the converseness property of directional relations.

(3) What do we do when the grid(A,B) is not 3x3?

If the grid is not 3 x 3, indexing can still be applied as indexes start -1 to 0 to 1. For example in Table 4, and then label the cells according to the indexing scheme as depicted in Tables 4. We illustrate this with two examples in Tables 4.

Example.
Suppose grid(A,B) is a 1x2 grid. In this case we can label the O cell arbitrarily.   We index the first or left cell as (-1,-1) , and the second cell is labeled (0,-1), then based on the same pattern as described for the 3x3 grid, see Table 4.

Table 4. 1x2 Grid For MBR(A,B) with Indexes

(-1,-1)	(0,-1)

Now A intersects the right cell (0,-1), B intersects the left cell (-1,-1), and dir(B,A) represents that direction A relative to B which is trivially east. We can use our definition to get

$$dir(B,A) = dir(A)-dir(B) = \{(0,-1)-(-1,-1)\} = \{(1,0)\}$$

It shows that A is to the east of B.

(4) How do we determine the composition of dir(A,O) and dir(O,B)?
Definition. The composition of dir(A,O) and dir(O,B) is defined as dir(A,O)+ dir(O,B).
It follows from the fact that

$$dir(A,B) = dir(B)-dir(A) = dir(O,B)-dir(O,A)= dir(A,O)+ dir(O,B)$$

The complete 9x9 composition table is outlined in the center of Table 5. Each entry is the direction of B relative to A; for example, the table entry SW means B is southwest of A.

4   Extension to 3D

For regions in 3D, all our exploration work in 2D can be seamlessly extended to 3D by adding a third component to the ordered pairs. For an object A, MBR(A) = $\{(x,y,z): x^A_m \leq x \leq x^A_M, y^A_m \leq y \leq y^A_M, z^A_m \leq z \leq z^A_M \}$, it is enclosed by the six grid lines $x=x^A_m$, $x=x^A_M$; $y=y^A_m$, $y=y^A_M$; $z=z^A_m$, $z=z^A_M$. This can be done similarly for an object B.

For grid transition from 2D to 3D, the 3x3 rectangular grid becomes a 3x3x3 voxel grid, see Fig. 8. The 3D cells are described by uniformly considering north, south, east, west, above, and below. The grid(A,B) has 27 cells, where all cells are bounded within the confines of minimum bounding volume, MBV(A,B), of A and B.

The center cell is labeled (0,0,0) and all other cells follow the pattern (u, v, w) where u = -1, 0, 1; v = -1, 0, 1; w = -1, 0, 1. In other words, the set of index labels now is

$$S = \{(u, v, w) : u, v, w \in \{0, 1, -1\} \}$$

In this scheme, u refers to east and west; v refers to north and south; and w refers to above and below the center voxel. These labels also become the directions of the cells with respect to O and indexes of the information storage array data structure. This set S is closed with respect to addition, subtraction, and negation as in the case of 2D. The directional relation of the object A relative to the center object O is denoted by dir(A) and is defined as the set of triples (u, v, w) where u, v, w refer to the cell index where the object A intersects the grid(A,B). If dir(A) = $\{(1,1,1)\}$, it means A intersects MBV(A,B) at the north-east-above cell of the grid(A,B). The grid(A,B) is at most 3x3x3; in general, it is pxqxr where $1 \le$ p, q, r ≤ 3, p rows, q columns and r planes. With our approach nothing is significantly different in 3D.   All the essential ideas occur in 2D.   Similar to the 2D case, for a pair of objects A and B, the following hold good.

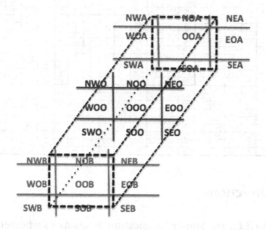

Fig. 8. 3D grid for A and consists of 27 voxel cells

Let dir(A) and dir(B) be the sets of indexes for cells where A and B intersect the grid(A,B).
(1) dir(O,A) = dir(A) = set of triples representing the cells where A intersects the grid(A,B). Formally, dir(A) ={ (a_1,a_2,a_3): (a_1,a_2,a_3) is the cell where A intersects the grid}.   Similarly, we can define direction dir(B) for object B.
(2) dir(A,O) = - dir(A) ={ $(-a_1,-a_2,-a_3)$: $(a_1,a_2,a_3) \in$ dir(A)}.
(3) dir(A,B) ={(sign(b_1-a_1), sign(b_2-a_2), sign(b_3-a_3)): $(a_1,a_2,a_3) \in$ dir(A) and $(b_1,b_2,b_3) \in$ dir(B)}

(4) If grid(A,B) is not 3x3x3, then cell indexes can be created as in the 2D case. If the grid has three cells in a row/column/plane, then we choose the middle cell coordinate for O; if it does not have three cells, then we can choose any available cell in that direction for O. From this knowledge, we can apply the 3D indexing scheme using the similar pattern scheme as for a 3x3 grid.

(5) The composition of atomic directions dir(A,O) and dir(O,B) is dir(A,B) = dir(A,O) + dir(O,B).

(6) In 3D, the object A may intersect grid(A,B) in any one of the cells. The composition direction of atomic object B relative to atomic object A will be any one of 27 directions. It would be a computationally intensive task to compute this 27x27 table with previously available methods (e.g., [11] and [6]). With our approach, it is quite simple, see Table 5.

Table 5. The Composition of Relations in 3D. Left Column Represents Object A Intersection with Grid(A,B), Top Row Identifies B Intersection with Grid, Table Entries Correspond to Orientation of B as Seen by A in 3D. Table Entries Are Read as North-South, East-West, Above-Below in 3D. SEB Means B is to the Southeast and Below A.

	NWA	NA	NEA	WA	A	EA	SWA	SA	SEA	NW	N	NE	W	O	E	SW	S	SE	NWB	NB	NEB	WB	B	EB	SWB	SB	SEB
NWA	O	E	E	S	SE	SE	S	SE	SE	B	EB	EB	SB	SEB	SEB	SB	SEB	SEB	B	EB	EB	SB	SEB	SEB	SB	SEB	SEB
NA	W	O	E	SW	S	SE	SW	S	SE	WB	B	EB	SWB	SB	SEB	SWB	SB	SEB	WB	B	EB	SWB	SB	SEB	SWB	SB	SEB
NEA	W	W	O	SW	SW	S	SW	SW	S	WB	WB	B	SWB	SWB	SB	SWB	SWB	SB	WB	WB	B	SWB	SWB	SB	SWB	SWB	SB
WA	N	NE	NE	O	E	E	S	SE	SE	NB	NEB	NEB	B	EB	EB	SB	SEB	SEB	NB	NEB	NEB	B	EB	EB	SB	SEB	SEB
A	NW	N	NE	W	O	E	SW	S	SE	NWB	NB	NEB	WB	B	EB	SWB	SB	SEB	NWB	NB	NEB	WB	B	EB	SWB	SB	SEB
EA	NW	NW	N	W	W	O	SW	SW	S	NWB	NWB	NB	WB	WB	B	SWB	SWB	SB	NWB	NWB	NB	WB	WB	B	SWB	SWB	SB
SWA	N	NE	NE	N	NE	NE	O	E	E	NB	NEB	NEB	NB	NEB	NEB	B	EB	EB	NB	NEB	NEB	NB	NEB	NEB	B	EB	EB
SA	NW	N	NE	NW	N	NE	W	O	E	NWB	NB	NEB	NWB	NB	NEB	WB	B	EB	NWB	NB	NEB	NWB	NB	NEB	WB	B	EB
SEA	NW	NW	N	NW	NW	N	W	W	O	NWB	NWB	NB	NWB	NWB	NB	WB	WB	B	NWB	NWB	NB	NWB	NWB	NB	WB	WB	B
NW	A	EA	EA	SA	SEA	SEA	SA	SEA	SEA	O	E	E	S	SE	SE	S	SE	SE	B	EB	EB	SB	SEB	SEB	SB	SEB	SEB
N	WA	A	EA	SWA	SA	SEA	SWA	SA	SEA	W	O	E	SW	S	SE	SW	S	SE	WB	B	EB	SWB	SB	SEB	SWB	SB	SEB
NE	WA	WA	A	SWA	SWA	SA	SWA	SWA	SA	W	W	O	SW	SW	S	SW	SW	S	WB	WB	B	SWB	SWB	SB	SWB	SWB	SB
W	NA	NEA	NEA	A	EA	EA	SA	SEA	SEA	N	NE	NE	O	E	E	S	SE	SE	NB	NEB	NEB	B	EB	EB	SB	SEB	SEB
O	NWA	NA	NEA	WA	A	EA	SWA	SA	SEA	NW	N	NE	W	O	E	SW	S	SE	NWB	NB	NEB	WB	B	EB	SWB	SB	SEB
E	NWA	NWA	NA	WA	WA	A	SWA	SWA	SA	NW	NW	N	W	W	O	SW	SW	S	NWB	NWB	NB	WB	WB	B	SWB	SWB	SB
SW	NA	NEA	NEA	NA	NEA	NEA	A	EA	EA	N	NE	NE	N	NE	NE	O	E	E	NB	NEB	NEB	NB	NEB	NEB	B	EB	EB
S	NWA	NA	NEA	NWA	NA	NEA	WA	A	EA	NW	N	NE	NW	N	NE	W	O	E	NWB	NB	NEB	NWB	NB	NEB	WB	B	EB
SE	NWA	NWA	NA	NWA	NWA	NA	WA	WA	A	NW	NW	N	NW	NW	N	W	W	O	NWB	NWB	NB	NWB	NWB	NB	WB	WB	B
NWB	A	EA	EA	SA	SEA	SEA	SA	SEA	SEA	A	EA	EA	SA	SEA	SEA	SA	SEA	SEA	O	E	E	S	SE	SE	S	SE	SE
NB	WA	A	EA	SWA	SA	SEA	SWA	SA	SEA	WA	A	EA	SWA	SA	SEA	SWA	SA	SEA	W	O	E	SW	S	SE	SW	S	SE
NEB	WA	WA	A	SWA	SWA	SA	SWA	SWA	SA	WA	WA	A	SWA	SWA	SA	SWA	SWA	SA	W	W	O	SW	SW	S	SW	SW	S
WB	NA	NEA	NEA	A	EA	EA	SA	SEA	SEA	NA	NEA	NEA	A	EA	EA	SA	SEA	SEA	N	NE	NE	O	E	E	S	SE	SE
B	NWA	NA	NEA	WA	A	EA	SWA	SA	SEA	NWA	NA	NEA	WA	A	EA	SWA	SA	SEA	NW	N	NE	W	O	E	SW	S	SE
EB	NWA	NWA	NA	WA	WA	A	SWA	SWA	SA	NWA	NWA	NA	WA	WA	A	SWA	SWA	SA	NW	NW	N	W	W	O	SW	SW	S
SWB	NA	NEA	NEA	NA	NEA	NEA	A	EA	EA	NA	NEA	NEA	NA	NEA	NEA	A	EA	EA	N	NE	NE	N	NE	NE	O	E	E
SB	NWA	NA	NEA	NWA	NA	NEA	WA	A	EA	NWA	NA	NEA	NWA	NA	NEA	WA	A	EA	NW	N	NE	NW	N	NE	W	O	E
SEB	NWA	NWA	NA	NWA	NWA	NA	WA	WA	A	NWA	NWA	NA	NWA	NWA	NA	WA	WA	A	NW	NW	N	NW	NW	N	W	W	O

4   Future Directions

If dir(A,B) and dir(B,C) are known, a question arises as to whether we can leverage this knowledge to determine dir(A,C). Such techniques are used in RCC8 [14] models for spatial reasoning. For example, if dir(A,B)=(0,-1) (i.e., object B is to the south of object A) and dir(B,C)=(0,1) (i.e., C is to the north of B), it is possible that A is to the north of C or south of C or in the same space as C. It is not clear how to *uniquely* determine dir(A,C) without the knowledge of distance. Qualitative distance information is required for resolving this query. Previous approaches do not address this or are too computation intensive and inefficient [6]. In some cases, even knowing distance is not sufficient to answer the query. For example, if A is close to B and B is far from C, we can't tell whether A is far from C. Distances coupled with directions help resolve the direction relation query: A is south of C. An alternative solution that builds upon the work presented herein will be presented in a future paper, thereby increasing the usefulness of this methodology.

5   Conclusion

In this paper, we have presented an approach to representing and determining cardinal directions between objects, which may be multiply connected, with heterogeneous dimensions. The analysis and development is very simple, mathematically sound, and can be implemented efficiently. This approach does not require complex computation to define additional object location matrices which are used in other models for computing complex cardinal directions. Converseness is preserved while the relation between gridded parts of the complex objects is determined. The extension to 3D is seamless; it needs no additional formulation or data structures for transition from 2D to 3D. This work is directly applicable to geographical information systems for location determination, robot navigation, and spatio-temporal network databases where direction changes frequently.

References

[1] Chen, T., Schneider, M., Viswanathan, G., Yuan, W.: The Objects Interaction Matrix for Modeling Cardinal Directions in Spatial Databases. In: Kitagawa, H., Ishikawa, Y., Li, Q., Watanabe, C. (eds.) DASFAA 2010. LNCS, vol. 5981, pp. 218–232. Springer, Heidelberg (2010)

[2] Sabharwal, C.L., Leopold, J.L.: Cardinal Direction Relations in Qualitative Spatial Reasoning. International Journal of Computer Science & Information Technology (IJCSIT) 6(1), 1–13 (2014), doi:10.5121/ijcsit.2014.6101

[3] Viswanathan, G., Schneider, M.: Querying Cardinal Directions between Complex Objects in Data Warehouses. Fundamenta Informaticae CVII, 1001–1026 (2011)

[4] Moratz, R.: Representing relative direction as binary relation of oriented points. In: Proceedings of the 17th European Conference on Artificial Intelligence (ECAI 2006), Riva del Garda, Italy (August 2006)

[5] Frommberger, L., Lee, J.H., Wallgrün, J.O., Dylla, F.: Composition in OPRAm, Report Series of the Transregional Collaborative Research Center SFB/TR 8 Spatial Cognition Universität Bremen / Universität Freiburg, SFB/TR 8 Report No. 013-02/2007

[6] Chen, J., Jia, H., Liu, D., Zhang, C.: Composing Cardinal Direction Relations Based on Interval Algebra. Int. J. Software Informatics 4(3), 291–303 (2010)

[7] Borrmann, A., van Treeck, C., Rank, E.: Towards a 3D Spatial Query Language for Building Information Models. In: Proc. Joint Int. Conf. of Computing and Decision Making in Civil and Building Engineering, ICCCBE-XI (2006)

[8] Goyal, R.K., Egenhofer, M.J.: Consistent Queries over Cardinal Directions Across Different Levels of Detail. In: 11th Int. Conf. on Database and Expert Systems Applications, pp. 876–880 (2000b)

[9] Skiadopoulos, S., Sarkas, N., Sellis, T., Koubarakis, M.: A Family of Directional Relation Models for Extended Objects. IEEE Trans. on Knowledge and Data Engineering 19(8), 1116–1130 (2007); Brageul, D., Guesgen, H.W.: A Model for Qualitative Spatial Reasoning Combining Topology, Orientation and Distance. ACM Artificial Intelligence, 653–658 (2007)

[10] Frank, A.U.: Qualitative Spatial Reasoning about Cardinal Directions. In: Int. Research Symp. on Computer-based Cartography and GIScience AutoCarto., vol. 6, pp. 148–167 (1991)

[11] Schneider, M., Chen, T., Viswanathan, G., Yuan, W.: Cardinal Directions between Complex Regions. ACM Transactions on Database Systems V(N), 1–45 (2011)

[12] Chang, S.-K., Li, Y.: Representation of Multi-resolution Symbolic/Binary Pictures using 2DH Strings. In: IEEE Workshop on Languages for Automation, pp. 190–195 (1988)

[13] Sabharwal, C.L., Bhatia, S.K.: Perfect Hash Table Algorithm for Image Databases Using Negative Associated Values. Pattern Recognition Journal 28(7), 1091–1101 (1995)

[14] Sabharwal, C.L., Leopold, J.L.: Evolution of Region Connection Calculus to VRCC-3D+. Journal of New Mathematics and Natural Computation 10(2), 1–39 (2014), doi:http://dx.doi.org/10.1142/S1793005714500069

Qualitative Spatial Reasoning in 3D: Spatial Metrics for Topological Connectivity in a Region Connection Calculus

Chaman L. Sabharwal and Jennifer L. Leopold

Missouri University of Science and Technology
{chaman,leopoldj}@mst.edu

Abstract. In qualitative spatial reasoning, there are three distinct properties for reasoning about spatial objects: connectivity, size, and direction. Reasoning over combinations of these properties can provide additional useful knowledge. To facilitate end-user spatial querying, it also is important to associate natural language with these relations. Some work has been done in this regard for line-region and region-region topological relations in 2D, and very recent work has initiated the association between natural language, topology, and metrics for 3D objects. However, prior efforts have lacked rigorous analysis, expressive power, and completeness of the associated metrics. Herein we present new metrics to bridge the gap required for integration between topological connectivity and size information for spatial reasoning. The new set of metrics that we present should be useful for a variety of applications dealing with 3D objects.

Keywords: Region Connection Calculus, Metrics, Spatial Reasoning, Qualitative Reasoning.

1 Introduction

Qualitative spatial reasoning is intrinsically useful even when information is imprecise or incomplete. The reasons are: (1) precise information may not be available or required, (2) detailed parameters may not be necessary before proceeding to decision making, and (3) complex decisions sometimes must be made in a relatively short period of time. However, qualitative reasoning can result in ambiguous solutions due to incomplete or imprecise quantitative information. In RCC8 [1], [2], the regions have a well-defined interior, boundary, and exterior. The RCC8 relations are bivalent with true and false values. Mathematically defined and computer drawn objects are crisp and well-defined, whereas hand-drawn regions tend to have a vague boundary [3]. When regions are vague, the relations between regions can be vague also. That makes the possible values for relations to be true, false, or even 'maybe.' We may have an application where regions and relations are vague; in RCC8, regions and relations are crisp. While topology is sufficient to determine the spatial connectivity relations, it lacks the capability to determine the degree (or extent) of connectivity of such relations.

R. Prasath et al. (Eds.): MIKE 2014, LNAI 8891, pp. 215–228, 2014.

For example, in Fig. 1, for two objects A and B, the RCC8 proper overlap relation, PO(A,B), evaluates to true, yet it does not provide any information about the degree of connectivity; we do not know how much is the overlap — are they barely overlapping or are they are almost equal? The usefulness of metrics lies in providing such additional information which can be quite critical for some applications.

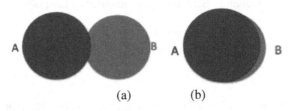

(a)   (b)

Fig. 1. RCC8 determines that there is an overlap between A and B, but it does not quantify the proper overlap whereas in (a) they are barely overlapping, and in (b) they are almost equal

Metrics are quantitative, whereas topology is qualitative and both together can supplement each other in terms of spatial knowledge. The metric refinements provide for quality of connectivity of each relation. The goal of this exposition is to bridge the gap between topology and size via metrics.

The paper is organized as follows. Section 2 provides a brief mathematical background relevant to subsequent discussions in the paper. Section 3 explains the motivation for metrics. Section 4 discusses the development of our metrics, as well as the association between size and topology. Section 5 explains the association between connectivity, size and metrics. Section 6 gives the conclusion and future directions, followed by references in Section 7.

2   Background

2.1   Spatial Relations in General

Historically, there are two approaches to topological region connection calculus, one is based on first order logic [1], and the second is based on the 9-intersection model [2]. Both of these approaches assume that regions are in 2D and the regions are crisp, and that relation membership values are true and false only. Metrics were used in 1D to differentiate relative terms of proximity like *very close, close, far*, and *very far* [4]. Metrics were used to refine natural language and topological relationships for line-region and region-region connectivity in 2D [5]. These approaches lack determining the strength of relation, the combination of the connectivity and size information. Recently more attention has been directed to these issues in 2D [6] and in 3D [7]. However, prior work has been deficient in rigorous analysis, expressive power, and completeness of the metrics. The complete set of metrics presented herein differs from the previous approaches in its completeness and enhanced expressiveness.

2.2   Mathematical Preliminaries

R^3 denotes the three-dimensional space endowed with a distance metric. Here the mathematical notions of *subset, proper subset, equal sets, empty set (∅), union, intersection, universal complement,* and *relative complement* are the same as those typically defined in set theory. The notions of *neighborhood, open set, closed set, limit point, boundary, interior, exterior,* and *closure* of sets are as in point-set topology. The interior, boundary, and exterior of any region are disjoint, and their union is the universe.

A set is *connected* if it cannot be represented as the union of disjoint open sets. For any non-empty bounded set A, we use symbols A^c, A^i, A^b, and A^e to represent the universal complement, interior (Int(A)), boundary (Bnd(A)), and exterior (Ext(A)) of a set A, respectively. Two regions A and B are equal if $A^i == B^i$, $A^b == B^b$, and $A^e == B^e$ are true. For our discussion, we assume that every region A is a non-empty, bounded, regular closed, connected set without holes; specifically, A^b is a closed curve in 2D, and a closed surface in 3D.

2.3   Region Connection Calculus Spatial Relations

Much of the foundational research on qualitative spatial reasoning concerns a region connection calculus (RCC) that describes 2D regions (i.e., topological space) by their possible relations to each other [1], [2]. Conceptually, for any two regions, there are three possibilities: (1) *One is outside the other*; this results in the RCC8 relation DC (disconnected) or EC (externally connected). (2) *One overlaps across boundaries*; this corresponds to the RCC8 relation PO (proper overlap). (3) *One is inside the other*; this results in topological relation EQ (equal) or PP (proper part). To make the relations jointly exhaustive and pairwise distinct (JEPD), there is a converse relation denoted by PPc (proper part converse), PPc(A,B) ≡ PP(B,A). For completeness, RCC8 decomposes proper part into two relations: TPP (tangential proper part) and NTPP (non-tangential Proper part). Similarly for PPc, RCC8 defines TPPc and NTPPc. RCC8 can be formalized by using first order logic [1] or using the 9-intersection model [2].

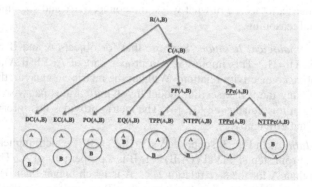

Fig. 2. RCC8 Relations in 2D

Region connection calculus was designed for 2D [1], [2]; it was extended to 3D [8], [6]. In [5], metrics were used for associating line-region and region-region connectivity in 2D to natural language. The metrics were adapted from [5] for qualitative study of the dependency between metrics and topological relations, and between metrics and natural-language terms; conclusions then were drawn for association between the natural-language terms and topological connectivity RCC8 terms [7]. However, the 2D metrics were adopted and adapted to 3D objects without any regard for viability or completeness. Herein we introduce new metrics and explore the degree of association between them in terms of strength of connectivity and relative size information.

3   Motivation for Metrics

In qualitative spatial reasoning, there are three distinct properties for reasoning about spatial objects: connection, dimension, and direction. Reasoning over *combinations* of these properties can provide additional useful knowledge. The prior efforts [5] have lacked rigorous analysis, expressive power, and completeness of the associated metrics. Revision of the metrics is required before we can begin to bridge the gap between topological connectivity and size information for automated spatial reasoning.

We start with following example for motivation to study the degree (or extent) of spatial relations. This example centers around one metric and one pair of objects; see Fig. 3 for concept illustration. Consider the interior volume of an object A, split by the interior volume of an object B; let this be denoted by metric, IVsIV(A,B). This metric calculates how much of A is part of B. Since sizes of objects can vary in units of measurement, it is more realistic to compare qualitative relative sizes for objects. Recall from section 2.2 that A^i represents the interior of A. We define the relative (i.e., normalized) part of A in B by the equation,

$$IVsIV(A, B) = \frac{volume(A^i \cap B^i)}{volume(A^i)}$$

With this metric, let us see in what ways, the connectivity and size information are useful in spatial reasoning.

(1) RCC8 Topological Relation: Suppose that for objects A and B in Fig. 3, we have IVsIV(B,A) = 1. This implies B is a proper part of A, PP(B,A), which is an RCC8 qualitative connectivity relation. Without the metric, in general, this relation is computed by using the 9-intersection model involving various pairwise intersections before arriving at this conclusion [2], [6]. The metric provides this information much more quickly and efficiently.

(2) Size Relations: In Fig. 3, suppose IVsIV(B,A) = 0.1, which implies that 10% of B is part of A. From step (1), IVsIV(A,B) = 1, B is a proper part of A. Therefore, B is much smaller than A for the size relation (i.e., A is much bigger than B). In general, if IVsIV(A,B) < IVsIV(B,A), then A is larger than B in size (i.e., or B is smaller than A). Thus the metric is a useful tool for qualitative size comparison of pairs of objects.

(3) Cardinal Direction Relations: We will concentrate on steps (1) and (2) in this paper. The detailed discussion of directions metrics is beyond the scope of this exposition; the reader may consult [9]. The direction metric in [9] determines that B is in the northeast of part of A. With this directional knowledge, it means that in addition to B being a tangential proper part of A, TPP(B,A), tangency is in the NE direction.

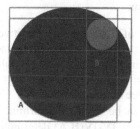

Fig. 3. Object B is a proper part of A, B is much smaller than A in size, and B is in the northeast relative to A. The grid is generated by grid lines for A and B, where the minimum-bounding rectangle is composed of horizontal and vertical gridlines.

Thus we see that B is a proper part of A, and B is much smaller than A. Moreover B is a tangential proper of A and is located in the northeast part of A.

For an example of the need and usefulness of the metrics, see Section 5, how metrics measure the degree of connectivity strengthening the toplogical classification tree.

4   Introduction to Metrics

Quantitative metrics are defined to determine the extent of connectivity of the topological relations between pairs of objects in 3D. The metrics are normalized so that the metric values are constrained to [0,1]. The metrics also allow for qualitative reasoning with the spatial objects in determining their topological relations between objects. As seen in Fig. 3., a metric can be used to derive the qualitative size of the overlap. The overlap relation, PO(A,B), is symmetric, but the overlap metric IVsIV(A,B) is anti-symmetric. The metric values are also sensitive to the location of the objects in addition to topological connectivity, see Fig. 1.

For the purposes of precisely defining the metrics herein, we will need two additional topological concepts in addition to the traditional interior, exterior, and boundary parts of an object (or region). The classical boundary of an object A is denoted by A^b; for fuzzy regions, the boundary interior neighborhood (Bin) is denoted by A^{bi} and the boundary exterior neighborhood (Bex) is denoted by A^{be}. We give the complete details of these concepts in Section 4.2; an application can selectively use the kind of boundary information available. The exterior and interior boundary neighborhoods even may be combined into one fuzzy/thick boundary which is denoted by A^{bt} and defined as $A^{bt} \equiv A^{bi} \cup A^{be}$.

Based on these five region parameters, the 9-Intersection table expands to a 25-Intersection table; see Table 1. For 9-intersection, there are $2^9=512$ possible combinations out of which only eight are physically realizable; see Fig. 2. Similarly out of 2^{25} possible combinations derivable from the five region parameters, only a few are physically possible. The possible relations using metrics are as crisp as for bivalent 9-intersection values, see Section 5.

4.1   Volume Considerations

For 3D regions, the volume of a region is a positive quantity, as is the volume enclosed by a cube or a sphere. The classical crisp boundary of a 3D object is 2D, the volume of a 2D region in a plane or space is zero. Topological relations are predicates that represent the existence of a relation between two objects; metrics measure the strength of the relation or degree of connectivity.

The metric IVsIV(A,B) can be used to determine the extent of overlap $A \cap B$ relative to A, whereas the metric IVsIV(B,A) determines the extent of overlap $A \cap B$ relative to B. For ease and consistency, the metrics are always normalized with respect to the first parameter of the metric function. Recall from section 3 that this metric IVsIV(A,B) is not symmetric. This metric represents the amount of overlap relative to first argument of the metric.

For practical applications, the first parameter is never the exterior volume of an object, because the exterior of a bounded object is unbounded with infinite volume. It is also observed that since volume(A) = volume($A \cap B$) + volume($A \cap B^e$), then IVsIV(A,B^e) = 1- IVsIV(A,B).

4.2   Boundary Considerations

The boundary neighborhood is the region within some small positive radius of the boundary. This is useful for regions with vague boundary. There are two types of neighborhoods, the boundary interior neighborhood, A^{bi}, and the boundary exterior neighborhood, A^{be}; see Fig. 4. By combining the two, we can create a thick boundary for vague regions.

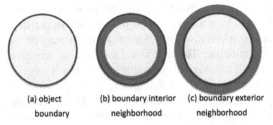

(a) object boundary   (b) boundary interior neighborhood   (c) boundary exterior neighborhood

Fig. 4. (a) A 3D object, (b) the exterior neighborhood of the boundary of the object, and (c) the interior neighborhood of the boundary of the object

Several metrics are designed for cases where the boundary is vague; these are discussed in Section 4.6.1 and 4.6.2. To compensate for an accurate crisp boundary, an application-dependent small neighborhood is used to account for the thickness of the boundary. For the 3D object shown in Fig. 4(a), let the boundary interior neighborhood of A^b of some radius $r>0$, be denoted by A^{bi} or $N_{Ir}(A^b)$, i.e., $A^{bi} \equiv N_{Ir}(A^b)$ (Fig. 4(b)), and let the boundary exterior neighborhood of A^b of some radius $r>0$, be denoted by A^{be} or $N_{Er}(A^b)$, i.e., $A^{be} \equiv N_{Er}(A^b)$; see Fig. 4(c). The smaller the value of r, the less the ambiguity in the object boundary. We denote the qualitative interior neighborhood by $\Delta_I A$ and exterior neighborhood by $\Delta_E A$ without specific reference to r, as $\Delta_I A \equiv A^{bi}$ and $\Delta_E A \equiv A^{be}$ in the equations that follow in this paper.

Many times in geographical information system (GIS) applications the region's exact boundary is not available. Thus the problem in spatial domains becomes that of how to identify and represent these objects. In such analyses, the external connectedness would be resolved by using metric BexsBex and examining whether the value $BexsBex(A,B) < \min(r_1, r_2)$ (instead of $BsB(A,B)=0$) where the objects have boundary exterior r_1- and r_2-neighborhoods for thick boundaries of objects.

In fact, some applications may need only one r-neighborhood (the combination of r_1-interior and r_2-exterior neighborhood along a vague boundary), while others may need two separate neighborhoods as in [5]. The value of $r = \min(r_1, r_2)$ is specified by the application. In general, for numerical calculations, it is approximately one percent of the sum of the radii of two spheres. Intuitively, r accounts for the minimum thickness of the boundary for the object.

4.3   Intersection Consideration in General

All the metrics and topological relations involve intersections (see Table 1) between a pair of objects. An intersection between a pair of objects may be interior to interior (i.e., 3D), or boundary to boundary (neighborhood), which may be turn out to be 2D, or 1D or even 0D. Metrics measure the quantitative values for topological relations. The intersection of 3D objects may remain 3D, as in the case of PO(A,B). If the intersection such as $A^i \cap B^i$ exists, then we can calculate the volume of the 3D intersection $A^i \cap B^i$, which is practical. But if the boundary is 2D, the volume of the boundary is zero, which does not provide any useful information. The intersection between two 3D objects may also be 3D, 2D, 1D, or even 0D. Since intersection is a significant component of topological relations, we can extract useful information from intersections of lower dimensional components also. We can calculate the area of a 2D object (e.g., $A \cap B^b$ may be a 2D surface), and surface area can provide essential information for relations EC(A,B), TPP(A,B), and TPPc(A,B). For example, if two cubes touch face to face, they intersect in a surface; the volume of intersection will be zero, but surface area will be positive, which can still provide a measure of how close the objects are to each other. So we will need metrics that accommodate 2D surface area also. Sometimes intersection is a curve or a line segment, in which case we can analyze the strength of the relation from the length of the segment. Consequently, we also need metrics that handle the length of edge intersection. For a single point intersection (degenerate line segment), the volume of a point is zero, as are the area and length of a single point.

4.4   Space Partitioning

Each object divides the 3D space into three parts: interior, boundary and exterior. The interior and exterior of the object are 3D parts of space, and the boundary of the object is 2D. The intersection between two 3D objects can be 3D, or a 2D surface, or a 1D curve, or a line segment, or even 0D (i.e., a point). In many geographical applications, regions may not have a well-defined boundary. For example, the shoreline boundary of lake is not fixed. If the lake is surrounded with a road, the road can serve as the boundary for practical purposes. We need to compensate for the blur in the boundary. Consequently we utilize two additional topological regions: Boundary inner neighborhood (Bin) and Boundary exterior neighborhood (Bex). They can be used to measure how close the objects are from boundary to boundary. The thick boundary becomes a 3D object rather than a 2D object, so the volume calculation for boundary becomes meaningful. For non-intersecting objects, it can be used to account for the distance between them, and for the tangential proper part relation between objects A and B, TPP(A, B), it can measure how close is inner object A is from the outer object boundary B^{be}. Thus the terms Boundary interior neighborhood (Bin) and Boundary exterior neighborhood (Bex) for an object A account for the fuzziness, $A^{bt} \equiv A^{bi} \cup A^{be}$, in the boundary description or thickness of the boundary; see Fig. 4.

4.5   25-Intersections

To keep full generality available to the end-user, an object space can be defined in terms of five parts: interior, boundary, exterior, boundary interior neighborhood, and boundary exterior neighborhood. As descriptive as we can be for symbols to be close to natural language: we use Int(A) for A^i the interior of A, Ext(A) for A^e the exterior of A, Bnd(A) for A^b the boundary of A, Bin(A) for A^{bi} the boundary interior neighborhood A, and Bex(A) for A^{be} the boundary exterior neighborhood of the boundary of A. This will lead to a 25-intersection table where the boundary can be a crisp boundary A^b, or a thick boundary $A^{bt} \equiv A^{bi} \cup A^{be}$; see Table 1 for all 25 combinations of intersections.

Table 1. 25-Intersection table

	Int	Bnd	Ext	Bin	Bex
Int	$A^i \cap B^i$	$A^i \cap B^b$	$A^i \cap B^e$	$A^i \cap B^{bi}$	$A^i \cap B^{be}$
Bnd	$A^b \cap B^i$	$A^b \cap B^b$	$A^b \cap B^e$	$A^b \cap B^{bi}$	$A^b \cap B^{be}$
Ext	$A^e \cap B^i$	$A^e \cap B^b$	$A^e \cap B^e$	$A^e \cap B^{bi}$	$A^e \cap B^{be}$
Bin	$A^{bi} \cap B^i$	$A^{bi} \cap B^b$	$A^{bi} \cap B^e$	$A^{bi} \cap B^{bi}$	$A^{bi} \cap B^{be}$
Bex	$A^{be} \cap B^i$	$A^{be} \cap B^b$	$A^{be} \cap B^e$	$A^{be} \cap B^{bi}$	$A^{be} \cap B^{be}$

Now Bnd(A) represents the crisp boundary of A, if any, whereas Bin(A) and Bex(A) account for the crisp representations of the vague boundary. There are 2^{25} possible 25-intersection vectors in all. However, all the vectors are not physically realizable. For example, all entries in any row in Table 1 cannot be true simultaneously, and all entries in any column in Table 1 cannot be true simultaneously. Another use of the metrics is to see, for the proper part relation between A and B, PP(A,B), how far the inner object A is from the inner boundary neighborhood of the outer object, B^{bi}. A commonly used predicate for determining connectivity between crisp regions is boundary-boundary intersection, $A^b \cap B^b$. We must be mindful that space now is portioned into five parts instead of three parts. It is clear that A^i, A^e are open sets, and A^b is a closed set. For spatial reasoning, when A^{bi} A^{be} are used, they are semi-open, semi-closed sets — open towards A^b and closed towards inside of A^{bi} and outside of A^{be}.

4.6 Metrics

Here we complete the development of the remaining metrics; an application may selectively use the metrics applicable to the problem at hand. Conventionally, a 4-intersection [6] (BndBnd, IntBnd, BndInt, IntInt) is sufficient for crisp 3D data. Some applications may need Bex and Bin separately [5], while fuzzy logic applications may need to combine Bex and Bin into one Bnd [6]. For all 25 intersections (see Table 1) the metrics are defined by normalizing the intersections. There are 25 possible pairswise intersections to be considered in the metrics. For one pair of objects, there are eight distinct versions {(A,B), (A,B^e), (A^e,B), (A^e,B^e), (B,A), (B,A^e), (B^e,A), (B^e,A^e)} as input arguments for which a metric value may be computed. That is, the domain for each metric consists of eight distinct pairs corresponding to each input pair of objects A and B. Since metrics are normalized, some metrics may not be realizable; for example, IVsIV cannot be defined for the combinations {(A^e,B), (A^e,B^e), (B^e,A), (B^e,A^e)} because the corresponding metrics involve infinity. In fact, five of the metrics are impossible (not realizable); see Table 2. Here we will identify the possible (realizable) 20 metrics.

Since the metrics are not symmetric, the converse metrics can be obtained by switching arguments A and B (e.g., the converse of IVsIV(A,B) is IVsIV(B,A)). To make the list of metrics exhaustive, we can append suffix c to the name to indicate the converse metric when needed. Table 2 lists directly possible and impossible metrics, which are developed in Sections 4.6.1 and 4.6.2.

Table 2. Complete list of metrics corresponding to 25 intersections in Table 1. 20 metrics are viable and 5 metrics are not possible.

Possible	Impossible
IVsIV, IVsEV	EVsIV, EVsEV
BinsIV, BinsEV, IVsBin	EVsBin
BexsIV, BexsEV, IVsBex	EVsBex
BinsBin, BinsBex,	
BexsBin, BexBex	
BsIV, BsEV, IVsB	
BsBin, BsBex,	EVsB
BinsB, BexsB	
BsB	

Next we define 20 viable metrics and show their connection with the RCC8 topological relations and size relations on 3D objects only. First we look at the two metrics together: IVsIV(A,B) and IVsEV(A,B) which measure how much space one object shares with the other object. We have already defined interior volume split by interior volume, IVsIV(A,B), earlier in the motivation discussion, Section 3.

4.6.1 Anatomy of Volume Metrics

Recall, interior volume splitting (IVsIV) computes the scaled (normalized) part of one object that is split by the interior of the other object. It measures how much of A is part of B. The boundary of a 3D object is 2D. Here boundary does not matter, as the volume of the boundary is zero. Exterior volume splitting (IVsEV) describes the proportion of one object's interior that is split by the other object's exterior. The exterior volume splitting (IVsEV) is defined by

$$IVsEV(A, B) = \frac{volume(A^i \cap B^e)}{volume(A^i)}$$

It measures how much A is away from B. Again, boundary does not matter. Observe that volume(A) = volume(A∩B) + volume(A∩B^e), and hence IVsEV(A,B) = 1- IVsIV(A,B). The metric value is between 0 and 1, inclusive. If the metric value IVsIV(A,B) = 0, the objects are disjoint or externally connected. If the metric value IVsIV(A,B) > 0, then this value indicates two things. First, A^i∩B^i ≠ ∅. Usually, the truth value of A^i∩B^i is established by considering the intersection of the boundaries of two objects (extensive computation takes place because the objects are represented with boundary information only). Here the metric value IVsIV(A,B) > 0, so we can quickly determine the truth value of A^i∩B^i. Secondly, the actual value of the metric IVsIV(A,B) measures what relative portion of object, A is common with object B; the larger the value of the metric, the larger the commonality and conversely. Let

$$x = \frac{volume(A^i \cap B^i)}{volume(A^i)} * 100 \quad y = \frac{volume(B^i \cap A^i)}{volume(B^i)} * 100$$

This can directly answer queries such as object A has x percent in common with B, whereas object B has y percent in common with A. If x=y=0, then the objects are either externally connected or disjoint, but this metric alone does not tell how far apart they are. In order to determine that, we simply compute the distance between the centers to differentiate between DC and EC. The metric does embody knowledge about which object is larger.

4.6.2 Anatomy of Boundary Metrics

Recall, for the 3D object shown in Fig. 4(a), A^{be} is the boundary exterior neighborhood of A^b with some radius (Fig. 4(b)), and A^{bi} is the boundary interior neighborhood of A^b with some radius; see Fig. 4(c). The value of the radius is application-dependent. We use the qualitative interior and exterior neighborhood without specific reference to r, as $\Delta_I A \equiv A^{bi}$ and $\Delta_E A \equiv A^{be}$ in the following equations.

Considering the interior neighborhood of an object, we define the closeness to interior volume (BinsIV) as follows:

$$BinsIV(A, B) = \frac{volume(\Delta_I A \cap B^i)}{volume(\Delta_I A)}$$

This metric contributes to the overall degree of relations of PO, EQ, TPP, and TPPc.

Similarly, we can consider the exterior neighborhood of an object, and can define a metric for exterior volume closeness (BexsIV) as by replacing $\Delta_I(A)$ by $\Delta_E(A)$. This metric is a measure of how much of the exterior neighborhood of A^b is aligned with the interior of B. This metric is useful for the degree of relations of PO, EQ, TPP, and TPPc.

Similarly the metrics for the exterior of B are defined for completeness as follows:

$$BinsEV(A, B) = \frac{volume(\Delta_I A \cap B^e)}{volume(\Delta_I A)}$$

BexsEV(A,B) is defined by replacing $\Delta_I(A)$ by $\Delta_E(A)$. Boundary-boundary intersection is an integral predicate for distinguishing RCC8 relations. Similarly, for quantitative metrics, it can be important to consider how much of the inside and outside of the boundary neighborhood of one object is shared with the boundary neighborhood of the other object.

BinsBin(A,B) is designed to measure how much of the Interior Neighborhood of A is split by the Interior Neighborhood of B. This metric is useful for fuzzy regions with fuzzy interior boundary.

$$BinsBin(A, B) = \frac{volume(\Delta_I A \cap \Delta_I B)}{volume(\Delta_I A)}$$

BexsBin(A,B) is designed to measure how much of the Exterior Neighborhood of A is split by the Interior Neighborhood of B. This metric may be useful when the region is vague around both sides of the boundary.

BinsBex(A,B) is defined by replacing $\Delta_I(A)$ by $\Delta_E(A)$ and is designed to measure how much of the Interior Neighborhood of A is split by the Exterior Neighborhood of B, It is useful to analyze topological relations DC and EC.

$$BinsBex(A, B) = \frac{volume(\Delta_I A \cap \Delta_E B)}{volume(\Delta_I A)}$$

BexsBex(A,B) is designed to measure how much of the Exterior Neighborhood of A is split by the Exterior Neighborhood of B. This metric is useful for fuzzy regions, if BexsBex(A,B) = 0 then we can narrow down the candidates of possible relations between A and B to DC, NTPP, and NTPPc.

$$BexsBex(A, B) = \frac{volume(\Delta_E A \cap \Delta_E B)}{volume(\Delta_E A)}$$

We define several splitting metrics to specifically examine the proportion of the boundary of one object that is split by the volume, boundary neighborhoods, and boundary of the other object; we denote these metrics accordingly for boundary splitting. It should be noted that there are five versions of the equations for this metric. First, the boundary may be the thick boundary composite neighborhood (interior and exterior), in which case it is a volume. If the boundary is a simple boundary, it's a 2D area. Therefore, for numerator calculations, we will be calculating $A^b \cap B$ as either a volume or an area. It also is possible that $A^b \cap B$ is an edge (a curve or a line segment). For example, for two cubes, a cube edge may intersect the face of the cube as a line segment or an edge of another cube in a line segment, or even as a single point (i.e., a degenerate line segment). If $A^b \cap B$ is an edge, we calculate edge length. For denominator, volume(A^b) and area(A^b) are self-evident depending on whether we have a thick or simple boundary. However, length(A^b) calls for an explanation. In the numerator, when length($A^b \cap B$) is applicable, then this intersection is part of an edge in A^b; length(A^b) is computed as the length of the enclosing edge. These metrics are defined and described below. The converses of the metrics can be derived similarly.

BsIV(A,B) measures the Boundary of A split by the Interior Volume of B.

$$BsIV(A,B) = \frac{volume(A^b \cap B^i)}{volume(A^b)} \; or \; \frac{area(A^b \cap B^i)}{area(A^b)} \; or \; \frac{length(A^b \cap B^i)}{length(A^b)}$$

BsEV(A,B) is defined by replacing B^i by B^e and measures the Boundary of A split by the Exterior Volume of B. BsBin(A,B) is defined by replacing B^i by $\Delta_I(B)$ and measures the Boundary of A split by the Interior Neighborhood of B. BsBex(A,B) is defined by replacing B^i by $\Delta_E(B)$ and measures the Boundary of A split by the Exterior Neighborhood of B. BsB(A,B) is defined by replacing B^i by B^b and measures the Boundary of A split by the Boundary of B.

This metric is again directly applicable to computing $A^b \cap B^b$ which is used to distinguish many of the RCC8 relations. This subsequently allows us to narrow down the candidates of possible relations between A and B to DC, NTPP, and NTPPc.

For crisp regions, we have an interior, boundary, and exterior. For vague regions, we have boundary interior and exterior neighborhoods. The smaller the radius for boundary neighborhoods, the smaller the ambiguity in the object boundary. For consistency, we can combine the interior and exterior neighborhoods into one, which we call a thick boundary. For a thick boundary, the object has three disjoint crisp parts: the interior, the thick boundary, and the exterior. Now we can reason with these parts similar to how we use crisp regions for determining the spatial relations.

5   Connectivity, Size and Metrics

If the regions are crisp, we can use the 9-intersection model for determining connectivity relations for 2D connectivity knowledge [2], and for relative size information we use the 3D metrics from Section 4. The relative size of objects and boundary is obtained by using volume metrics IVsIV, IVsB, and boundary-related BsB metrics. Metrics measure the degree of connectivity; for example, for the proper overlap relation PO(A,B), IVsIV metrics help to determine the relative extent of overlap of each object. In Section 4 we discussed which metrics are specific to each of the connectivity relations. If one or both regions are vague, we can use metrics to create a thick boundary, $A^{bt} \equiv A^{bi} \cup A^{be}$, by using the interior and exterior neighborhoods. Again we have, crisp interior A^i, exterior A^e, and thick boundary A^{bt}. By using the 9-Intersection model on A^i, A^e, and A^{bt}, we can derive the connectivity, degree of connectivity, and relative size information for vague regions. Other applications such as natural language and topological association [5] can use appropriate combinations of these topological parts. Fig. 5 provides a visual summary of: (1) what metrics are required to classify each topological relation, and (2) the contribution (0/+) each metric has with regards to the overall quality of the relation. This tree can be used to classify crisp relations. Similarly, a tree could be generated for vague regions with appropriate metrics from the set of 20 metrics.

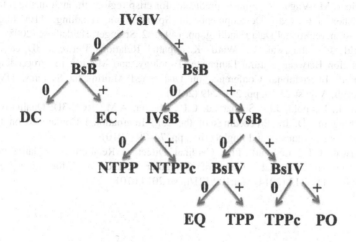

Fig. 5. Tree for the metrics required for classification and the contribution (0/+) of the respective metrics to the overall quality of classification

6   Conclusion and Future Directions

Herein we presented an exhaustive set of metrics for use with both crisp and vague regions, and showed how each metric is linked to RCC8 relations for 3D objects. Our metrics are systematically defined and are more expressive (consistent with natural language) than previously published efforts. Further, we showed the association between our metrics and the topology and size of objects. This work should be useful for a variety of applications dealing with automated spatial reasoning in 3D. In the future, we plan to use these metrics to associate natural language terminology with 3D region connection calculus including occlusion considerations. Also we will explore the applications of these metrics between heterogeneous dimension objects, $O_m \in R^m$ and $O_n \in R^n$ for m, n $\in \{1,2,3\}$.

References

[1] Randell, D.A., Cui, Z., Cohn, A.G.: A Spatial Logic Based on Regions and Connection. In: Proceedings of the 3rd International Conference on Knowledge Representation and Reasoning, Cambridge, MA, pp. 156–176 (1992)

[2] Egenhofer, M.J., Herring, J.: Categorizing binary topological relations between regions, lines, and points in geographic databases. In NCGIA Technical Reports 91-7 (1991)

[3] Schockaert, S., De Cock, M., Kerre, E.E.: Spatial reasoning in a fuzzy region connection calculus. Artificial Intelligence 173(2), 258–298 (2009)

[4] Hernandez, D., Clementini, E.C., Di Felice, P.: Qualitative Distances. In: Kuhn, W., Frank, A.U. (eds.) COSIT 1995. LNCS, vol. 988, pp. 45–57. Springer, Heidelberg (1995)

[5] Shariff, R., Egenhofer, M.J., Mark, D.M.: Natural-Language Spatial Relations Between Linear and Areal Objects: The Topology and Metric of English-Language Terms. International Journal of Geographical Information Science 12, 215–246 (1998)

[6] Schneider, M.: Vague topological predicates for crisp regions through metric refinements. In: Fisher, P.F. (ed.) Developments in Spatial Data Handling, 11th International Symposium on Spatial Data Handling, pp. 149–162. Springer, Heidelberg (2005)

[7] Leopold, J., Sabharwal, C., Ward, K.: Spatial Relations Between 3D Objects: The Association Between Natural Language, Topology, and Metrics. In: Proceedings of the Twentieth International Conference on Distributed Multimedia Systems, DMS 2014, Pittsburgh, August 27-29, pp. 241–249 (2014)

[8] Albath, J., Leopold, J.L., Sabharwal, C.L., Maglia, A.M.: RCC-3D: Qualitative Spatial Reasoning in 3D. In: Proceedings of the 23rd International Conference on Computer Applications in Industry and Engineering, pp. 74–79 (2010)

[9] Sabharwal, C.L., Leopold, J.L.: Cardinal Direction Relations in Qualitative Spatial Reasoning. International Journal of Computer Science & Information Technology (IJCSIT) 6(1), 1–13 (2014), doi:10.5121/ijcsit.2014.6101

Context-Aware
Case-Based Reasoning

Albert Pla, Jordi Coll, Natalia Mordvaniuk, and Beatriz López

University of Girona, Institut d'Informàtica i Aplicacions,
Campus Montilivi - Edifici P4, 17041 Girona, Catalonia, Spain
{albert.pla,jordicoll,beatriz.lopez}@udg.edu,
nat606@gmail.com
http://exit.udg.edu

Abstract. In the recent years, there has been an increasing interest in ubiquitous computing. This paradigm is based on the idea that software should act according to the context where it is executed in what is known as context-awareness. The goal of this paper is to integrate context-awareness into case-based reasoning (CBR). To this end we propose thee methods which condition the retrieval and the reuse of information in CBR depending on the context of the query case. The methodology is tested using a breast-cancer diagnose database enriched with geospatial context. Results show that context-awareness can improve CBR.

Keywords: Context Management, Case-based reasoning, Stacking, Pervasive computing.

1 Introduction

Context-aware applications aim to take advantage of contextual data in order to improve user experience. These kind of applications are based on the concept that a particular set of information can have different meanings or implications depending on the context it is placed or from where it has been obtained [13]: for instance, in the medical domain, patients can obtain different blood pressure measures depending on whether the pressure test has been taken by a doctor or not (this is known as the white coat effect) [20]. Therefore, keeping track of contextual information can be useful to improve data analysis and reasoning processes.

Case-based reasoning has been proved a useful tool to reason about context [8,11]. On the other hand, it could be interesting to analyze how context-reasoning relates to case-based reasoning. Our research concerns such study: to assess how case-based reasoning (CBR) can integrate and use contextual information. For that purpose we present three different methods for endowing case-based reasoning with contextual information in what we called context-aware case-based reasoning. Two of the methods are intended for classical case-based reasoning whilst the third one is designed for ensemble case-based reasoning systems.

R. Prasath et al. (Eds.): MIKE 2014, LNAI 8891, pp. 229–238, 2014.

In this work, context-aware case-based reasoning is tested using data extracted from the medical domain. Particularly, we use a database containing information regarding cancer patients from different cities and hospitals. In such scenario, we use geographical information to establish the context for each patient and to take benefit from it during the reasoning process.

This paper is organized as follows: first we introduce some basic notions regarding context-aware computing; next we present some related work in order to contextualize our research. In Section 4 we propose 3 different approaches to integrate context to case-based reasoning. Those approaches are then studied using a breast cancer database which contains geospatial context information. Finally, the paper ends with some conclusions and the proposal of further research.

2   Background

In pervasive computing context-awareness is the property of changing the behavior of a computing system depending on the context where it is being executed or the context where the user is placed [21]. This implies that a same application can present different outputs and behaviors depending on where and when it is used and who is using it. Context management is the whole process of context modeling, context gathering and context reasoning; nevertheless, in this paper we are focused in context reasoning and, specifically, on how it can be applied in case-based reasoning.

The definition of context is strongly task and domain dependent[14]. Nevertheless, in context-aware applications, different kinds of context are typically considered: temporal context, taking advantage of when the application is being executed (e.g. at which time of the day the application is being executed) or when the data used in the system was collected; source-related context, which considers data regarding who or what is using the application or collected the data (e.g. differentiating if the data used in the system was automatically or manually gathered); environmental context, referring to environmental conditions such as the weather (for instance, an activity recommender system should take into account whether it is a sunny or a rainy day); and, finally, geospatial context which involves the location of the user, the application or the data [16].

The methodology presented in this paper involves the context where the case-base was gathered. Such context can refer to any of the above listed context typologies albeit this work has been developed and tested within the medical domain where geospatial context is of special relevance [2].

3   Related Work

The increasing importance of ubiquitous computing and the arising of smartphones have favored the appearance of context-aware applications in several fields. For instance, context has been used to improve the outputs of recommender systems [5,3], to improve the monitoring of patients in hospitals [1] and to combine the outputs of different recommenders [6]. An exhaustive survey regarding context management and context-aware application can be found in [12].

Regarding the relation between CBR and ubiquitous computing, many authors have used CBR to identify context in context-aware applications [11,8]. On the other hand, to the best of our knowledge, not many approaches have tried to improve CBR performance by means of context awareness. In this direction, Zimmerman [23] poses the question if separating case-bases by context might improve the outcomes of CBR. Answering such challenge, [9] presents a dual CBR for recommending music depending on the listener's context. In it a first context-aware CBR generates a database of users with a similar context to the query case, then, a second CBR process recommends songs using such users database as case-base. Our proposals differ from those approaches as we pretend to include context-awareness directly to the CBR reasoning process.

4 Methodology

This section presents three different approaches for including contextual information into the reasoning process of a CBR. Such approaches mainly affect the retrieve and the reuse stages of the CBR, however, they also require certain preliminary work before starting the CBR process in order to identify the context of the query case and case-base instances.

The first two presented methods, plain context CBR and case-base filtering, aim to improve CBR by modifying the way cases and attributes are handled during the reasoning process. The third approach (context stacking) tackles the problem of using contextual information in ensemble CBR by modifying the way the information comes from the different CBR systems is aggregated.

We assume a plain CBR system where the cases are represented by attribute-value pairs. Nevertheless, we consider that the presented methodology can be useful for any other case representation.

4.1 Plain Context CBR

The first approach we propose is the most intuitive way to tackle the inclusion of context into CBR: extending the cases with the contextual information being treated as attributes (Figure 1).

This approach can be described as follows:

- Context representation: Before creating the case-base containing information regarding different cases, it is important to identify the different context information which can be used for each case. For instance, geospatial context information as where the data has been recorded or temporal information such as when the data has been gathered. This information, then, needs to be mapped into the case representation as a simple attribute.

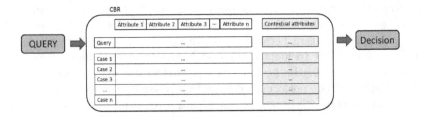

Fig. 1. Adding plain context to CBR

Each individual of the case-base is extended with the available contextual information. As happens with the rest of attributes, contextual attributes can be weighted according to their significance and importance.

$$\begin{aligned}
\text{attributes:} \quad & \langle at_1, \ldots, at_n \rangle \\
\text{context:} \quad & \langle c_1, \ldots, c_n \rangle \\
\text{case:} \quad & \langle at_1, \ldots, at_n, c_1, \ldots, c_n \rangle
\end{aligned}$$

– Query case representation: As well as it is done with the case-base, when a new query arrives to CBR system, it is extended with the available contextual information. For instnce:

$$\begin{aligned}
\text{query case:} \quad & \langle at_1, \ldots, at_n \rangle \oplus \langle c_1, \ldots, c_n \rangle = \langle at_1, \ldots, at_n, c_1, \ldots, c_n \rangle \\
\text{query case:} \quad & \langle \text{age,weight} \rangle \oplus \langle \text{city,location} \rangle = \langle \text{age,weight,city,location} \rangle
\end{aligned}$$

– Context CBR: The CBR system is executed with the desired retrieve, reuse, retain and revise parameters as it would have been done without the contextual information.

The plain approach can be considered as the natural approach of CBR methodology to handle the concept of context: it considers contextual data as valuable as any other data of the case, therefore, contextual data is represented as a normal attribute within the cases. Nevertheless, this approach cannot be included into the concept of "ubiquitous computing" as the inclusion of context is not modifying the behavior of the CBR or the way the data is interpreted.

4.2   Case-Base Filtering

The second approach uses context to filter which parts of the case-base are used by the CBR. In this way, context is used to narrow the case-base of the CBR and to ensure that the CBR only compares cases which have happened under similar or the same context. For that purpose, the different instances of the database and the query cases are labeled with their context(s). Context labels define against who a new case is compared to (Figure 2):

– Context representation: Using only the contextual information of the cases, the different types of context available into the case-base are identified. The process of identifying the existing contexts can be performed by an expert

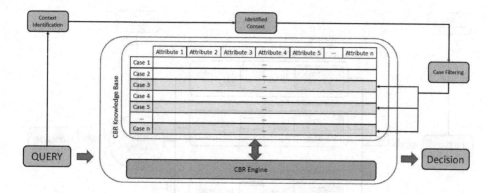

Fig. 2. Case-base contextual filtering schema

or by context classification algorithms (e.g. *ConText* [4], data fusion techniques [19] or subgroup discovery [22,7]). The different types of detected contexts act as labels l which identify the different cases of the case-base.

$$\text{context: } \langle c_1, \ldots, c_n \rangle \rightarrow \text{label: } l$$
$$\text{case } i: \quad (\langle at_1^i, \ldots, at_n^i, \rangle, l_{i_j})$$

- Query case representation: Every time a new case is introduced to the CBR its contextual information is analyzed in order to determine which are their context labels. This analysis needs to be performed following the same methodology used for the case-base.
- Context CBR: The CBR system is executed with the desired retrieve, reuse, retain and revise methods as it would have been done without the contextual information. However, during the retrieve stage, the CBR considers only instances which share context labels with the query case:

$$sim(c, i_j) = \begin{cases} f(c, i_j) & \text{if } l_c = l_{i_j} \\ 0 & \text{otherwise} \end{cases} \tag{1}$$

where $sim(c, i_j)$ is the similarity between the query case c and the case i_j of the case-base; $f(c, i_j)$ the metric which defines its similarity; and l_c and l_{i_j} their context labels.

This approach can be useful for cases where the context differentiates the interpretation of the cases' information. For instance, in a case where the case-base contains information regarding patients recovering at a hospital and patients recovering at their home the geospatial context information of the patient (e.g. whether the patient is being treated at home, at an ambulatory or at a hospital) can condition the interpretation of the data. Conversely to the previous approach, in this methodology context is conditioning the way the CBR retrieves the data by only considering certain cases from the knowledge-base. Following the model of knowledge containers of CBRs described in [18], it can be said that context is conditioning the case-base container by deciding which case-base should be considered in the reasoning.

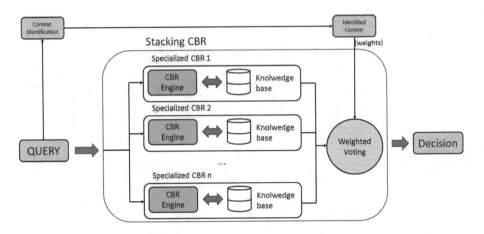

Fig. 3. Contextual CBR stacking schema

4.3 Context Stacking

Finally, the last approach is intended for stacking CBR approaches [17]. In such methodologies different case-based reasoning systems cooperate in order to deliver a final solution. Stacking means that the results of different CBR systems are aggregated following one criteria or another according to certain rules (as a meta-reasoner). For instance, the different solutions can be combined into a single one using a weighted multi-criteria decision method (e.g. a weighted sum or a rated ranking [10]) and the rules can be applied to select the appropriate set of weights for the aggregating function. In context stacking we propose to use the contextual information as the rules which decide the set of weights to be used during the aggregation process (see Figure 3).

As in previous approaches, the available contexts within the database should be identified. When this is done, the CBR system needs to be trained in order to learn the set of weights of the aggregation functions which correspond to each of the contexts learned. Then, depending on the context of the query case, one or another set of weights will be used:

- Context representation: First of all the available types of contexts within the data base needs to be identified (manually or automatically). Conversely to the previous approaches, in stacking the case-base can be distributed among different CBRs systems (which may have different cases among them), therefore the context identification process needs to be done for each existing case-base. Thus, each of the cases is labeled with its context(s).
- Stacking CBR weighting: For each of the identified contexts, the attributes used to weight the different CBRs systems need to be learned. To this end, each set of weights are learned using only cases of a particular context. Therefore, there are as many sets of weights as different contexts, and each set is composed by the same number of weights than the number of CBR systems:

$$WS^{l_m} = \left\langle ws_1^{l_m}, ..., ws_k^{l_m} \right\rangle \tag{2}$$

where WS^{l_m} is the set of weights for the context label l_m when using and stacking approach and $ws_k^{l_m}$ is the weight for the output of the CBR n inside the stacking schema.

- Query case representation: As previous approaches, the context l_c of the new query case c needs to be identified.
- Context Stacking: After the context of the case has been identified, the set of weights corresponding to the case context is selected. Then, the query case query is submitted to each of the stacking CBRs. Every CBR component within the stacking schema studies the received case and proposes a solution according to their knowledge, next, they submit their different solutions. The different solutions are then combined using the desired aggregation function but taking into account the set of weights corresponding to the case of the query case c. In this way, the solution of the CBR is obtained in the following way:

$$r^c = mcdm(S, WS^{l_m}) \tag{3}$$

where r^c is the result of the query case c, WS^{l_m} is the set of weights for the context label l_m, $S = \langle sol_1^c \ldots sol_k^c \rangle$ the solutions to the case c provided by each of the stacking CBRs and $mcdm(S, WS^{l_m})$ the multi-criteria decision method used to combine the provided solutions.

This approach is useful to aggregate the outputs of CBRs with differentiated characteristics. For instance, a stacking CBR where the CBRs' case-bases contain information of cases from different geographical areas (in this case context might give more weight to the case-bases containing information from the same geographical area than the query case); note that in this case, the learning process will weight more the CBRs having a relevant relation with the specified context.

5 Results

In this section we evaluate the methods presented in this paper using a database for breast-cancer prognosis which contains information regarding patients from different cities.

To test the influence of context, we consider the city of residence of patients as the geographical context. The database consists of 502 cases from 3 cities off different size and climate (434 from *CityA*, 37 from *CityB* and 31 from *CityC*). Each case is characterized by 1197 attributes and the distribution of cases is 202 healthy women and 300 women with breast cancer.

The experimentation compares the three presented methods with a CBR without contextual information. In each experiment, the city of each patient has been used as a context label and the methodology has been tested using a cross validation approach. All the tested approach use the same retrieve reuse and revise

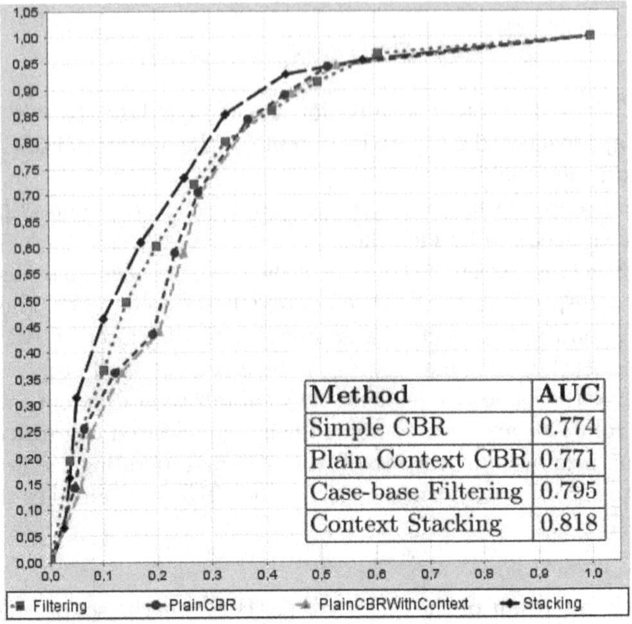

Fig. 4. ROC curve of the plain context CBR method

configuration[1]. Results are evaluated using Receiver Operator Characteristics (ROC) curves and the area under the ROC curve (AUC).

The results of the performed experiments (Figure 4) state that context-awareness can be useful to enhance case-based reasoning[2]. The results of case-base filtering and context stacking outperform the results of the simple CBR without context (which acts as baseline method), being Context stacking the method which obtained the best outcomes. Conversely, the results of modeling context as a regular case attribute (plain context CBR) have not improved the results of the CBR without context; particularly, a two-sample KolmogorovSmirnov test points that there are no significative difference between the simple CBR and the plain context CBR with a confidence of the 99%. These results are in line with the assumptions of ubiquotious computing that pose that context affects the way a data should be interpreted: whilst adding context as an attribute has not affected the outputs of the CBR, changing the way the data retrieval and reuse is done depending on the context has improved the CBR classification.

6 Conclusions

Pervasive computing poses that a software must adapt its behavior depending on the context where it is executed or to the context of its user. Therefore, we

[1] Each CBR configured as the other methods but containing information regarding only one city.

[2] Additional experiments can be found at [15].

propose integrating case-based reasoning and context-handling in what we called context-aware case-based reasoning.

We have proposed three different methods. In all of them it is considered that the context of both the query case and the cases within the CBR case base can be identified. The first one, plain context CBR, includes context into the reasoning process by treating it as another attribute(s) within a case. Case-base filtering proposes to delimit the space search within the case-base depending on the context of the query case. Finally, context stacking, proposes to aggregate the outputs of different CBR systems under a stacking schema; in such approach context determines how the different outputs should be aggregated in order to provide an ultimate solution to the query case.

The methodology has been tested using a breast-cancer diagnose database which contained information of patients from different cities. The city of each patient has been used as geographical context. Experiments have shown that contextual information can improve the outputs of case-based reasoning. CBR without contextual information has been used as a baseline method. Results have shown that adding context as a case attribute does not improve the CBR output. Conversely, both case-base filtering and context stacking, which handle context in a different way, outperformed the other approaches.

As a future work, context might be handled by using different sets of attribute weights for each existing context; we have performed some preliminary in such direction, however, the characteristics of the test database used have prevented obtaining concluding results. In addition, further researches should include studying how context can affect the revise and retain steps of CBR.

Acknowledgements. The work described in this paper was carried out as part of the MoSHCA project. The project has been funded by the Ministerio de Economa y Competitividad of the Spanish Government (Ref. EUREKA ITEA 2 no 11027 - IPT-2012-0943-300000) according to article 31 in General Grants Law 38/2003, November 17th, approved by Real Decreto 887/2006, July 21st. The project is also cofounded by the European Union through the European Regional Development Fund (ERDF). Grup de recerca consolidat de la Generalitat de Catalunya (Ref. SGR 2014 -2016).

References

1. Alonso, R.S., Tapia, D.I., García, Ó., Sancho, D., Sánchez, M.: Improving context-awareness in a healthcare multi-agent system. In: Corchado, J.M., Pérez, J.B., Hallenborg, K., Golinska, P., Corchuelo, R. (eds.) Trends in Practical Applications of Agents and Multiagent Systems. AISC, vol. 90, pp. 1–8. Springer, Heidelberg (2011)
2. Bergquist, R., Rinaldi, L.: Health research based on geospatial tools: a timely approach in a changing environment. Journal of Helminthology 84(01), 1–11 (2010)
3. Bose, I., Chen, X.: A framework for context sensitive services: A knowledge discovery based approach. Decision Support Systems 48(1), 158–168 (2009)
4. Chapman, W.W., Chu, D., Dowling, J.N.: Context: An algorithm for identifying contextual features from clinical text. In: BioNLP 2007: Biological, Translational, and Clinical Language Processing, pp. 81–88 (2007)

5. Chen, A.: Context-aware collaborative filtering system: Predicting the user's preference in the ubiquitous computing environment. In: Strang, T., Linnhoff-Popien, C. (eds.) LoCA 2005. LNCS, vol. 3479, pp. 244–253. Springer, Heidelberg (2005)

6. Hayes, C., Cunningham, P.: Context boosting collaborative recommendations. Knowledge-Based Systems 17(2), 131–138 (2004)

7. Herrera, F., Carmona, C.J., González, P., del Jesus, M.J.: An overview on subgroup discovery: foundations and applications. Knowledge and Information Systems 29(3), 495–525 (2011)

8. Kwon, O.B., Sadeh, N.: Applying case-based reasoning and multi-agent intelligent system to context-aware comparative shopping. Decision Support Systems 37(2), 199–213 (2004)

9. Lee, J.S., Lee, J.C.: Context awareness by case-based reasoning in a music recommendation system. In: Ichikawa, H., Cho, W.-D., Satoh, I., Youn, H.Y. (eds.) UCS 2007. LNCS, vol. 4836, pp. 45–58. Springer, Heidelberg (2007)

10. López, B., Pous, C., Gay, P., Pla, A.: Multi criteria decision methods for coordinating case-based agents. In: Braubach, L., van der Hoek, W., Petta, P., Pokahr, A. (eds.) MATES 2009. LNCS, vol. 5774, pp. 54–65. Springer, Heidelberg (2009)

11. Ma, T., Kim, Y.-D., Ma, Q., Tang, M., Zhou, W.: Context-aware implementation based on cbr for smart home. In: IEEE Wireless and Mobile Computing, Networking And Communications (WiMob 2005), vol. 4, pp. 112–115. IEEE (2005)

12. Makris, P., Skoutas, D.N., Skianis, C.: A survey on context-aware mobile and wireless networking: On networking and computing environments' integration. IEEE Communications Surveys & Tutorials 15(1), 362–386 (2013)

13. Nurmi, P., Przybilski, M., Lindén, G., Floréen, P.: A framework for distributed activity recognition in ubiquitous systems. In: IC-AI, pp. 650–655 (2005)

14. Öztürk, P., Aamodt, A.: A context model for knowledge-intensive case-based reasoning. International Journal of Human-Computer Studies 48(3), 331–355 (1998)

15. Pla, A., Coll, J.: Context-aware cbr tests with a breast cancer dataset. Technical Report IIiA 14-01-RR, IIiA, Universitat de Girona (August 2014)

16. Pla, A., Lopez, B., Coll, J., Mordvaniuk, N., Lopez-Bermejo, A.: Context management in health care apps. In: Proceedings of the 25th European Medical Informatics Conference (MIE 2014), Turkey, p. 1207 (2014)

17. Pla, A., López, B., Gay, P., Pous, C.: exit*cbr v2: Distributed case-based reasoning tool for medical prognosis. Decision Support Systems 54(3), 1499–1510 (2013)

18. Richter, M.M., Aamodt, A.: Case-based reasoning foundations. The Knowledge Engineering Review 20(03), 203–207 (2005)

19. Steinberg, A.N., Bowman, C.L.: Adaptive context discovery and exploitation. In: 2013 16th International Conference on Information Fusion (FUSION), pp. 2004–2011. IEEE (2013)

20. Stergiou, G.S., Zourbaki, A.S., Skeva, I.I., Mountokalakis, T.D.: White coat effect detected using self-monitoring of blood pressure at home comparison with ambulatory blood pressure. American Journal of Hypertension 11(7), 820–827 (1998)

21. Varshney, U.: Pervasive healthcare computing: EMR/EHR, wireless and health monitoring. Springer (2009)

22. Vavpetič, A., Lavrač, N.: Semantic subgroup discovery systems and workflows in the sdm-toolkit. The Computer Journal 56(3), 304–320 (2013)

23. Zimmermann, A.: Context-awareness in user modelling: Requirements analysis for a case-based reasoning application. In: Ashley, K.D., Bridge, D.G. (eds.) ICCBR 2003. LNCS, vol. 2689, pp. 718–732. Springer, Heidelberg (2003)

Determining the Customer Satisfaction
in Automobile Sector Using the Intuitionistic Fuzzy
Analytical Hierarchy Process

S. Rajaprakash[1], R. Ponnusamy[2], and J. Pandurangan[3]

[1] Department of computer Science and Engineering
[3] Department of Mathematics
Aarupadai Veedu Institute of Technology
Vinayaka Mission University, Chennai, India
srajaprakash_04@yahoo.com
[2] Department of Computer Science and Engineering
Rajiv Ganthi College of Engineering, Chennai, India
rponnusamy@acm.in

Abstract. Customer satisfaction is an important factor sustaining the business and its further development of the organization. To retain the customer is one of the important task in production industries. In these days of high competition customer satisfaction is very much essential, but uncertainty creeps. Analytical hierarchy process (AHP) is an important theory in the decision making problem. In this work we are combining Intuitionistic Fuzzy Analytical Process (IFAHP).The intuitionistic fuzzy set is able to give a very good outcome on uncertainty, and vagueness. Therefore the objective of the work is using Intuitionistic fuzzy analytical hierarchy process (IFAHP) to determine the customer satisfaction.

1 Introduction

The automobile industry plays a major role in the Indian economy. The Indian market opening its wings to MNCS, and the competition has become in imminent product quality and service. In recent years, changes in the business environment have made it harder for firms to maintain long-term sales growth and probability levels. Global competition has increased dramatically. A larger selection of products and services is available to the same set of buyers, with little growth in overall markets. Thus satisfied customers are important to companies in the production industries. Here the big manufacturing company is the customer and Supplier Company is the vendor. Based on the experts' suggestion about customer satisfaction in the production industries, the customer satisfaction is classified into several attributes in the form of hierarchy.

1.1 Fuzzy Set Theory

Fuzzy sets have been introduced by Lotfi A. Zadeh (1965).The extension of classical set theory is fuzzy sets used in fuzzy logic. In classical set theory the membership of

R. Prasath et al. (Eds.): MIKE 2014, LNAI 8891, pp. 239–255, 2014.

elements in relation to a set is assessed in binary terms according to a crisp condition an element either belongs or does not belong to the set. By contrast, fuzzy set theory permits the gradual assessment of the membership of elements in relation to a set; this is described with the aid of a membership function valued in the real unit interval [0, 1]. Fuzzy sets are an extension of classical set theory since, for a certain universe, a membership function may act as an indicator function, mapping all elements to either 1 or 0, as in the classical notion. The membership function $\mu_A(x)$ quantifies the grade of membership of the elements x to the fundamental set X. An element mapping to the value 0 means that the member is not included in the given set, 1 describes a fully included member. Values strictly between 0 and 1 characterize the fuzzy members.

1.2   Intuitionistic Fuzzy Set

Let U be the universe and A be the subset of U then the intuitionistic fuzzy set (IFS) is defined by $A^* = \left\{ \left\langle x, 0 \le \mu_A(x) + v_A(x) \le 1 \right\rangle \mid x \in U \right\}$ Where $0 \le \mu_A(x) + v_A(x) \le 1$.

The function $\mu_A(x):U \rightarrow [0,1]$ and $v_A(x):U \rightarrow [0,1]$ represent the Membership and Non Membership function respectively. $\pi(x) = 1 - \mu_A(x) + v_A(x)$ represent the degree of indeterrminacy or hesitation degree. Where $\pi(x):U \rightarrow [0,1]$. Obviously every Fuzzy set has the form $\left\{ \left\langle x, \mu_A(x), 1 - \mu_A(x) \right\rangle \mid x \in U \right\}$. The distance between the two Intuitionistic fuzzy set (IFS) has been studied by Szmidt and Kacprzyk [6]. They mention in their work we cannot omit $\pi_A(x)$ in the IFS. So we consider the $\alpha = (\mu_\alpha, v_\alpha, \pi_\alpha)$ is an intuitionistic fuzzy values where $\mu_\alpha \in [0,1]$ and $v_\alpha \in [0,1]$ and $\mu_\alpha + v_\alpha \le 1$. Moreover they studied about the mathematical form

$$\rho(\alpha) = 0.5(1 + \pi_\alpha)(1 - \mu_\alpha) \tag{1}$$

Smaller the value of $\rho(\alpha)$, the grater is the intuitionistic fuzzy value. The α cut of normal distribution contain all positive information included. Therefore intuitionistic fuzzy set mainly based on membership function and non membership function and inderminacy or hesitation degree.

1.3   Intuitionistic Relation

The fuzzy t-norm and t-conorm of intuitionistic fuzzy set has been studied by Deschrijver et al. [1]. The intuitionistic relation was studied by Xu [2] and also about the pair wise comparison judgment matrix using the Intuitionistic Fuzzy Values. In that he mention the that Let R be the Intuitionistic relation on the set $X = X = \left\{ x_1, x_2, x_3, ... x_n \right\}$ is represented by matrix $R=(M_{ik})_{nxn}$, where $M_{ik} = \left\langle (x_i, x_k), \mu(x_i, x_k), v(x_i, x_k) \right\rangle$ i,k=1,2,3...n. Let Assume that $M_{ik} = (\mu_{ik}, v_{ik})$ and $\pi(x_i, x_k) = 1 - \mu(x_i, x_k) - v(x_i, x_k)$ is interpreted as an indeterminacy degree.

1.4   Analytic Hierarchy Process

The AHP is a multi-criteria decision making approach and it was introduced by Satty (1994 and 1997). It used to solve complex decision problems. It uses a multi level hierarchical structure and each level have comparison matrix. These comparisons are used to obtain the weights of important of the decision criteria. The comparisons are checked by the consistent test and this check is used to improve the comparison matrix consistency.

Table 1. Relative Importance Scale [3]

Numerical values	Linguistics scale	Explanation
1	Equal importance of both elements	Two elements contribute equally
3	Moderate importance of one element over another	Experience and judgment favour one element over another
5	Strong importance of one element over another	An element is very strongly dominant
7	Very strong importance of one element over another	An element is favoured by at east an order of magnitude
9	Extreme importance of one element over another	An element is favored by at least more than an order of magnitude
2,4,6,8	Intermediate values	Used to compromise between two judgments

1.5   Fuzzy Analytical Hierarchy Process (FAHP)

An AHP reflect the human thinking but in some area it is not reflecting fully since the decision makers usually feel more confident to interval judgments rather than expressing their judgment in the form of single number value. But FAHP use the membership value and based on that weight is calculated. Thus the FAHP is the combination of AHP and Fuzzy set theory is called Fuzzy Analytical Hierarchy Process (FAHP). It was proposed by Laahoven and Pedrycs (1983). The fuzzy comparison are able to tolerate vagueness more than AHP.

Table 2. FAHP Triangular Fuzzy scale

Linguistic Scale	Triangular Fuzzy scale	Fuzzy Number	Intensity importance
Equally important	(1,1,1)	$\tilde{1}$	1
Moderately more	(1,3,5)	$\tilde{3}$	3
Strongly more important	(3,5,7)	$\tilde{5}$	5
Very strongly more	(5,7,9)	$\tilde{7}$	7
Extremely more	(7,9,11)	$\tilde{9}$	9

1.6   Intuitionistic Fuzzy Analytical Hierarchy Process (IFAHP)

In FAHP generally triangular membership function and Trapezodial membership function are used. It is combination of Intuitionistic fuzzy set and AHP. The Intuitionistic fuzzy set theory gives a batter output over fuzzy set theory suggested values.

Table 3. Comparison Scale [4]

Linguistic	1-9 scale	Linguistic (0.1 to 0.9) scale
9	0.9	Extreme Important
7	0.8	Very Strong Important
5	0.7	Strong Important
3	0.6	Moderately Important
1	0.5	Equal preference
1/3	0.4	Moderately not Important
1/5	0.3	Strong not Important
1/7	0.2	Very Strong not Important
1/9	0.1	Extreme not Important

2   Literary Survey

The notion of intuitionistic fuzzy sets was introduced by Atanassov [5]. The new definition of distance between two intuitionistic fuzzy sets apart from that Atanassov studied by Elualia szmidt et al. [6]. The study of human capital indicator and ranking by using IFAHP to evaluate the four main indicators of Human capital indicator concurrently viva expert judgment by Lazim Abdullah et al [7]. Among the several candidates the selection of DBMS in the Turikish National Identity Card Management project using the Fuzzy AHP done by F.Ozgur Catak et al. [8]. Using the FAHP and the students expectation in the present education system in Tamilnadu , India. In the work the authors taken a sample work on Engineering education studied by S.Rajaprakash et al. [9]. The study was done about the quality, and price should aspect the customer satisfaction and the impact of the quality in addition that market trends by Eugene W.Anderson et al. [10]. The new method of customer satisfaction and loyalty measurement has been developed for company in Poland by Joanna Waligora et al. [11] Determination of customer satisfaction is important for consultants since customer satisfaction in which the service business depends on for repeat business by Anne-Mette Sonne [12]. Intuitionistic fuzzy pair, intuitionistic fuzzy couple, intuitionistic fuzzy value studied by Krassimir Atanassov et al. [13]. S,M.Reza Nasserzadeh et al [14] uses the Fuzzy Cognitive Maps to dealing with customer satisfaction in the Banking Industry by. The customer satisfaction can be increased without necessarily improving operation by Andy [15].

3   Methodology

Step-1: Identify the experts related to the particular domain and get the suggestion about the problem. Based on the suggestion identify the attributes of the problem and make into hierarchy order, confirm from the experts that hierarchy is correct. If it is not correct then use suggestion from the experts and make the hierarchy list suitably.

Step-2: Based on the attributes, frame the questionnaire and collect opinions from the experts by using the linguistic variable (Table-3). Calculate the geometric mean of each answer of the question.

Step-3: After finalizing the expert suggestion the values are converted to intuitionistic value (table-3) and the pair wise comparison Matrix M is constructed.

Step-4: Before deriving the priorities and of the alternatives and criteria we have to check the consistency of the intuitionistic preference relation of the matrix M. According to Xu et al. [4] the consistent interval fuzzy preference relations are as follows.

R= $(M_{ik})_{nxn}$ with $M_{ik}=(\mu_{ik},v_{ik})$ (i , k= 1,2,3...n) is multiplicative consistent if

$$
\mu_{ik} = \begin{cases} 0, & if\,(\mu_{it},\mu_{tk})\in\{(0,1),(1,0)\} \\ \dfrac{\mu_{it}\mu_{tk}}{\mu_{it},\mu_{tk}+(1-\mu_{it})(1-\mu_{tk})}, & otherwise \end{cases}
$$
(2)

$$
and\,v_{ik} = \begin{cases} 0, & if\,(v_{it},v_{tk})\in\{(0,1),(1,0)\} \\ \dfrac{v_{it}v_{tk}}{v_{it},v_{tk}+(1-v_{it})(1-v_{tk})}, & otherwise \end{cases}
$$
(3)

Xia and Xu [2] have proved the fuzzy preference relation.

Theorem [2]: In the fuzzy preference relation the following statement are equivalent:

1) $\quad b_{ik}=\dfrac{b_{ik}b_{tk}}{b_{ik}b_{tk}+(1-b_{ik})(1-b_{tk})}\quad i,t,k=1,2,3...n$

2) $\quad b_{ik}=\dfrac{\sqrt[n]{\prod\limits_{s=1}^{n}b_{is}b_{sk}}}{\sqrt[n]{\prod\limits_{s=1}^{n}b_{is}b_{sk}}+\sqrt[n]{\prod\limits_{s=1}^{n}(1-b_{is})(1-b_{sk})}}\quad i,k=1,2,3...n$
(4)

From on the above result and theorem, Zeshuri Xu *et al* [6] developed two algorithms in the intuitionistic fuzzy analytic hierarchy process. But we used only one algorithm based on that algorithm we can have the following formula.

For k > i+1, and let $\overline{M}_{ik}=(\overline{\mu}_{ik},\overline{v}_{ik})$ where

$$
\overline{\mu}_{ik}=\dfrac{\sqrt[k-i-1]{\prod\limits_{t=i+1}^{k-1}\mu_{it}\mu_{tk}}}{\sqrt[k-i-1]{\prod\limits_{t=i+1}^{k-1}\mu_{it}\mu_{tk}}+\sqrt[k-i-1]{\prod\limits_{t=i+1}^{k-1}(1-\mu_{it})(1-\mu_{tk})}}\quad k>i+1
$$
(5)

$$
\overline{v}_{ik}=\dfrac{\sqrt[k-i-1]{\prod\limits_{t=i+1}^{k-1}v_{it}v_{tk}}}{\sqrt[k-i-1]{\prod\limits_{t=i+1}^{k-1}v_{it}v_{tk}}+\sqrt[k-i-1]{\prod\limits_{t=i+1}^{k-1}(1-v_{it})(1-v_{tk})}}\quad k>i+1
$$
(6)

And for k= i+1 let $\overline{M}_{ik} = M_{ik}$ and k< i. Let $\overline{M}_{ik} = (\overline{v}_{ki}, \overline{\mu}_{ki})$. From the above equations we can get the lower triangular elements of the matrix. These values are computed and they are multiplicative consistent and the intuitionistic relation is obtained.

Step-5: The distance between intuitionistic relation [16]   is calculated by the following way

$$d(\overline{M}, M) = \frac{1}{2(n-1)(n-2)} \sum_{i=1}^{n} \sum_{k=1}^{n} (|\overline{\mu}_{ik} - \mu_{ik}| + |\overline{v}_{ik} - v_{ik}| + |\overline{\pi}_{ik} - \pi_{ik}|) \quad (7)$$

If $\qquad\qquad\qquad\qquad d(\overline{M}, M) < \tau \qquad\qquad\qquad\qquad (8)$

Then the relation matrix is consistence in the intuitionistic in relation. Suppose equation (8) is not satisfied then go to setp.1.

Step-6: The priority of the intuitionistic preference relation Zeshuri Xu [2] is calculated by the following method.

$$w_i = \frac{\sum_{k=1}^{n} M_{ik}^1}{\sum_{i=1}^{n} \sum_{k=1}^{n} M_{ik}^1} = \frac{\sum_{k=1}^{n} [\mu_{ik}, 1-v_{ik}]}{\sum_{i=1}^{n} \sum_{k=1}^{n} [\mu_{ik}, 1-v_{ik}]} \qquad w_i = \left(\frac{\sum_{k=1}^{n} [\mu_{ik}]}{\sum_{i=1}^{n} \sum_{k=1}^{n} [1-v_{ik}]}, 1 - \frac{\sum_{k=1}^{n} [1-v_{ik}]}{\sum_{i=1}^{n} \sum_{k=1}^{n} [\mu_{ik}]} \right) \quad (9)$$

Step-7: After finding the weights for two level based on the weights ranking the weight by using the formula (1) then calculate the preference ranking of the attributes. The flow diagram is provided in figure1

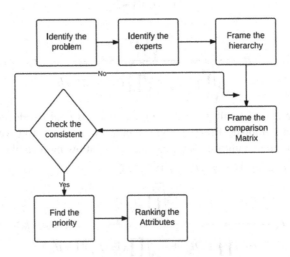

Fig. 1. Flow diagram

4   Observation from the Expects

In the automobile sector the customer satisfaction plays the major role. To satisfy the customer in all aspects is very difficult. In this work the customer is the big automobile company where as the supplier is the vendor. The objective of this work is finding the customer satisfaction from supplier point of view.

Here customer satisfaction initially classified into three parts namely quality, delivery and commercial dealings in the level-1.

4.1   Quality

The suppler inspect the all the products before it is delivered to the customer. The purpose of the quality department is on based "zero defects" philosophy. The supplier is full responsible for the quality of the product. In that we have identified into eleven attributes. The attributes are as follows

1. **Product quality:** The Suppliers are fully responsible for the quality of their products including their sub-suppliers. Both are responsible for providing products that meet all e requirements, specifications, and drawings as identified on the purchase order and that the products are free from defects as warranted in General Purchasing Conditions. Zero-defect products are expected from all suppliers.

2. **Response to Queries:** Based on quickly the queries from the customer end are addressed about the quality of product or any defect of the product make the customer fully satisfied about the product.

3. **Concern of your (customer) problem:** The defect about the quality of the product from the customer end to supplier, if it is not solved about the defect of the product it will give stressful and frustration to the customer.

4. **Understanding your (customer) needs:** The supplier understanding the customer needs about the quality of the product. They can understanding by using simple information sheet about the product. It may contain simple questions about the product quality.

5. **Support in crisis:** The supplier should support the customer at the critical time in the product.

6. **Resolving of Complaints:** Based on the complaint about the quality of the product from the customer end, the supplier should take suitable immediate action and ensure that similar complaints about the product should not come again.

7. **Visit frequency and quality:** The customer visit to the supplier industries visit in certain interval for the quality audit for the Q1 certification.

8. **Line stop do to quality:** The production should not stop due to any defect in the quality of the product in the customer end. The supplier is fully responsible for quality of the product.

9. **Field return:** If any major damage or defective in the quality of the product is found. The customer will return the product. Thus it will give major problem to the suppliers.

10. **Notification:** Q1 Certification it means that based on marks scored in customer audit and supplier performance in systematic intervals.

11. **Implementation of Improvement Action:** Based on the customer end complaints the supplier should improve the quality of the product.

4.2   Delivery

All suppliers to provide 100% on-time delivery performance with the correct quantity and pricing agreed upon.. The quantity shipped per order or release cannot vary from specified quantity without the consent of the planner of customer.

1. **On Time Deliveries:** The supplier should not deliver the goods late or unauthorized and partial or over shipments
2. **Packaging Aspects:** The suppliers should Pack the product with safe, secure, efficient and effectively. The supplier will check whether it has reached the customer safely without any damage of the product.
3. **Schedule Adherence:** its means that calculate the timeline of deliveries from supplier end. The customer will check the schedule adherence of the suppliers. It is calculated by

$$Schedule\ Adherance = \frac{Number\ of\ ontime\ delivery\ in\ period}{Total\ Number\ delivery} \times 100$$

4. **Clarity and completeness of communication:** supplier should give clarity in the communication of the delivery of the product and on time delivery.
5. **Honoring Commitments:** The supplier should honor the commitment which is given earlier to the customer.
6. **Line Stoppage due to Delivery Issues:** production line is shut down due to poor quality, late delivery, or incorrect quantity on any shipment, the supplier will be responsible for all costs incurred including expediting shipments and claim for damages at the customer end.

4.3   Commercial Dealings

1. **Submission of Quotation:** The supplier should submit the quotation to the customer in time. The Data Analyst is responsible for the quotation. He will interact with commercial and technical key contacts to support the relationship with the customers.
2. **Confirmation of order:** The relationship manager is responsible to get the confirmation of order from the customer.

Customer requirement hierarchy diagram figure-2

3. **Documentation Accuracy:** The customer should provide requirements document with all necessary details and the supplier should responsible for the production plan documentation according the customer requirements.
4. **Price competitiveness:** In the quotation the data analyst should give the competitive price compare to the other suppliers or in the market.
5. **Business integrity:** How honestly fulfill the customer requirement in the quality, delivery, cost, service etc. On the other end, how honestly accept the fact of the supplier?

Development of New Product: The supplier readiness to develop new product based on the requirement of customer or modify the existing product to improving its quality. In the Level-2 in Quality have eleven attributes Q1,Q2,Q3,Q4,Q5,Q6,Q7, Q8,Q9,Q10,Q11 represent product quality, response to queries, concern of your (customer) problem, Understanding your (customer) needs, Support in crisis, Resolution of Complaints, Visit frequency and quality, Line stop do to quality, Field return, Notification, Implementation of Improvement Action respectively. From the suggestion the initial Intuitionistic fuzzy comparison Matrix given below

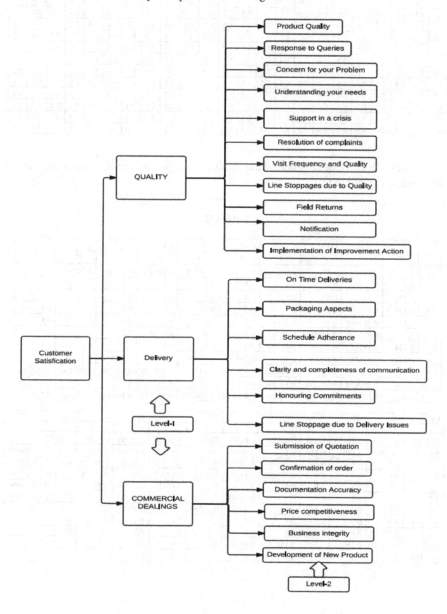

Table Q1

M	Q1 μ	Q1 ν	Q2 μ	Q2 ν	Q3 μ	Q3 ν	Q4 μ	Q4 ν	Q5 μ	Q5 ν	Q6 μ	Q6 ν	Q7 μ	Q7 ν	Q8 μ	Q8 ν	Q9 μ	Q9 ν	Q10 μ	Q10 ν	Q11 μ	Q11 ν
Q1	0.5	0.5	0.5	0.5	0.5	0.5	0.2	0.5	0.4	0.5	0.2	0.6	0.4	0.5	0.2	0.7	0.2	0.8	0.4	0.5	0.4	0.5
Q2	0.5	0.5	0.5	0.5	0.5	0.4	0.5	0.5	0.4	0.5	0.5	0.5	0.4	0.5	0.1	0.7	0.1	0.8	0.5	0.5	0.5	0.5
Q3	0.5	0.5	0.4	0.5	0.5	0.5	0.4	0.5	0.4	0.5	0.4	0.5	0.5	0.5	0.3	0.7	0.1	0.8	0.4	0.5	0.4	0.6
Q4	0.5	0.2	0.5	0.5	0.5	0.4	0.5	0.5	0.4	0.5	0.4	0.5	0.5	0.5	0.3	0.7	0.3	0.7	0.5	0.5	0.5	0.5
Q5	0.5	0.4	0.5	0.4	0.5	0.4	0.5	0.4	0.5	0.5	0.5	0.5	0.5	0.5	0.4	0.5	0.4	0.5	0.5	0.5	0.5	0.5
Q6	0.6	0.2	0.5	0.5	0.5	0.4	0.5	0.5	0.5	0.5	0.5	0.5	0.5	0.5	0.4	0.5	0.3	0.7	0.4	0.5	0.5	0.5
Q7	0.5	0.4	0.5	0.4	0.5	0.5	0.5	0.5	0.5	0.5	0.5	0.4	0.5	0.5	0.5	0.4	0.4	0.6	0.5	0.5	0.2	0.8
Q8	0.7	0.2	0.7	0.1	0.7	0.3	0.6	0.3	0.5	0.4	0.7	0.3	0.5	0.4	0.5	0.5	0.4	0.6	0.4	0.6	0.5	0.5
Q9	0.8	0.2	0.8	0.1	0.8	0.1	0.7	0.3	0.5	0.4	0.7	0.3	0.6	0.4	0.6	0.4	0.5	0.5	0.5	0.5	0.5	0.6
Q10	0.5	0.4	0.5	0.5	0.5	0.4	0.5	0.5	0.5	0.5	0.5	0.4	0.5	0.5	0.6	0.4	0.5	0.5	0.5	0.5	0.5	0.5
Q11	0.5	0.4	0.5	0.5	0.6	0.4	0.5	0.5	0.5	0.5	0.5	0.5	0.8	0.2	0.5	0.5	0.6	0.5	0.5	0.5	0.5	0.5

To check the consistent generate the intuitionistic preference relation Matrix using the equation (5) and (6)

Table Q2

	Q1 μ	Q1 ν	Q2 μ	Q2 ν	Q3 μ	Q3 ν	Q4 μ	Q4 ν	Q5 μ	Q5 ν
Q1	0.5	0.5	0.5	0.5	0.5	0.44949	0.44949	0.44949	0.307692	0.5
Q2	0.5	0.5	0.5	0.5	0.5	0.4	0.5	0.44949	0.35247	0.5
Q3	0.449	0.5	0.4	0.5	0.5	0.5	0.4	0.5	0.5	0.5
Q4	0.449	0.44949	0.44949	0.5	0.5	0.4	0.5	0.5	0.4	0.5
Q5	0.5	0.307692	0.5	0.35247	0.5	0.5	0.5	0.4	0.5	0.5
Q6	0.5	0.307692	0.44949	0.35247	0.5	0.4	0.44949	0.5	0.5	0.5
Q7	0.634	0.385953	0.44949	0.44949	0.4	0.5	0.5	0.4	0.5	0.5
Q8	0.814	0.07675	0.608696	0.16	0.555006	0.225381	0.44949	0.307692	0.5	0.5
Q9	0.552	0.005605	0.897231	0.048441	0.777778	0.134591	0.777778	0.134591	0.696168	0.262726
Q10	0.821	0.057032	0.707194	0.170632	0.740781	0.16	0.740781	0.16	0.55051	0.307692
Q11	0.853	0.025521	0.859355	0.04625	0.823529	0.104458	0.823529	0.104458	0.666667	0.213939

Q6 μ	Q6 ν	Q7 μ	Q7 ν	Q8 μ	Q8 ν	Q9 μ	Q9 ν	Q10 μ	Q10 ν	Q11 μ	Q11 ν
0.30769	0.5	0.38595	0.633	0.07675	0.81361	0.00560	0.55230	0.05703	0.82087	0.02552	0.85286
0.35247	0.44948	0.4494	0.449	0.16	0.60869	0.04844	0.89723	0.17063	0.70719	0.04625	0.85935
0.4	0.5	0.4	0.5	0.225381	0.55500	0.13459	0.77777	0.16	0.74078	0.10445	0.82359
0.44949	0.5	0.4	0.5	0.307692	0.44949	0.13459	0.77777	0.16	0.74078	0.10445	0.82359
0.5	0.5	0.5	0.5	0.5	0.5	0.26272	0.69616	0.30769	0.55051	0.21393	0.66666
0.5	0.5	0.5	0.5	0.5	0.44949	0.35247	0.6	0.30383	0.65166	0.15122	0.75339
0.5	0.4	0.5	0.5	0.5	0.4	0.44949	0.5	0.44949	0.55051	0.4	0.5
0.44949	0.5	0.5	0.4	0.5	0.5	0.4	0.6	0.44949	0.55051	0.4	0.6
0.6	0.35247	0.5	0.449	0.6	0.4	0.5	0.5	0.5	0.5	0.5	0.5
0.651669	0.30383	0.55051	0.449	0.55051	0.44949	0.5	0.5	0.5	0.5	0.5	0.5
0.753394	0.15122	0.5	0.4	0.6	0.4	0.5	0.5	0.5	0.5	0.5	0.5

Calculate the distance between intuitionistic fuzzy values using the above two matrix by equation (7) and (8). The $d(\overline{M},M) = 0.1119115$. Which is less than τ. Here we are fix the threshold value $\tau = 0.15$. Therefore the above matrix is consistent.

The next step is calculating the weight of all attributes using the equation (9). From the equation (1) the weight ($\rho(\alpha)$) is calculated and Ranking the attribute based on the calculated values. We get the following Table-Q4 and Diagram-Q.

Table Q3

Weight	μ	ν
W(Q1)	0.0596	0.873
W(Q2)	0.062	0.867
W(Q3)	0.0622	0.872
W(Q4)	0.0608	0.866
W(Q5)	0.0719	0.857
W(Q6)	0.0672	0.854
W(Q7)	0.0721	0.856
W(Q8)	0.0727	0.835
w(Q9)	0.0865	0.813
w(Q10)	0.1705	0.818
W(Q11)	0.1838	0.809

Table Q4

Attribute	$\rho(\alpha)$	Preference
Product Quality	0.93665238	1
Concern for your problem	0.93351615	2
Response Queries	0.93602175	3
Understanding the problem	0.932932	4
Support in crisis	0.92827737	5
Visit Frequency and Quality	0.92687275	6
Resolution of Complaints	0.92801914	7
Line stoppages due to Quality	0.91774617	8
Notification	0.9065264	9
Field Returns	0.90921681	10
Implementation of Improvement action	0.90452988	11

Diagram Q

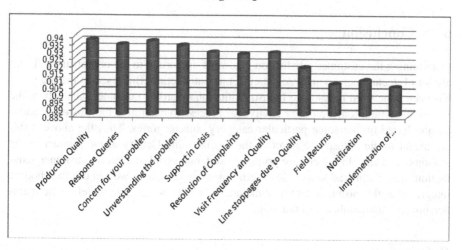

From the above Table Q4 and Diagram-1 the first preference product quality, second concern for your problem, third response queries, forth is understanding the problem, fifth one support in crisis, sixth is visit frequency and quality, seventh is resolution of complaints, eight is line stoppages due to quality, ninth is notification, tenth is field returns and the last one is implementation of improvement action.

5   Empirical Result

Based on the opinion of the experts the customer satisfaction initially identified into important three categories. In that using the IFAHP we calculated the preference relation that is the supplier has to concentrate first on quality, second is delivery and the third is commercial dealing (Diagram CS and Table CS-4). In particularly in the quality is classified into eleven categories (level-2). From the Table-Q4 and Diagram-Q the supplier has to concentrate product quality first , second is concern for your problem, third is response to queries, forth is understanding the problem, fifth is support in crisis, sixth is visit frequency and quality, seventh is resolution of complaints, eighth is line stoppages due to quality, and the last one is notification. In the delivery part, it is subdivided into six categories. Based on the Table D-4 and the Diagram-D the supplier has to give importance to on time deliveries, the next part is packaging aspects, third is clarity and completeness of communication, forth is schedule adherence, fifth is area of honoring commitments and last is line stoppage due to delivery issues. In the next category is commercial dealing. In that supplier have to concentrate to submission of quotation first, the next sub category is confirmation of order, third is documentation accuracy, forth is price competitiveness, the fifth sub category is business integrity, and last sub category is development of new product (Table CD-4 and Diagram-CD).

6   Conclusion

Determining the customer satisfaction is always important to the organization .There are lot of techniques to analyze the customer satisfaction .But it seems that no study has been done from Industry to Industry situation regarding customer satisfaction, the we are used Intuitionistic Fuzzy Analytical Process to determine the Customer Satisfaction based on the some particular category, subcategories. Thus the above result are useful to the industry to improve the business and Quality of the product in the automobile sector. In the end we propose a suitable model to evaluate customer satisfaction in the industrial sector. For further work in this paper we can use the choquet integral over this problem. But in some areas we may not be able to define the membership and nonmember ship function.

References

1. Deschrijver, G., Cornelis, C., Kerre, E.: On the representation of intuitionistic fuzzy t-norms and t-conorms. Notes on Intuitionistic Fuzzy Sets 8(3), 1–10 (2002)
2. Xu, Z.: Intuitionistic preference relations and their application in group decision making. Inf. Sci. 177(11), 2363–2379 (2007)
3. Saaty, T.: The Analytic Hierarchy Process, Planning, Priority Setting, Resource Allocation. McGraw-Hill, New York (1980)

4. Xu, Z., Liao, H.: Intuitionistic fuzzy analytic hierarchy process. IEEE Transactions on Fuzzy Systems 22(4), 749–761 (2014)
5. Atanassov, K.T.: Intuitionistic fuzzy sets. Fuzzy Sets Syst. 20(1), 87–96 (1986)
6. Szmidt, E., Kacprzyk, J.: Distances between intuitionistic fuzzy sets. Fuzzy Sets Syst. 114(3), 505–518 (2000)
7. Abdullah, L., Jaafar, S., Imran: Intuitionistic fuzzy analytic hierarchy process approach in ranking of human capital indicator. Journal of Applied Science 3(1), 423–429 (2013)
8. Catak, F.O., Karabas, S., Yildirim, S.: Fuzzy analitical hierarchy based dbms selection in turikish national identity card management project. IJIST 4(2), 212–224 (2012)
9. Rajaprakash, S., Ponnusamy, R.: Determining students expectation in present education system using fuzzy analytic hierarchy process. In: Prasath, R., Kathirvalavakumar, T. (eds.) MIKE 2013. LNCS, vol. 8284, pp. 553–566. Springer, Heidelberg (2013)
10. Anderson, E.W., Fornell, C., Lehmann, D.R.: Customer satisfaction, market share, and probability: Findings from Sweden. Journal of Marketing 58(4), 53–66 (1994)
11. Waligora, J., Walligora, R.: Measuring customer satisfaction and loyalty in the automotive industry, aarhus school of business, Denmark, Faculty of business performance management. Journal of Marketing 43(6), 55–69 (2007)
12. Sonne, A.M.: Determinants of customer satisfaction with professional service-a study of consultant services. Journal of Marketing 41(2), 159–171 (1999)
13. Atanassov, K., Szmidt, E., Kacprzyk, J.: On intuitionistic fuzzy pairs. Notes on Intuitionistic Fuzzy Sets 19(3), 1–13 (2013)

A   Appendix

A.1   Customer Satisfaction

In customer satisfaction initially three categories C1, C2, C3 represent product quality, delivery, commercial dealing respectively. From the opinion of experts, intuitionistic fuzzy relation matrix M is constructed (Table CS-1).

Table CS-1

	C1		C2		C3	
M	μ	v	μ	v	μ	v
	0.5	0.5	0.2	0.7	0.2	0.7
C2	0.7	0.2	0.5	0.5	0.23	0.7
C3	0.7	0.2	0.7	0.23	0.5	0.5

Table CS-2

	C1		C2		C3	
$\overline{M}$	μ	v	μ	v	μ	v
C1	0.5	0.5	0.2	0.7	0.214619	0.7
C2	0.6	0.2	0.5	0.5	0.23	0.7
C3	0.7	0.214619	0.7	0.23	0.5	0.5

Using the equation (5) and (6) the intuitionistic fuzzy relation Matrix formed to check the consistency

$d(\overline{M}, M) = 0.1119115$ using the equation (7) and (8) which is less than $\tau = 0.15$

and the weight is calculated using the equation (7). $\rho(\alpha)$ is calculated using the equation (1) we can get the Ranking of the attributes (Table CS-4 & Diagram CS)

Table CS-3

Weight	μ	v
Weight(C1)	0.192334	0.734596
Weight(C2)	0.279683	0.613957
weight(C3)	0.399547	0.504085

Table CS-4

Attribute	$\rho(\alpha)$	Ranking
Quality	0.867298	1
Delivery	0.806979	2
Commercial Dealing	0.752042	3

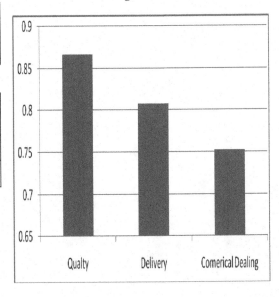

Diagram CS

A.2   Delivery

In Delivery part have six attributes are identified six sub categories D1,D2, D3,D4,D5,D6 represents On time delivery, packing aspects, schedule adherence, clarity and completeness of communications, honoring commitments, line stoppage due to delivery. In that initial intuitionistic fuzzy matrix is constructed based on the opinion of the experts.

Table D-1

M	D1 μ	v	D2 μ	v	D3 μ	v	D4 μ	v	D5 μ	v	D6 μ	v
D1	0.5	0.5	0.3	0.7	0.4	0.5	0.5	0.5	0.4	0.6	0.4	0.6
D2	0.7	0.3	0.5	0.5	0.4	0.6	0.2	0.8	0.2	0.8	0.2	0.8
D3	0.5	0.4	0.6	0.4	0.5	0.5	0.3	0.5	0.2	0.5	0.5	0.5
D4	0.5	0.5	0.8	0.2	0.5	0.3	0.5	0.5	0.4	0.6	0.4	0.5
D5	0.6	0.4	0.8	0.2	0.5	0.2	0.6	0.4	0.5	0.5	0.4	0.5
D6	0.6	0.4	0.8	0.2	0.5	0.5	0.5	0.4	0.5	0.4	0.5	0.5

Using the equation (5) and (6) the matrix $\overline{M}$ is calculated.

Using the equation (5) and (6) the matrix $\overline{M}$ is calculated.

Table D-2

$\overline{M}$	D1 μ	v	D2 μ	v	D3 μ	v	D4 μ	v	D5 μ	v	D6 μ	v
D1	0.5	0.5	0.3	0.7	0.4232	0.5767	0.3956	0.5635	0.4142	0.5505	0.1270	0.7891
D2	0.7	0.3	0.5	0.5	0.4	0.6	0.4232	0.5250	0.3956	0.5635	0.1197	0.8304
D3	0.7	0.3	0.6	0.4	0.5	0.5	0.3	0.5	0.449	0.5	0.1791	0.5
D4	0.57676	0.42323	0.52506	0.42323	0.5	0.3	0.5	0.5	0.4	0.6	0.4	0.5505
D5	0.5505	0.41421	0.56350	0.39564	0.5	0.449	0.6	0.4	0.5	0.5	0.4	0.5
D6	0.7891	0.12700	0.83047	0.11978	0.5	0.1791	0.550	0.4	0.5	0.4	0.5	0.5

$d(M,\overline{M})$ = 0.137831 using the equation (7) and (8) which is less than τ =0.15 and the weight is calculated using the equation (7) (Table D-3).

$d(\overline{M},M)$ = 0.137831 using the equation (7) and (8) which is less than τ =0.15 and the weight is calculated using the equation (7) (Table D-3).

Table D-3

Weight	μ	v
Weight(D1)	0.132699	0.864427
Weight(D2)	0.155955	0.843341
weight(D3)	0.167624	0.807168
weight(D4)	0.199798	0.812834
Weight(D5)	0.1913	0.804792
Weight(D6)	0.225461	0.750248

Table D-4

Attributes	$\rho(\alpha)$	Preference
On Time Deliveries	0.932213	1
Packaging Aspects	0.921671	2
Clarity and completeness of communication	0.903584	3
Schedule Adherence	0.906417	4
Honoring Commitments	0.902396	5
Line Stoppage due to Delivery Issues	0.875124	6

Using the equation (1) we can get the Ranking of the attributes (Table D-4 & Diagram D)

Diagram D

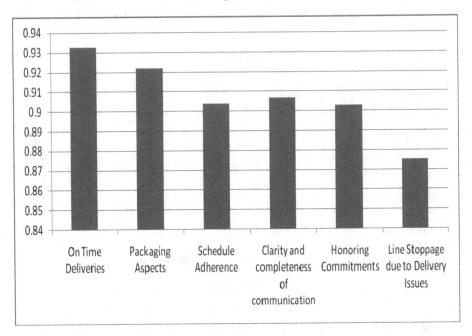

A.3   Commercial Dealing

In this category we are identified six sub categories based on the suggestion from the experts, CD1,CD2,CD3,CD4,CD5,CD6 represent Submission of Quotation , conformation of order, Documentation Accuracy , Price competitiveness, Business integrity , Development of new product respectively. Based on the opinion from the expects the matrix M is formed (Table CD-1).

Table CD-1

M	CD1		CD2		CD3		CD4		CD5		CD6	
	μ	v	μ	v	μ	v	μ	v	μ	v	μ	v
CD1	0.5	0.5	0.3	0.5	0.4	0.5	0.5	0.5	0.5	0.5	0.4	0.5
CD2	0.5	0.3	0.5	0.5	0.4	0.5	0.2	0.8	0.2	0.8	0.2	0.8
CD3	0.5	0.4	0.5	0.4	0.5	0.5	0.1	0.8	0.2	0.8	0.1	0.8
CD4	0.5	0.5	0.8	0.2	0.8	0.1	0.5	0.5	0.1	0.8	0.4	0.5
CD5	0.5	0.5	0.8	0.2	0.8	0.2	0.8	0.1	0.5	0.5	0.4	0.5
CD6	0.5	0.4	0.8	0.2	0.8	0.1	0.5	0.4	0.5	0.4	0.5	0.5

$d(\overline{M},M) = 0.135947$ using the equation (7) and (8) which is less than $\tau = 0.15$ and the weight is calculated using the equation (7) (Table CD-3). Using the equation (1) we can get the Ranking of the attributes (Table CD-4 & Diagram CD).

Table CD-2

	CD1		CD2		CD3		CD4		CD5		CD6	
$\overline{M}$	μ	v	μ	v	μ	v	μ	v	μ	v	μ	v
CD1	0.5	0.5	0.3	0.5	0.4232	0.5	0.3660	0.5767	0.3860	0.5767	0.0560	0.8
CD2	0.5	0.3	0.5	0.5	0.4	0.5	0.3660	0.5767	0.3501	0.6188	0.0433	0.888
CD3	0.5	0.3	0.5	0.4	0.5	0.5	0.1	0.8	0.3956	0.5767	0.1	0.8
CD4	0.5	0.4232	0.5767	0.3660	0.8	0.1	0.5	0.5	0.1	0.8	0.2139	0.666
CD5	0.5767	0.3860	0.6188	0.3501	0.5767	0.3956	0.8	0.1	0.5	0.5	0.4	0.5
CD6	0.8	0.0560	0.8888	0.0433	0.8	0.1	0.666	0.2139	0.5	0.4	0.5	0.5

Table CD-3

Weight	μ	v
Weight(CD1)	0.12604711	0.846647
Weight(CD2)	0.13400259	0.84249
weight(CD3)	0.13003608	0.842024
weight(CD4)	0.23848765	0.810657
Weight(CD5)	0.21546428	0.773073
Weight(CD6)	0.25785495	0.717763

Table CD-4

Attribute	$\rho(\alpha)$	Preference
Submission of Quotation	0.92332341	1
Confirmation of order	0.92124509	2
Documentation Accuracy	0.92101185	3
Price competitiveness	0.90532872	4
Business integrity	0.88653669	5
Development of New Product	0.85888156	6

Diagram CD

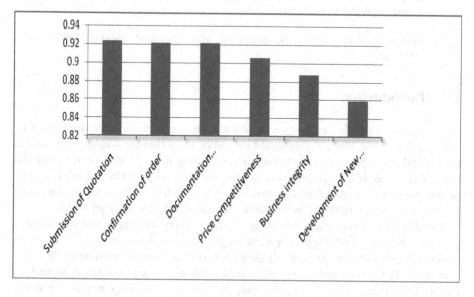

Pattern Based Bootstrapping Technique for Tamil POS Tagging

Jayabal Ganesh[1], Ranjani Parthasarathi[1], T.V. Geetha[2], and J. Balaji[2]

[1]Department of Inforamation Science and Technology, College of Engineering,
Anna University, Chennai. 25, Tamilnadu, India
[2]Department of Computer Science and Engineering, College of Engineering,
Anna University, Chennai. 25, Tamilnadu, India

Abstract. Part of speech (POS) tagging is one of the basic preprocessing techniques for any text processing NLP application. It is a difficult task for morphologically rich and partially free word order languages. This paper describes a Part of Speech (POS) tagger of one such morphologically rich language, Tamil. The main issue of POS tagging is the ambiguity that arises because different POS tags can have the same inflections, and have to be disambiguated using the context. This paper presents a pattern based bootstrapping approach using only a small set of POS labeled suffix context patterns. The pattern consists of a stem and a sequence of suffixes, obtained by segmentation using a suffix list. This bootstrapping technique generates new patterns by iteratively masking suffixes with low probability of occurrences in the suffix context, and replacing them with other co-occurring suffixes. We have tested our system with a corpus containing 20,000 Tamil documents having 2,71,933 unique words. Our system achieves a precision of 87.74%.

Keywords: POS Tagging, Bootstrapping, semi-supervised Tagging, Tamil Language.

1 Introduction

Part of speech (POS) is a category to which a word is assigned in accordance with its syntactic functions. The category could be one of the following – noun, verb, adjective, adverb, post-position, etc. Part of speech tagging is the process of assigning the POS label to words in a given text depending on the context. POS tagging is an important process in natural language processing applications such as natural language parsing, information retrieval, information extraction, machine translation etc.

Tamil is a morphologically rich language. It has many inflections and grammatical features. Therefore, POS tagging is a challenging task. However, because of the morphological richness, the suffixes and their context have enough information to determine the POS tag. The suffix context is rich enough to be used for other tasks such as morphological analysis as well [20]. But, the number of suffixes available and the combinations available is very large. Hence, it is difficult to determine or frame the rules to handle all possible combinations to identify the POS. It is for this reason that

R. Prasath et al. (Eds.): MIKE 2014, LNAI 8891, pp. 256–267, 2014.
© Springer International Publishing Switzerland 2014

initial attempts at POS tagging which have used a rule-based approach [1][2], cannot tag all words. It is for the same reason that machine-learning approaches are better suited for this task. Machine learning techniques that have been used earlier for POS tagging of Tamil documents are HMM, SVM, and other statistical techniques [15][16][17]. However, all the techniques require considerable amount of resources such as a tagged corpus, parallel corpus, or exhaustive dictionary.

Therefore, in this paper, we propose a pattern-based boot-strapping approach to POS tagging, which does not require a large tagged corpus or a dictionary. With a very small set of labeled words, and a suffix list, the boot-strapping approach learns new patterns and tags the words appropriately, based on the suffix context, in an iterative manner. The rest of the paper is organized into 5 sections. Section 2 presents the related work, section 3 describes the bootstrapping techniques used for POS tagging in detail, section 4 describes the evaluation and results and section 5 gives the conclusion.

2   Related Work

Building a POS tagging system for natural languages is a basic preprocessing technique for most of the NLP applications. While POS tagging for English and other European languages has been well researched, POS tagging of Indian languages is recently receiving significant attention. Many approaches - including rule based and statistical - have been proposed for POS tagging of Indian languages.

A rule based POS tagger has been proposed by NavneetGarg et al., [1] for Hindi. Here the words are first tokenized, and each word is searched in the database. Database contains the tag category for the words. If the tag is not in the database, a set of rules based on the suffix pattern, are applied to identify the tag. Similarly, later Pallavi Bagual et al., [2] has developed a Rule based POS tagger for Marathi text. This paper uses Wordnet to assign the tag. Nisheeth Joshi et al., [3] experimented with the Hidden Markov Model based POS tagger for Hindi. Treating the POS tagging as a sequence labeling task, this work uses the POS of the preceding and succeeding words to determine the POS of the given word. Similarly Manju K et al. [4] have proposed a POS for Malayalam and Navanath Saharia et al., [5] have proposed POS for Assamese language using the same HMM technique. A trigram based POS tagger has been proposed by Jyoti Singh et al., [6] for Marathi language. This method uses the tags of the previous two words to determine the next tag in sequence.

A POS tagger based on CRF and SVM has been proposed by Thoudam Doren Singh et al., [7]. This work uses different contextual and orthographic word-level features, with 63,200 manually annotated tokens and 26 POS tags. Similarly Pallavi et al., [8] have experimented with CRF model to build POS tagging for Kannada language and Chirag Patel et al., [9] for Gujarati language. An SVM based POS tagger has been proposed by Antony P.J et al. [10]. The objective is to identify the ambiguities in lexical items and develop an accurate POS tagger. This supervised machine learning POS tagging approach requires a large amount of annotated training corpus for correct tagging. Here they have used their own tagset for training and testing the POS-tagger generators. Similarly G. Sindhiya Binulal et al., [11] has developed a

POS tagger based on SVM approach for Telugu language. Eenadu Telugu news paper corpus has been used for training and testing.

D.Chandrakanth et al. [12] have reviewed various POS tagging approaches and discussed how different approaches solve the challenges in Tamil POS tagging. S. Lakshmana Pandian and T. V. Geetha [13] have proposed a Statistical language model for assigning part of speech tags using morphological analyzer for separating the Stem and the Suffixes. Similarly R.Akilan [14] has proposed a rule based model of part of speech (POS) tagger for Classical Tamil Texts. Arulmozhi Palanisamy and Sobha Lalitha Devi [15] have proposed a HMM based POS tagger for a relatively free word order language. This approach uses the POS of the previous and next words to determine the POS of the given word. This is demonstrated for Tamil. Later Sobha Lalitha Devi et al., [16] have developed a Hybrid approach which consists of HMM and Rule based approach. Dhanalakshmi V et al. [17] have used SVM based on linear programming method for assigning Tamil POS tags. The authors have designed and used their own 32 tags for preparing the annotated corpus for Tamil. Kavi Narayana Murthy and Srinivasu Badugu [18] have proposed a new approach for automatic tagging of Indian languages without machine learning techniques and training data. Instead of these techniques they have used the morphological analyzer for assigning the POS tags. S. Lakshmana Pandian et al. [19] have proposed a language model for POS tagging and chunking using CRFs for Tamil. Here the morphological information which is a very important feature for designing CRF, is obtained from a morphological analyzer. Anand Kumar M et al. [20] have proposed a Sequence Labeling Approach to Morphological Analyzer for Tamil Language. In this paper they used a suffix list for morpheme segmentation. This approach uses the machine learning techniques of SVM, MBT and CRF taggers.

While various machine learning techniques have been used, bootstrapping approach has not been used for Tamil POS. However bootstrapping approach has been proposed by Silviu Cucerzan and David Yarowsky[21] for part of speech tagger for any language. Bilingual dictionary, basic library reference grammar and monolingual text corpus has been used for their approach. Similarly Stephen Clark et al., [22] have developed a bootstrapping approach for POS using co-training. After tagging they retrain the results of newly labeled data. Similarly Wen Wang et al. [23] have proposed a similar co-training bootstrapping approach for POS tagging for Mandarin language. It uses mandarin broadcast news data.

3   Bootstrapping Approach for POS (Part of Speech) Tagging

In this section we discuss the details of the proposed pattern based bootstrapping technique.

Bootstrapping is a semi-supervised algorithm. It requires a small amount of seed patterns and learns new patterns from the test data so the seed patterns increase automatically. Through the new patterns we can tag more words in the test data. Bootstrapping consists of two phases. In the first phase shown in Fig. 1(a), the features for the seed pattern are identified, and a set of seed patterns are extracted from the corpus. These seed patterns are used to tag the test data through exact matching.

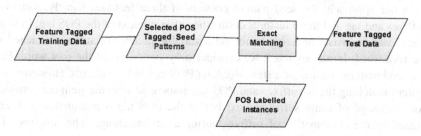

(a) Phase I – Exact matching

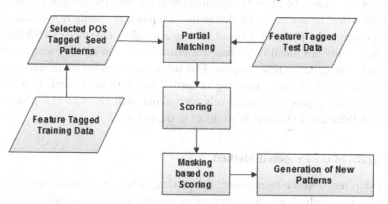

(b) New pattern generation

Fig. 1. Bootstrapping

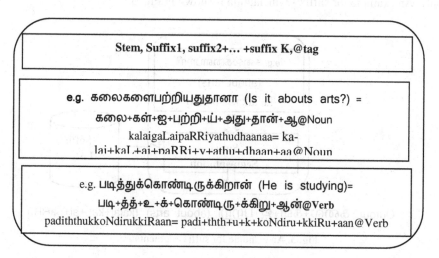

Fig. 2. Seed Pattern Representation

In our approach, the seed pattern consists of three features, namely, stem word, suffixes and tag and these features define the suffix context of the POS tag. In general Tamil word consists of a root word and a variable number of suffixes attached with the root word. In our approach we consider only stem and not the root word. Hence our seed pattern consists of a stem word, its POS tag and K suffixes. However during pattern matching the K suffixes and POS tag associated with the stem are considered. An advantage of using suffix context is that the POS tag is automatically disambiguated by the combination of suffixes during exact matching. The structure of our seed pattern is given in Fig. 2.

In the second phase the important component is the new pattern generation shown in Fig. 1(b), where instances not labeled in the first phase are partially matched. Using frequency count and a probability based scoring algorithm, the suffixes with least and highest probability are identified. These patterns are used to identify features which can be masked to generate new patterns. This forms the extended set of seed patterns. Then the first phase of exact matching is carried out with the extended set of seed patterns. This whole process is iteratively carried out until no more new patterns are generated or there are no unlabeled instances in the test data.

3.1   Details of the Proposed Method

The first step required for representing the pattern, is segmentation of words into its stem and suffixes, which we call suffix segmentation. An existing suffix list [20] of about 262 suffixes is used for this purpose. Using the suffix list, words are segmented by identifying the suffix of maximum length first and then looking for other suffixes. Finally the word is split into its stem word and suffixes. We do not consider the sandhi changes during segmentation since we are using the surface form of the suffixes only. An example for suffix segmentation is shown in Fig. 3.

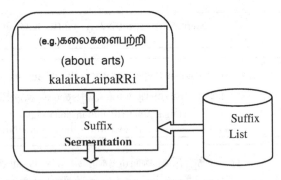

Output: கலை+கள்+ஐ+பற்றி (about arts) (kalai+kaL+ai+paRRi)

Fig. 3. An example for suffix segmentation

The next step is the identification of seed patterns which is an important task in our approach. The selection of the correct set of initial seed patterns decides the effectiveness of the bootstrapping method.

We use two approaches for creating the initial r seed patterns for the POS tagging task. One approach is based on frequently occurring words. The other is based on frequently occurring suffix patterns. In the **word frequency based** approach, to create "r" seed patterns, we take a corpus consisting of "N" unique words. Then we calculate the frequency of occurrence of all the words, and the average frequency. A threshold value based on this average value is chosen, and we select the seed patterns which are above the threshold value so as to contain "r" seed patterns covering nouns, verbs, adjectives and adverbs.

In the **suffix pattern frequency based** approach, we use suffix segmentation, and segment the unique words in the corpus. Then we calculate the frequency of occurrence of the suffixes in the test document, independent of the stem words and the average frequency of occurrence of the suffixes. Once again, based on a threshold value derived from this average, and the length of the suffix, r seed patterns are selected.

3.1.1 New Pattern Generation

The learning part of the bootstrapping algorithm takes place when we generate new patterns. The identification of least probable suffix in a given context of a given pattern, and the dominant suffixes in a given context is the core of the new pattern generation process. As already mentioned, our POS identification pattern is based on suffix context. We use a conditional probability based scoring mechanism (Maskable Suffix Selection Score) to find the suffix that is least probable in a given suffix context of a given pattern. This least probable suffix is masked and replaced by a suffix that can occur in that context to generate the new pattern.

Maskable Suffix Selection Score

Scoring is an important task for new pattern generation. Scoring is based on probability of a suffix conditioned on its suffix context. Scoring computes the average suffix probabilities of all suffixes in a given suffix context of a given seed pattern. For a given pattern, this scoring helps to select a suffix S with least probability for the purpose of masking. This least probability indicates that the occurrence of the suffixes in the given suffix context is least likely and can be replaced.

We consider each seed pattern consisting of m suffixes, Si, i=1 to m, and determine the probability of occurrence of Si in the context of its neighbor Sj, in the corpus, as shown in Eqn. (1).

$$P(S_i|S_j) = \frac{P(S_i \text{ and } S_j)}{P(S_j)}, \quad i,j = 1 \text{ to } m, i \neq j$$

(1)

In Equation (1), the probability of occurrence of suffix Si given the neighborhood suffix context Sj is defined as the probability of occurrence of Si followed by Sj among all occurrences of Sj. In this way the probability of each suffix given its neighbor in the suffix context is determined for all suffixes in the given pattern P. The suffix in the pattern with the least probability score is chosen as the maskable suffix.

Suffix Replacement

The next step is to select a suffix to replace the maskable suffix. In this case we match the suffix context of the maskable pattern with all instances in the test corpus, to find all suffixes that can occur at the masked position.

Selection of New Patterns

There may be many suffix replacements for the masked suffix resulting in the generation of many new patterns. Here we use two methods to select the new patterns. In the first method, we assume that all the generated patterns can be new patterns. In the second method, we filter the new patterns based on the frequency of occurrence of the patterns in the test data corpus. By fixing a threshold based on the average frequency value, new patterns whose frequency of occurrence are above the threshold value are selected.

An example for new pattern generation is given in Table 1.

Considering the word, படிக்கின்றார்(padikkindRaar (He (Honorific) reading), the stem and suffixes are identified as (படி+க்+கின்ற+ஆர்(He (Honorific) reading) @verb padi+k+kin+aar@verb). The suffix parts are split into single and context of suffixes. Table 1(a) shows the suffix split. Table 1(b) shows the suffix frequency. By using the Maskable Suffix Selection Score we find which suffix is to be masked in the pattern in Table 1 (c).

Table 1. (a) Suffix context

	க்(k)	க்+கின்ற் (k+kinR)
Example	கின்ற்(kinR)	க்+ஆர் (k+aar)
	ஆர்(aar)	கின்ற்+ஆர் (kinR+aar)

Table 1. (b) Suffix frequency

க் (k)	கின்ற்(kinR)	ஆர் (aar)	க்+கின்ற் (k+kinR)	கின்ற்+ஆர் (kinR+aar)	க்+ஆர் (k+aar)
47	55	65	30	20	42

In can be seen that for this example, ஆர்(aar) has the minimum probability value. So we mask the suffix, ஆர்(aar), and get an input pattern 'க்(k) + கின்ற்(kinR) + masked suffix' and search for this pattern in the test data to find new patterns. For this example, the new patterns obtained are: படி+க்+கின்ற+ஆன் (padi+k+kinR+aan) (He is reading), படி+க்+கின்ற்+ஆள் (padi+k+kinR+aaL) (She is reading), படி+க்+கின்ற்+ஆய்(padi+k+kinR+aay)(You are reading), படி+க்+கின்ற்+அது (padi+k+kinR+athu) (It Reads).

Table 1. (c) Maskable Suffix Selection Score

$P(க்(k)\|கின்ற்(kinR)) = \dfrac{P(க்(k) \, and \, கின்ற்(kinR))}{P(கின்ற்(kinR))}$	0.54
$P(கின்ற்(kinR)\|க்(k)) = \dfrac{P(கின்ற்(kinR) \, and \, க்(k))}{P(க்(k))}$	0.64
$P(ஆர்(aar)\|கின்ற்(kinR)) = \dfrac{P(ஆர்(aar) \, and \, கின்ற்(kinR))}{P(கின்ற்(kinR))}$	0.36
$P(கின்ற்(kinR)\|ஆர்(aar)) = \dfrac{P(கின்ற்(kinR) \, and \, ஆர்(aar))}{P(ஆர்(aar))}$	0.30
$P(ஆர்(aar)\|க்(k)) = \dfrac{P(ஆர்(aar) \, and \, க்(k))}{P(க்(k))}$	0.89
$P(க்(k)\|ஆர்(aar)) = \dfrac{P(க்(k) \, and \, ஆர்(arr))}{P(ஆர்(aar))}$	0.76

4   Evaluation and Results

In this section, we discuss the performance of the POS tagger with respect to the two methods for identification of seed patterns and the two methods for new pattern selection. The final POS tagging results are evaluated using two measures, namely, recall and precision.

We first discuss the effect of the two methods of seed pattern identification, namely, word frequency based and suffix pattern based selection during the exact matching phase of the bootstrapping method. For both types of seed pattern selection, a corpus consisting of 2,71,933 unique words (Corpus I) is used as input, and 30 seed patterns are selected. The percentage of words tagged by exact matching with these seed patterns is shown in Table 2. There are two observations we make from this result. One is that exact matching does not handle most of the instances, pointing to the need for bootstrapping. The second is that suffix pattern based selection gives better seed patterns. This is as expected since our POS tagging system depends on suffix information.

Table 2. Exact Matching Results with given seed patterns

Method	Unique Words	Percentage Handled
Word frequency based	2,71,933	20.5359%
Suffix Pattern frequency based	2,71,933	30.0533%

4.1 New Pattern Generation

Next we discuss the effectiveness of our pattern selection method. In our work the new patterns generated are chosen in two ways. In the first method (Method I), we consider all the generated patterns as possible new patterns. By using this method, the accuracy of POS tagging is reduced due to presence of some incorrect patterns that occur in the generated new patterns as shown in Table 3. In the second method (Method II) we find the frequency of occurrence of each new pattern in the test data. We consider only patterns whose frequency of occurrence is above the average frequency of all new patterns generated. The results show that there is an increase in accuracy of the new patterns when the average threshold is used for selection of new patterns. The correct and incorrect patterns are evaluated manually.

$$Accuracy = \frac{no.\,of\,true\,correct\,patterns\,generated + no.\,of\,true\,incorrect\,patterns\,generated}{Total\,no.\,of\,patterns\,generated}$$

Table 3. New Pattern Generation Table

Method	New patterns generated	New Pattern Considered	True Correct	True Incorrect	False Correct	False Incorrect	Accuracy
I	160	160	46	48	29	37	58.75%
II	160	77	67	72	11	10	86.87%

Hence we considered only patterns selected by Method II, where initially we started with a set of 30 seed patterns and at the end of 20th iteration of bootstrapping for new pattern generation, a total of 77 new patterns were generated. At the end of 20th iteration the process stopped since no more new patterns were generated

4.2 Recall and Precision

Finally the overall performance of the system was measured using precision, recall and F-measure. In our case, precision is defined as the number of words tagged correctly against the number of words tagged.

$$Precision = \frac{No.\ of\ words\ tagged\ correctly}{No.\ of\ words\ tagged}$$

Recall is defined as the proportion of the target items that the system selected.

$$Recall = \frac{No.\ of\ words\ tagged\ correctly\ by\ the\ system}{No.\ of\ words\ to\ be\ tagged}$$

Correspondingly, F-measure the harmonic mean of recall and precision is defined as

$$F\text{-}Measure = 2PR/P+R \quad where\ P\ is\ Precision\ and\ R\ is\ Recall$$

Here we used an existing morphological analyzer tool for finding the correctly and incorrectly tagged words from the results of all tagged words for measuring precision. The system was tested with a test data set. The test data set consists of untagged data as well as data used for selection of seed patterns. The results of POS tagging for the two methods of seed selection using Method II for new pattern selection, with the corpus of 2,71,933 words (Corpus I) are given in Table 4. We achieve an overall recall of 92.14%, since the new patterns generated allowed good coverage of the corpus. The overall precision of 87.74 % was achieved since the patterns generated for each POS tag was based on seed patterns selected by a suffix pattern frequency method.

Table 4. Recall-Precision F-measure Table

Selection of Seed Pattern	Recall	Precision	F-measure
word frequency based	53.52%	46.32%	49.66%
suffix pattern frequency based	92.14%	87.74%	89.88%

After bootstrapping and the generation of new patterns the system was again tested with an entirely new corpus consisting of 2,32,164 unique words (Corpus II). The result (Table 5) shows that 74.79% of the words are handled, when the additional new patterns generated with Corpus I are used. This shows that still more new patterns need to be added to handle the new corpus. Hence, with appropriate selection of seed patterns from different domains and generating new patterns from corpora of different domains, the performance of the system can be further improved.

Table 5. Performance of new pattern generation in exact matching

Matching	Pattern	Unique Words	Percentage Handled
Exact Matching	Seed Pattern	2,32,164	28.13%
	New pattern + Seed Pattern	2,32,164	74.79%

5 Conclusion and Future Work

We have developed a Part of speech tagger system for Tamil, using a specialized machine learning technique named bootstrapping. The input for a POS tagger system is a string of words. The suffix context is used in determining the POS of a word. The bootstrapping method makes use of a conditional probability approach for masking and generating new patterns with new suffix contexts. The overall precision of the POS tagger is 87.74%. Even on an entirely new corpus, just an exact matching with the new patterns gives a precision of 74.79%. Even though other POS tagging systems for Tamil have achieved precision of above 90%, the uniqueness of this approach is that a dictionary is not required and light weight segmentation is sufficient.

This system can be improved further by adding more suffix patterns to the Seed Pattern by using words from different domains. The same approach can be applied for any morphologically rich language by use of word based context. Further improvements include POS tagging with a richer set of tags and using a richer suffix context which considers more than one neighbor.

References

1. Garg, N., Goyal, V., Preet, S.: Rules Based Part of Speech Tagger. In: The Proceedings of COLING, pp. 163–174 (2012)
2. Bagul, P., Mishra, A., Mahajan, P., Kulkarni, M., Dhopavkar, G.: Rule Based POS Tagger for Marathi Text. The Proceedings of International Journal of Computer Science and Information Technologies (IJCSIT) 5(2), 1322–1326 (2014)
3. Joshi, N., Darbari, H., Mathur, I.: Hmm Based Pos Tagger For Hindi. In: The Proceedings of the Computer Science Conference Proceedings, CSCP (2013)
4. Manju, K., Soumya, S., Idicula, S.M.: Development of a Pos Tagger for Malayalam-An Experience. In: Proceedings of the International Conference on Advances in Recent Technologies in Communication and Computing (2009)
5. Saharia, N., Das, D., Sharma, U., Kalita, J.: Part of Speech Tagger for Assamese Text. In: The Proceedings of ACL-IJCNLP Conference Short Papers, pp. 33–36 (2009)
6. Singh, J., Joshi, N., Mathur, I.: Part of Speech Tagging of Marathi Text Using Trigram method. Proceedings of the International Journal of Advanced Information Technology (IJAIT) 3(2) (April 2013)
7. Singh, T.D.: Manipuri POS Tagging using CRF and SVM: A Language Independent Approach. In: Proceedings of the International Conference on Natural Language Processing, ICON (2008)
8. Pallavi, A.S.P.: Parts Of Speech (POS) Tagger for Kannada Using Conditional Random Fields (CRFs). In: Proceedings of the National Conference on Indian Language Computing, NCILC (2014)
9. Patel, C., Gali, K.: Part-Of-Speech Tagging for Gujarati Using Conditional Random Fields. In: Proceedings of the IJCNLP Workshop on NLP for Less Privileged Languages, pp. 117–122 (2008)
10. Antony, P.J., Mohan, S.P., Soman K.P.: SVM Based Part of Speech Tagger for Malayalam. In: Proceedings of the International Conference on Recent Trends in Information (2010)

11. Sindhiya Binulal, G., Anand Goud, P., Soman, K.P.: A SVM based approach to Telugu Parts of Speech Tagging using SVMTool. Proceedings of the International Journal of Recent Trends in Engineering 1(2) (2009)
12. Chandrakanth, D., Anand Kumar, M., Gunasekaran, S.: Part-Of-Speech Tagging For Tamil Language. Proceedings of the International Journal of Communications and Engineering 06(6(1)) (March 2012)
13. Lakshmana Pandian, S., Geetha, T.V.: Morpheme based Language Model for Tamil Part-of-Speech Tagging. Proceedings of the Research Journal on Computer Science and Computer Engineering with Applications, 19–25 (July-December 2008)
14. Akilan, R., Naganathan, E.R.: Pos Tagging for Classical Tamil Texts. Proceedings of the International Journal of Business Intelligent 1(01) (January-June 2012)
15. Palanisamy, A., Devi, S.L.: HMM based POS Tagger for a Relatively Free Word Order Language. Proceedings of the Research in Computing Science (18), 37–48 (2006)
16. Arulmozhi, P., Pattabhi R K Rao, T., Sobha, L.: A Hybrid POS Tagger for a Relative Free Word Order Language. In: Proceedings of the MSPIL 2006 (2006)
17. Dhanalakshmi, V., Anand Kumar, M., Rajendran, S., Soman, K.P.: POS Tagger and Chunker for Tamil Language. In: Proceedings of Tamil Internet Conference (2009)
18. Murthy, K.N., Badugu, S.: A New Approach to Tagging in Indian Languages. Proceedings of the Research in Computing Science (70), 45–56 (2013)
19. Lakshmana Pandian, S.: Language models developed for POS tagging and chunking. In: Proceedings of 22nd International Conference, ICCPOL 2009 (2009)
20. Anand Kumar, M., Dhanalakshmi, V., Soman, K.P., Rajendran, S.: A Sequence Labeling Approach to Morphological Analyzer for Tamil Language. Proceedings of International Journal on Computer Science and Engineering International Journal on Computer Science and Engineering (IJCSE) 02(06), 1944–1951 (2010)
21. Cucerzan, Yarowsky, D.: Bootstrapping a Multilingual Part-of-speech Tagger in One Person-day. In: Proceedings of the Sixth Conference on Natural Language Learning (CoNLL), pp. 132–138 (2002)
22. Clark, S., Curran, J.R., Osborne, M.: Bootstrapping POS taggers using Unlabelled Data. In: Proceedings of the Seventh CoNLL Conference (2003)
23. Wang, W., Huang, Z., Harper, M.: Semi-Supervised Learning for Part-of-Speech Tagging of Mandarin Transcribed Speech. In: Proceedings of the ICASSP, vol. 4 (2007)

Anaphora Resolution in Tamil Novels

A. Akilandeswari and Sobha Lalitha Devi

AU-KBC Research Centre, MIT Campus of Anna University, Chennai
sobha@au-kbc.org

Abstract. We have presented a robust anaphora resolution system for Tamil, one of Indian languages belonging to Dravidian language family. We have used Conditional Random Fields (CRFs), a machine learning technique with linguistically motivated features. We have performed exhaustive experiments using data from different genres and domains and evaluated for portability and scalability. We have obtained an average accuracy of 64.83% across texts of different domains/genres. The results obtained are encouraging and comparable with earlier reported works.

Keywords: Anaphora resolution, Tamil, Machine learning, CRFs, Different genres/domains.

1   Introduction

Web data has become the source of information for people in every field. Unimaginary growth of web data poses a challenge in accessing these data. The various data mining and extraction tools perform fairly well in culling out information from structured data. But only an insignificant portion of the web data is in structured form and rest of the data are in unstructured form. These data contains natural language text such as news articles, blogs, reports, travel logs, anecdotes etc. Various inherent features of the natural languages such as focus shift, re-ordering, coherence in text, use of co-referring entities turnout to be the real challenge in extracting information from the unstructured data. To overcome these hurdles and come up with tools such as information extraction and question answering systems we require various sophisticated natural language processing tools such as parser, lexical disambiguator, coherence identifier, coreference/anaphora resolver etc. In this paper, we present the algorithm and experiments done on anaphora resolution engine for Tamil language, a south Dravidian language. Anaphora resolution is the task of identifying the referent (antecedent) for the corresponding pronoun (anaphor). Here we have developed our engine using machine learning technique, CRFs with linguistically motivated features and evaluated it with various source of text to explore the robustness of the resolution engine.

The research in anaphora resolution was initiated in early 80's. The early systems were classified as knowledge intensive approaches, where various knowledge resources such as case frames, world knowledge, semantic information etc were used in designing these systems. AR systems were built using Centering theory and salience factor/indicator based approaches. Various machine learning techniques such as

R. Prasath et al. (Eds.): MIKE 2014, LNAI 8891, pp. 268–277, 2014.

decision tree, memory based learning, support vector machine, conditional radon fields, integrated linear programming were used to come up with AR systems. A detailed survey on anaphora resolution engines is presented in [8, 14].

In Indian languages automatic resolution of anaphors are developed for Hindi, Bengali, Tamil and Malayalam. One of the early works in this area was the development of system named "Vasisht" [11]. Vasisht was a rule-based engine developed as a common platform for Indian languages and demonstrated with Hindi and Malayalam. A tool contest on anaphora resolution for Indian languages was conducted as part of ICON 2011. This tool contest has been a catalyst in generating interest among researchers in anaphora resolution in Indian languages and especially in Hindi, Bengali and Tamil languages [15]. Prasad et al. [6] presented a Hindi anaphora resolution system based on the centering theory approach. This work was further taken by Upalappu [16] and Dekawale [4] where they used dependency tree features in resolving pronouns in Hindi and Dutta [5] customised Hobb's algorithms for Hindi.

In Tamil an anaphora resolution system using morphological features as salience measures was developed by Sobha[13]. Akilandeswari [2] had presented preliminary studies on pronominal resolution using CRFs, a machine learning approach. Balaji et al., [3] had presented anaphora resolution using UNL representation and boot strapping approach. A system using Tree CRFs, a machine learning approach was developed by Ram et. al., [8], for anaphora resolution in Tamil. ICON 2011 tool contest had been the starting point for the development of Bengali anaphora resolution systems. Sikdar et al., [10] had developed a pronominal resolution system by customizing BART framework for Bengali. Similarly Senapati [9] had customised GUITAR for Bengali.

In the present work we have described the development of a robust anaphora resolution system for Tamil, where we have used linguistically motivated features in the machine learning. We have performed experiments using different data sets from different sources and genres and have evaluated the system performance to measure its portability and scalability across different texts.

The paper is further organised as follows. In the next section we describe the characteristics of Tamil language in general and pronominals in Tamil in particular. In section 3 we describe our approach and methodology. In section 4, we describe our experiments, results and our observations. Section 5 presents the conclusion of the work and future directions.

2 Nature of Tamil Language and Characteristics of Pronominals in Tamil

Tamil belongs to Dravidian family of languages. It is a morphologically rich and highly agglutinated language. It is an accusative-dative language. As the language is morphologically rich, the suffix carries lots of information. Nouns are suffixed with plural and case markers and verbs are suffixed with tense, aspect, modal and person, number gender suffixes. Verbs are also suffixed with conditional, relative participle markers. Tamil is relative free word order language and the structures with in the clauses are rigid. Subject and Object agree in person, number and gender. In this language, there is gender and number distinction in most of the pronouns. First person pronouns and second person pronouns does not have gender distinction and has number distinction such as 'nii' (you), 'naan' (I), 'niingkal' (you-plural). Third person pronouns have number and gender distinction such as 'avan' (he), 'aval' (she), 'atu' (it). In third person pronoun, plural pronoun 'avarkal' refers to both masculine and feminine genders and also represents honorific. These pronouns can be anaphoric and non-anaphoric. In anaphoric pronouns, the antecedent noun phrase can be in the same sentences or in the previous sentences. These antecedents agree in person, number and gender with the pronouns. Following are the set of examples, which explains anaphor-antecedent relation in Tamil.

1. (a) muthalai,   maraththin   verkaLukku   maththiyilirunthu veLiye varuvathaRku
       Crocodile (N) tree (N+gen) root+PL+dat middle+abl   outside come+dat
       ciRithu   neram   aayiRRu.
       some (ADJ) time (N) took+past

(It took some time for the crocodile to come out from the middle of the tree's root.)

1. (b) ippothu   athu   veLiye vanthuvittathu.
       Now (ADV) that(PN) outside came+past
       (Now that came outside.)

In above Example 1, the anaphoric pronoun is "athu" (3rd person neuter pronoun with nominative case) and its antecedent is "muthalai" that occurs in previous sentence. In 1(a) antecedent is the subject of the sentence and is present in the initial position of the sentence. The person, number and gender of the anaphor match with the antecedent.

2. athaRkup poongkuzhali   maRumozhi onRum collavillai, aanaal aval azhath
   that+dat pungkuzhali(N) comment   any   say+neg but   she cry
   thotangkinaaL.
   start+past+3SF

(For that pungkuzhali did not say any comment, but started to cry.)

    Example 2 is a compound sentence, where anaphor and antecedent occurs intra sententially. "avaL" (3rd person, singular and feminine pronoun) is the anaphoric pronoun. The anaphor "avaL" refers to the noun phrase "poongkuzhali" which is the subject in nominative case.

3. uththama chozhar thammutaiya kulaththin   pazhaiya perumaiyai
   Uthama chola (N) he (PN)+gen family(N)+gen old (ADJ) pride (N)+acc
   maRakkaveyillai.
   forget+past+neg
(uthama chola did not forget his family's old pride)

Considering Example 3, the anaphoric pronoun is "thammutaiya" (possessive anaphor) that refers to the noun phrase "uththama chozhar", which is the subject of the sentence. Here, the anaphor and antecedent occurs in same sentence. If anaphor is possessive, then noun phrase is the subject of the clause which contains the anaphora.

4. innum paranjothiyinutaiya thuNi maNikaLum, avan koNtu vanthiruntha
   still   paranjothi(N)+gen clothes(N)+um(conj) he   bring   came+past
   coRpap   paNamum   muutaikkuLLethaan irunthana.
   meagre(Adj) money(N)+and(conj) bag+inside+emp   is
   (Still, paranjothi's clothes and the meager money she brought were inside the bag.)

In Example 4, "avan" is the anaphoric pronoun and "paranjothi" is its antecedent in genitive case. The antecdent and anaphor is present in the same clause. The anaphor refers to genitive noun phrase, 'paranjothiyinutaiya' Here, the antecedent occurs as the subject of the sentence.

Following is an example where the pronoun occurs as non-anaphoric.

5. atu oru iniya   maalai neeram.
   It one beautiful evening time
   (It is one beautiful evening time)

In the above sentence has occured. In that sentence the pronoun 'atu' is not anaphoric. In anaphora resolution task, we need to identify the anaphoric and non anaphoric pronouns and resolve the antecedents for the anaphoric pronouns.

3   Our Approach

In this work we have developed a robust, scalable anaphora resolution system for Tamil. We have tested this using different corpus of different genres. We have used a machine learning based approach. Conditional Random Fields (CRFs) is the machine learning technique used here. CRFs have been used successfully for different natural language processing tasks. CRFs is an undirected graphical model, for labelling and

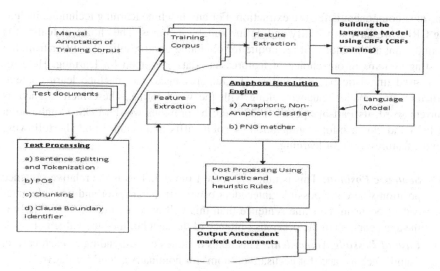

Fig. 1. System Architecture

segmenting structured data, such as sequences, trees and lattices, where the conditional probabilities of the output are maximized for a given input sequence (Lafferty et al., 2001). The task of identifying antecedents to anaphors is a syntactic and semantic task, which is non-linear. Though CRFs have been predominantly used to model sequential data, but we could also use it to model non-linear data. CRFs could be modelled as a binary classifier in this particular task. The other reason for us to choose CRFs is the ability to incorporate linguistic features in the learning. CRFs can contain number of feature functions.

We have two phases training phase and testing phase. In the training phase the system is provided with annotated data and the features for learning. After the system learns, a model file is generated as output. In the testing phase any unseen text is given for the automatic anaphora resolution. There are mainly three components in this system viz., i) Text Processing module ii) Feature identification and extraction module and iii) Anaphora Engine. Figure 1 shows the system architecture diagram during the training and testing phase.

In the text processing module we first perform the basic task of sentence splitting and tokenization. After that we annotate the text with Part-of-Speech (POS). For the POS tagging we have used the system developed by Arulmozhi et al., [1]. And next we perform chunking to identify noun phrases. For this purpose we have used the system developed by Sobha and Ram [12]. Here we have processed the text with clause boundary tags. Identification of clause boundaries is very much helpful in antecedent identification. As said in the earlier section, Tamil is a relatively free word order language and within a clause it is normally fixed structure and order. Thus identification of clause boundaries helps in structural learning. For the purpose of clause identification we have used system developed by Ram and Sobha [7]. The text processing module is completely automated.

3.1 Features for Learning

The next component is feature extraction. For any machine learning technique including CRFs, we have to identify and define suitable features. The choice of features is as important as the choice of technique for obtaining good results. We differ from the existing systems in our choice of features and data presented for learning. Here we have used linguistic based features. The features make the machine learn context, structure, grammatical nuances along with the word forms. These include such as Current word, its Part-of-Speech (POS) and the previous word's POS; Combination of POS and chunk info; Combination of noun suffix and chunk info. The following are the features used for learning.

i)   *Sentence Position:* This is numeric value from 0-5. This is the relative sentence position where the possible antecedent occurs. If the anaphor and possible antecedent occur in the same sentence then this will have a value of 0. Similarly if the antecedent is in the previous sentence of the anaphor then the value is 1.

ii) *Case of Possible Antecedent:* This takes 8 values corresponding to each case in Tamil, such as 'acc' for accusative, 'nom' for nominative, 'dat' for dative.

iii) *Case of the Anaphor:* This takes 8 values corresponding to each case in Tamil.

iv) ***Anaphor Type:*** This takes the following values "possessive", "reflexive", "honorific", "other" depending on the type of anaphors.

v) ***Anaphor Position:*** This is the relative position of the anaphor in the sentence. This has three values 'initial', 'middle' and 'final'. If the anaphor is in the initial few words of the sentence it takes value 'initial, and if in the middle of the sentence then 'middle' and if in the last few words of the sentence then 'final'.

vi) ***Antecedent Position:*** This is the relative position of the possible antecedent in the sentence. The values are similar to anaphor position

vii) ***Antecedent Clause Position:*** This is the clausal position of the possible antecedent with respect to the anaphor. If the possible antecedent occurs in the same clause as anaphor occurs then the value is '0', if it is in the next immediate clause then the value is "1".

viii) ***PNG (Person, Number and Gender) of the Anaphor:*** This is the PNG value of the anaphor. For example if the anaphor is 3rd person, masculine, singular then the value is 3sm.

ix) ***PNG of Possible Antecedent:*** This is the PNG of the possible antecedent. The values are similar to the one explained in the previous feature.

x) ***Possible Antecedent Type:*** This is whether the possible antecedent is existential, possessive or prepositional complement.

xi) ***POS Tag of Antecedent:*** This the POS tag of the possible antecedent.

Before we identify and extract the features, all the possible candidate antecedents for a given anaphor are collected and pairs of anaphor, antecedent is formed. And the features for each pair is identified and extracted in the feature extraction module.

3.2   Anaphora Engine – Training

In this module we perform three sub-tasks viz., i) Anaphoric and Non Anaphoric Pronominal discrimination ii) PNG matching and iii) Antecedent identification. As explained in section 2, anaphors in some instances do not have any referents or antecedents. We term such anaphors for whom there exists no antecedent in the text as non-anaphoric. Thus it is essential that we first identify such non-anaphoric anaphors before proceeding to identify antecedents. So we first build a language model using CRFs, which would identify actual anaphors and non-anaphoric. In the training phase using the manually annotated corpus we train CRFs with the features explained in previous section and develop a language model which distinguishes anaphors and non-anaphors. After this using a heuristic rules we perform a PNG match between the anaphor and possible antecedent PNG features. In the anaphor – antecedent pairs, which do not have a matching PNG characteristics of the anaphors are removed from the further training data. This pruned training data is used for training to develop a language model which will identify the antecedent. And present a ranked list of antecedents suitable to the particular anaphor.

3.3   Anaphora Engine – Testing

After training we have language models built and ready for use for any new test document. A new test document is first processed using text processing module and then features are extracted as explained in previous sections. With the aid language models developed in training phase we perform non-anaphoric identification and then do PNG match and then identify correct antecedent. We can select the top most ranked antecedent as the actual antecedent.

4   Experiments, Results and Discussion

We have performed our experiments using different types of corpora. We have collected corpora from various online sources belonging to different genres. The different genres in our corpus are tourism, health, general news articles, historical and contemporary novels. We have collected widely popular Tamil historical novel titled 'Ponniyan Selvan'. This novel has a very unique style of writing; it is a mix of contemporary style and old classical style. The other novel is titled "Akalvillaku", this is a contemporary novel. These novels are relatively huge text and manually annotating them completely is quiet labour intensive and time consuming. With our limited resources and time, we have manually annotated only a portion of total corpus collected. Manually annotated text consists of 22,000 sentences, 2,19,691 word tokens. In Table 1 we have provided the pronoun distribution statistics for each type/genre of corpora.

In our present work we have primarily focussed on resolution of 3rd person pronominals 'avan' (he), 'aval' (she), 'avar' (he/she honorific), 'atu' (it). As can be seen from the distribution table 1., the pronoun 'avar' has more frequency and also resolution of this pronoun is more challenging. The pronouns 'atu' and 'avar' are found to also occur as non-anaphoric. In the earlier section we had explained about two phases training and testing. For the training, we use online news articles and a part of 'Ponninyan Selvan' corpus. For the testing we use 'Akalvillaku' corpus, unseen general new articles and unseen 'Poniyyan Selvam' test. The general news articles and 'ponniyan selvan' data is partitioned in the ratio of 80:20 for training and testing. Whereas the 'Akalvillaku' corpus is not partitioned and whole of the 'Akalvillakku' is used for testing.

This is done to test the system's portability on a completely unseen text. We have obtained very encouraging results on this. Table 2 shows the training and testing partitions of the data and Table 3 below shows the results obtained for various types of anaphors in different genres.

Though the 'Akalvillaku' corpus was not used in training, we observe that the results obtained for this corpus is comparable with other genres. This shows us that the system is portable and scalable. We observe that in the 'Ponniyan Selvan' corpus, the pronoun 'avaL' (she) obtained relatively less score than other genres. The reasons for this could be attributed to the sentences in this corpus are relatively lengthier and have more than two or three female characters in the discourse. So in some of the instances this was resolved to wrong entity.

Table 1. Pronominal Corpus Statistics in Each Corpora

pronoun	Online news articles – genres – Toursim, Health, Sports			Akalvilakku – a Contemporary Novel			Ponniyinselvan – a Historical Novel			Total
	anaphoric	Non-anaphoric	total	anaphoric	Non-anaphoric	total	anaphoric	Non-anaphoric	total	
avan	106	0	106	457	3	460	349	8	357	888
avaL	35	1	36	149	1	150	233	6	239	425
avar	202	16	799	352	24	823	507	19	526	2148
atu	273	22	295	248	36	284	553	89	642	1221
Total	616	39	1236	1206	64	1717	1642	122	1764	4717

Table 2. Training and testing Partition of data

Language	Training Partition No. of Pronouns	Testing Partition No. of Pronouns	Total
Online general news articles	989	247	1236
Akalvillaku	0	1717	1717
Ponniyan Selvan	1411	353	1764

Table 3. System Evaluation results – across genres and different anaphor types

pronoun	Online news articles – genres – Toursim, Health, Sports Accuracy %	Akalvilakku – a Contemporary Novel – Accuracy %	Ponniyinselvan – a Historical Novel Accuracy %	Average
avan	72.46	69.82	71.73	71.34
avaL	69.32	68.49	63.29	67.03
avar	62.92	67.37	64.57	64.95
atu	56.13	54.46	57.37	55.98
Average	65.21	65.03	64.24	64.83

Similarly in the 'Akalvillaku' corpus, the pronoun 'atu' (it) in some of the instances was resolved to wrong entity. And in general the pronominal 'atu' is having less accuracy in resolution. This can be attributed to two reasons a) we can observe from the pronominal distribution in Table 1, there are more number of non-anaphoric instances of 'atu'. Though we had used an anaphoric – Non-anaphoric disambiguator, not all non-anaphoric instances were properly identified. A deeper analysis of data has to be made to improve it. b) The other reason being the pool of possible candidates having same characteristic features is more and hence wrong resolution to incorrect entities.

Pronoun 'avar' is another challenging issue in resolution. This pronoun is used for honorific reference, and also can be used to refer to both male and female. In some instances we observe that in the set of candidate antecedents there are both male and female genders and it is difficult to distinguish by the system which entity is being referred by 'avar'. A more world knowledge is required for resolving such instances.

5　Conclusion

In the present work we have presented a robust, scalable, portable third person pronominal resolution system for Tamil using machine learning. In this work exhaustive experiments were performed to experimentally study the behaviour of different 3rd person pronouns in Tamil. The earlier reported works in Tamil have not exhaustively tested their system across domains/genres. Another main contribution of our work is the development of the 200K words manually annotated corpus. This is a resource poor language and creation of such a huge manually annotated corpus of anaphor-antecedent tags is a first of its kind. We have used CRFs, machine learning technique, though most of the earlier reported works have used rule based approach or memory based approaches. We have obtained comparable results which are very encouraging.

References

1. Akilandeswari, A., Sobha, L.: Conditional Random Fields Based Pronominal Resolution in Tamil. International Journal on Computer Science and Engineering 5, 601–610 (2013)
2. Balaji, J., Geetha, T.V., Ranjani, P.R., Karky, M.: Two-Stage Bootstrapping for Anaphora Resolution. In: COLING 2012, pp. 507–516 (2012)
3. Dakwale, P., Mujadia, V., Sharma, D.M.: A Hybrid Approach for Anaphora Resolution in Hindi. In: International Joint Conference on Natural Language Processing, Nagoya, Japan, pp. 977–981 (2013)
4. Dutta, K., Prakash, N., Kaushik, S.: Resolving Pronominal Anaphora in Hindi using Hobbs algorithm. Web Journal of Formal Computation and Cognitive Linguistics (10) (2008)
5. Prasad, R., Strube, M.: Discourse Salience and Pronoun Resolution in Hindi. In: Penn Working Papers in Linguistics, vol. 6(3), pp. 189–208 (2000)
6. Ram, R.V.S., Bakiyavathi, T., Sindhujagopalan, R., Amudha, K., Sobha, L.: Tamil Clause Boundary Identification: Annotation and Evaluation. In: 1st Workshop on Indian Language Data: Resources and Evaluation, Istanbul (2012)

7. Ram, R.V.S., Sobha, L.: Pronominal Resolution in Tamil Using Tree CRFs. In: 6th Language and Technology Conference, Human Language Technologies as a challenge for Computer Science and Linguistics, Poznan, Poland (2013)
8. Senapati, A., Garain, U.: GuiTAR-based Pronominal Anaphora Resolution in Bengal. In: 51st Annual Meeting of the Association for Computational Linguistics, Sofia, Bulgaria, pp. 126–130 (2013)
9. Sikdar, U.K., Ekbal, A., Saha, S., Uryupina, O., Poesio, M.: Adapting a State-of-the-art Anaphora Resolution System for Resource-poor Language. In: International Joint Conference on Natural Language Processing, Nagoya, Japan, pp. 815–821 (2013)
10. Sobha, L., Patnaik, B.N.: Vasisth: An Anaphora Resolution System for Indian Languages. In: International Conference on Artificial and Computational Intelligence for Decision, Control and Automation in Engineering and Industrial Applications, Monastir, Tunisia (2000)
11. Sobha, L., Ram, R.V.S.: Noun Phrase Chunking in Tamil. In: MSPIL 2006, pp. 194–198. Indian Institute of Technology, Bombay (2006)
12. Sobha, L.: Resolution of Pronominals in Tamil. In: Computing Theory and Application, pp. 475–479. The IEEE Computer Society Press, Los Alamitos (2007)
13. Sobha, L., Ram, R.V.S., Pattabhi, R.K.R.: Resolution of Pronominal Anaphors using Linear and Tree CRFs. In: Proc. 8th DAARC, Faro, Portugal (2011)
14. Sobha, L., Bandyopadhyay, S., Ram, R.V.S., Akilandeswari, A.: NLP Tool Contest @ICON2011 on Anaphora Resolution in Indian Languages. In: ICON (2011)
15. Uppalapu, B., Sharma, D.M.: Pronoun Resolution For Hindi. In: 7th Discourse Anaphora and Anaphor Resolution Colloquium (DAARC 2009), pp. 123–134 (2009)

An Efficient Tool for Syntactic Processing of English Query Text

Sanjay Chatterji, G.S. Sreedhara, and Maunendra Sankar Desarkar

Samsung R&D Institute India - Bangalore
{sanjay.chatt,m.desarkar}@samsung.com, sreedhargs89@gmail.com

Abstract. A large amount of work has been done on syntactic analysis of English texts. But, for analyzing the short phrases without any structured contexts like capitalization, subject-object-verb order, etc. these techniques are not yet proved to be appropriate. In this paper we have attempted the syntactic analysis of the phrases where contextual information is not available. We have developed stemmer, POS tagger, chunker and Named Entity tagger for English short phrases like chats, messages, and queries, using root dictionary and language specific rules. We have evaluated the technique on English queries and observed that our system outperforms some commonly used NLP tools.

Keywords: Stemming, Parts-of-Speech, Chunk, Named Entity, Trie, Short text analysis.

1 Introduction

Most of the recent Natural Language Processing (NLP) engines analyze texts based on contexts. From an annotated training corpus, supervised techniques learn the contexts in which the words or phrases of a category are used and it prepares a model. Then, for a test word or phrase it disambiguates the contexts and identify its most appropriate category. Some language dependent structures like, capitalization, Subject-Verb-Object/Subject-Object-Verb structure, etc. are often used in such systems. Such techniques are proved to be appropriate for different text analysis tasks like Parts-of-Speech (POS) tag identification, chunk boundary and tag identification (chunking), Named Entity boundary and tag identification (Named Entity Recognition(NER)), etc. for sentences. But, for analyzing the short texts these techniques are not yet proved to be appropriate. Short texts such as queries given to search engines are often different from "Natural Language Texts", in terms of their lengths, availability of language context or structure information, etc. This makes the analysis of short texts a challenging task. Supervised techniques are also probably not appropriate for this task as training corpus is not available in sufficient volume.

Search engines need to quickly analyze the search queries to identify the inherent concepts, the intent of the user and the category of the document being searched. The analysis of short messages and chat is also a nontrivial task. We use dictionary and rules in a hybrid framework for this task. Multiple techniques

R. Prasath et al. (Eds.): MIKE 2014, LNAI 8891, pp. 278–287, 2014.

have been used in a common framework for analyzing short and free English texts. We use Trie [1] data structure for storing all English root words. Specifically, we have used java implementation [2] of Patricia trie [3] and refer to it as trie in this paper. Then, different rules of English language are used for different analysis tasks. Finally, we implement multiple analysis modules without using any annotated corpus. These analysis modules may be used to extract the topic of a conversational short phrases like chat log, message log, etc. and to parse short queries into chunks and phrases which may be used to retrieve the relevant documents.

The first analysis task is the stemming where the roots for each of the words are identified. The task of POS tagging is to identify the POS category of the words in a phrase. These two tasks are carried out at word level. A chunk consists of a content word surrounded by a constellation of function words. The task of chunking is to group such words and identify their categories like, noun chunk, verb chunk, etc. Similarly, the task of Named Entity identification is to group the words which denote same person, location etc. and identify their categories like, person, location, etc. We compare the performances of these modules with the performances of one of the existing standard automatic stemmer, POS tagger, chunker and NE tagger which work well for full sentences. Experimental results show that our hybrid technique provides more accurate results for English query phrases. Our system is also efficient in terms of response time and memory consumption.

2   Related Work

A large amount of work has been done on performing syntactic analysis of words and phrases in English sentences. For example, Stanford tools for POS tagging [4], Named Entity Recognition [5], Penn TreeBank Tokenizer (PTBTokenizer) [6], Parsing [7], etc. perform significantly well for the analysis of English full sentences. The Carnegie Mellon University tools for POS tagging based on a universal tagset [8], parsing (TurboParser [9] and MSTParserStacked [10]), Named Entity Recognition [11] etc. are also very popular in NLP community. Most of these tools use statistical techniques and require annotated corpus for training the models. They analyze the words and phrases of full sentences by observing the surrounding words and phrases and by observing the analysis of surrounding words and phrases. Porter stemmer [12] is the most popular existing English stemmer which separates root and suffixes using a set of rules for English language.

There are different techniques of stemming. Frakes[13] have presented a review on this. The basic but non-practical technique is to rely on the dictionary of all words of a language. Another technique identifies n-gram similarity between the words and then decide their stem. The another most familiar technique is Porter algorithm in which the affixes are removed in 5 steps based on a set of rules. Finally, the Peak-and-Plateau technique [14] is also a familiar technique for the stemming of words where either the root dictionary is not available or where the time efficiency is given more importance than the accuracy. This technique is similar to our technique and hence given a short description below.

Suppose, p is a term and p_i is its substring with the first i characters. Again, suppose, there are D words in a corpus and $D(p_i)$ is the number of words having first i characters same as p_i. Then, we count the number of different characters after i^{th} character in $D(p_i)$. This count become the successor variety of p_i. In simpler words, number of possible characters after p_i in a corpus is called its successor variety. In Peak-and-Plateau technique, p_i may be decided as a stem of p if successor variety of p_i is highest for all i's. In simpler words, the substring p_i (from 1st to i^{th} character) may be decided as stem of p if it's next character may have maximum variety in the corpus. This position is called peak. The common way of storing the words in a corpus is Trie. The Trie helps in finding the number of varieties in each position. Though, this technique is not highly accurate, but it returns consistent result for the words having same stem. For example, both try, tries and tried are stemmed as tr.

Different approaches are also explored for identifying the chunks or linguistic phrases from short free texts. One statistical technique [15] considers every pair of contiguous content words as candidate chunks and retain those having frequency higher than a threshold value. Gey and Chen[16] have calculated mutual information between the words using co-occurrence frequency statistics and used it for identifying the English phrases. Syntactic phrases are also identified based on tagging and linguistic rules of a language [17]. De and Pedersen[18] have used probabilistic parsing for identifying the most likely linguistic phrases for a previously unseen user query. The method then identifies the exact linguistic phrases and expand them and subsequently generate formal query. Aida[19] and Stanford NER[5] are some of the entity detection and disambiguation modules which work well for natural language texts.

Patricia [3] is a compact representation of trie structure where the only child of a node is merged with the node. This algorithm enables us to store, index, and retrieve information in a large file in a flexible way. We do not need to re-index when a new data is added to the dataset. Knuth[2] have provided a Java implementation of Patricia trie following the algorithm of [20].

3   Our Approach

We focus on the problem of given a query identify the Named Entities present in it, root and POS of each of the words, and break the query into some chunks. We have collected a list of English root words, suffixes and prefixes from the dictionaries in "Natural Language ToolKit", Web and manual process. The POS categories of the root words are also collected. We have manually tagged the suffixes and prefixes with different POS categories in which they commonly occur. For example, 'ing' is tagged as verb and ''s' (apostrophe s) is tagged as noun. We have also collected a list of Named Entities (NEs) and their types from similar sources. There are 6 types of NEs namely, person, organization, location, measure, time/date, and others.

We have created a trie (TRIER) structure for the members of the Root Word List (RWL). In TRIER, we also store the POS categories of the root words. An example TRIER for four English words, 'get', 'go', 'good', and 'government' is shown in Figure 1.

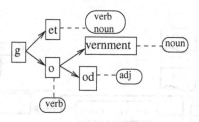

Fig. 1. Trie structure for the root words 'get' (verb or noun), 'go' (verb), 'good' (adjective), and 'government' (noun)

Similarly, we have created another trie (TRIEN) structure for the members of the Named Entity List (NEL). When a NE is identified in TRIEN, we store its NE category in TRIEN. One example TRIEN, for four NEs, 'subho', 'subhodevpur', 'subhodeep', and 'subhiksha', is shown in Figure 2.

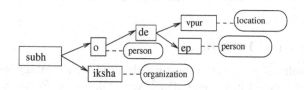

Fig. 2. Trie structure for the names subho (person), subhodevpur (location), subhodeep (person), and subhiksha (organization)

For the tasks of stemming and POS tagging, we rely on the RWL. If a word passes the TRIER, we immediately identify its root and POS category. The words which are not passed in TRIER, are tagged as unknown words. However, while traversing the TRIER for matching a word we also apply English orthographic rules. Similarly, we traverse the TRIEN to identify the NEs. If a named entity contains one word we identify it from TRIEN. We have manually identified a set of rules (RuleSet1) for chunking and another set of rules (RuleSet2) for identifying multword NEs and their tags.

The task of POS tagging depends on the stemming and the rules in chunking module use stem and POS tags of the words. Finally, the NE identification task depends on the stem, POS category and chunk tags of the words. Therefore, the modules are arranged in a pipeline architecture as presented in Figure 3. The first block does the stemming and POS tagging. Both these tasks use the

TRIER. The second block does chunking. It checks each pair of words in the phrase, their stems and POS tags in Ruleset1 and decide whether they may belong to same chunk. The third block identifies NE tag from stem, POS tag and chunk features with the help of TRIEN and Ruleset2.

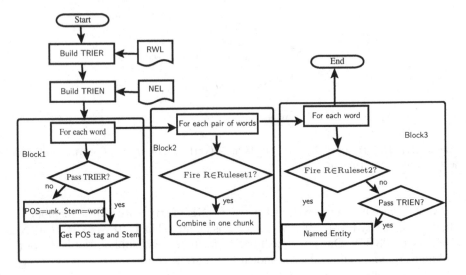

Fig. 3. Pipeline Architecture of (i) Stemming and POS tagging, (ii) chunking, and (iii) NE tagging

3.1 Stemmer and POS Tagger

Root words, suffixes and prefixes are widely used for stemming such as in the stemmers implemented in Lucene [21], YASS [22], and in Porter Stemmer [12]. For stemming we have also discarded the longest suffix. After discarding the suffix part we pass the remaining part through TRIER. When the root matches with this remaining part of the word we assign the POS tags of the root word as the POS tags of the input word. However, instead of searching for the exact match we have implemented orthographic rule based matching. For example, in English when 'ies' suffix is matched we replace it with 'y'. The stemming and POS tagging is carried out for each word without looking at the features of the surrounding words. If a word is passed through TRIER, we return its stem and POS tags. A word may have multiple matches. The longest stem matched in TRIER is considered as the actual stem. The POS tags of the corresponding stem are considered as the actual POS tags of the word. If there are multiple POS tags of this stem, we retain all of them as possible POS tags.

3.2 Chunker

The chunking task is carried out using a set of rules referred to as Ruleset1. Some example rules in Ruleset1 are presented below.

Preposition followed by Noun/Pronoun⇒Noun Chunk
Adjective/Determiner followed by Noun⇒Noun Chunk
Verb/Adverb followed by Verb/Adverb ⇒ Verb chunk

Each pair of words in the query phrase and their POS tags are searched in the Ruleset1. If a rule in Ruleset1 satisfies the features then we chunk that pair. However, for multiple matches we keep all the chunk possibilities.

3.3 Named Entity Tagger

We have developed a hybrid NER for short phrases. In the first step we have created a set of rules referred to as Ruleset2. The most important rule here is that the unknown words (not identified by the POS tagger) are Named Entities. The categories of such names are unknown. Ruleset2 is used to identify the categories of unknown Named Entities. Some rules from Ruleset2 are presented below.

A NE followed by relationship(son, wife, etc)⇒ Person
near/from/etc. followed by a NE ⇒ Location
from/to/etc. followed by a NE ⇒ Time
A NE having suffix pur/nagar/etc. ⇒ Location

Some other rules in Ruleset2 are used to identify more named entities and their categories. Some of such rules in Ruleset2 are presented below.

Number followed by Inch/Foot/etc. ⇒ Measure
Number followed by th/rd/etc. Jan./Feb./etc. ⇒ Date
Number followed by th/rd/etc. Jan./Feb./etc. 4 digit number ⇒ Date
Place followed by Univ./College/etc. ⇒ Organization

In the second step we have used NEL which contains the names having at least one of the root words of RWL. TRIEN is used to store those entries and for searching. The reason for using such a list is explained below.

The complete list of Named Entities, if at all possible to collect, is huge in size. So, it will take a large amount of memory space for loading it. Instead, it is easier to collect and use an exhaustive list of root words of a language which are used as noun, verb, etc. We consider the words which are in NEL and those which are not in RWL, i.e., those which are tagged as unknown by the POS tagger, as Named Entities.

4 Evaluation

We have evaluated our approach on English queries. The performances of our POS tagger, chunker and NER are measures in terms of F-measure and compared with some widely used systems.

4.1   Experimental Setup

We have listed 4054 English root words in RWL and 52 English suffixes and prefixes. Our NEL contains 618,823 English Names. The number of roots for each category is listed in Table 1. The numbers of English rules in the Rulesets are shown in Table 2. English specific 20 orthographic rules are hardcoded in the stemmer module.

Table 1. Number of Noun(NN), Pronoun(PN), Adjective(Adj), Verb(VB), Determiner(DT), and other root words

	NN	PN	Adj	VB	DT	Other
Root Words	1017	77	1126	736	3	87

Table 2. Number of rules in Rulesets

	Ruleset1	Ruleset2 for categorization	Ruleset2 for identification
No. of Rules	52	14	11

4.2   Experimental Result

We have tested the performances of the proposed query processing modules on 109 test English queries (643 words, 278 chunks). In these queries, there are 64 inflected words and 105 names.

The performance of our stemmer is compared with the Porter stemmer [12] demo available in NLTK. We have compared the performances of our POS tagger, chunker and NE tagger with the online demo systems developed by the CCG group of UIUC [23–25]. The comparison of performances of the modules in F-measure[26] are shown in Table 3.

The results show improvement in both stemmer, chunker and NE tagger over the standard NLP systems. The POS tagger has not crossed the accuracy of the available system due to short size of root word list.

Table 3. F-measure comparison of our and existing systems

	Stemmer	POS Tagger	Chunker	NE tagger
F-measure of our system	0.96	0.91	0.87	0.92
F-measure of existing system	0.82	0.92	0.69	0.44

4.3   Discussion with Some Example Outputs

From our 109 test queries, some example queries and the outputs of our stemmer, POS tagger, chunker and NE modules on these queries are presented below. The words which are stemmed are underlined. Chunk and named entities in the sentence are separated by ','. The wrong POS tags and chunks are shown using Boldface character. The words which are missed by the stemmer are shown using Boldface characters and the words which are wrongly stemmed are shown using Italic characters.

1. Input Query: standard sizes of pocket pictures in inches
   Stems: standard <u>size</u> of pocket <u>picture</u> in <u>inch</u>
   POS tags: Adjective Noun Preposition Noun Noun Preposition Noun
   Chunks: standard sizes, of pocket pictures, in inches
   NEs: inches
2. Input Query: daughters of ned stark game of thrones
   Stems: <u>daughter</u> of ned stark game of <u>throne</u>
   POS tags: Noun Preposition Noun Noun Noun Preposition Noun
   Chunks: daughters, of ned stark, game of thrones
   NEs: ned stark, game of thrones
3. Input Query: khaleesi game of thrones
   Stems: khaleesi game of <u>throne</u>
   POS tags: Noun Noun Preposition Noun
   Chunks: khaleesi, game of thrones
   NEs: khaleesi, game of thrones
4. Input Query: preprocessing of facial image using opencv
   Stems: <u>preprocess</u> of <u>face</u> image <u>use</u> opencv
   POS tags: Verb Preposition Adjective Noun **Verb** Noun
   Chunks: preprocessing, of facial image, **using, opencv**
   NEs: opencv
5. Input Query: rebuild precompiled header visual studio 2010
   Stems: rebuild **precompiled** <u>head</u> *vision* studio 2010
   POS tags: Verb Adjective Noun **Adjective** Noun Noun
   Chunks: rebuild, precompiled header, visual studio 2010
   NEs: visual studio 2010

In Example 4, incorrect POS tag of the word 'using' [it need to be preposition, but is tagged as verb], leeds to incorrect chunking of the phrase 'using opencv' [it need to be a single chunk but is tagged as two separate chunks]. In example 5, 'visual' is stemmed as 'vision' and it leeds to incorrect POS tag of the word 'visual' [it need to be noun, but is tagged as adjective]. But it does not lead to incorrect chunking of the phrase 'visual studio 2010'. In example 5, 'precompiled' could not be stemmed, but it did not affect the POS tagging and chunking of the word.

5   Conclusion

The approach presented here is efficient in terms of F-measure. As we have used optimized implementation using TRIE structure, the modules are fast and space efficient. Time and space efficiency will be reported in extended paper. More rules and dictionary words will be able to further increase the accuracy. The approach with the corresponding language specific rules and dictionary entries may also be used for other languages and for other short phrases like message, chat texts, etc.

References

1. Folk, M.J., Zoellick, B.: File structures. Addison-Wesley, Reading (1992)
2. Knuth, D.E.: The art of computer programming, 3rd edn. Sorting and Searching, vol. iii. Addison & Wesley, Reading (1998)
3. Morrison, D.R.: Patricia-practical algorithm to retrieve information coded in alphanumeric. Journal of the ACM (JACM) 15(4), 514–534 (1968)
4. Toutanova, K., Manning, C.D.: Enriching the knowledge sources used in a maximum entropy part-of-speech tagger. In: Proceedings of the 2000 Joint SIGDAT Conference on Empirical Methods in Natural Language Processing and Very Large Corpora: Held in Conjunction with the 38th Annual Meeting of the Association for Computational Linguistics, vol. 13, pp. 63–70. Association for Computational Linguistics (2000)
5. Finkel, J.R., Grenager, T., Manning, C.: Incorporating non-local information into information extraction systems by gibbs sampling. In: Proceedings of the 43rd Annual Meeting on Association for Computational Linguistics, pp. 363–370. Association for Computational Linguistics (2005)
6. Marcus, M., Santorini, B., Marcinkiewicz, M., Taylor, A.: Treebank-3 (tech. rep.). Linguistic Data Consortium, Philadelphia (1999)
7. Socher, R., Bauer, J., Manning, C.D., Ng, A.Y.: Parsing with compositional vector grammars. In: Proceedings of the 51st Annual Meeting on Association for Computational Linguistics. Citeseer (2013)
8. Petrov, S., Das, D., McDonald, R.: A universal part-of-speech tagset. arXiv preprint arXiv:1104.2086 (2011)
9. Martins, A.F., Almeida, M., Smith, N.A.: Turning on the turbo: Fast third-order non-projective turbo parsers. In: ACL (2), pp. 617–622 (2013)
10. Martins, A.F., Das, D., Smith, N.A., Xing, E.P.: Stacking dependency parsers. In: Proceedings of the Conference on Empirical Methods in Natural Language Processing, pp. 157–166. Association for Computational Linguistics (2008)
11. Mohit, B., Schneider, N., Bhowmick, R., Oflazer, K., Smith, N.A.: Recall-oriented learning of named entities in arabic wikipedia. In: Proceedings of the 13th Conference of the European Chapter of the Association for Computational Linguistics, pp. 162–173. Association for Computational Linguistics (2012)
12. Porter, M.F.: An algorithm for suffix stripping. Program: Electronic Library and Information Systems 14(3), 130–137 (1980)
13. Frakes, W.B.: Stemming algorithms. In: Frakes, W.B., Baeza-Yates, R. (eds.) Information Retrieval: Data Structures and Algorithms, pp. 131–160. Prentice Hall, Englewood Cliffs (1992)
14. Hafer, M.A., Weiss, S.F.: Word segmentation by letter successor varieties. Information Storage and Retrieval 10(11), 371–385 (1974)
15. Mitra, M., Buckley, C., Singhal, A., Cardie, C., et al.: An analysis of statistical and syntactic phrases. In: RIAO, vol. 97, pp. 200–214 (1997)
16. Gey, F.C., Chen, A.: Phrase discovery for english and cross-language retrieval at trec 6. NIST SPECIAL PUBLICATION SP, 637–648 (1998)
17. Strzalkowski, T., Lin, F., Perez-Carballo, J., Wang, J.: Natural language information retrieval trec-6 report. In: TREC, pp. 347–366. Citeseer (1997)
18. De Lima, E.F., Pedersen, J.O.: Phrase recognition and expansion for short, precision-biased queries based on a query log. In: Proceedings of the 22nd Annual International ACM SIGIR Conference on Research and Development in Information Retrieval, pp. 145–152. ACM (1999)

19. Hoffart, J., Altun, Y., Weikum, G.: Discovering emerging entities with ambiguous names. In: Proceedings of the 23rd International Conference on World Wide Web, pp. 385–396. International World Wide Web Conferences Steering Committee (2014)
20. Sedgewick, R.: Algorithms in Java, Parts 1-4. Addison-Wesley Professional (2002)
21. Hatcher, E., Gospodnetic, O., McCandless, M.: Lucene in action (2004)
22. Solanki, K., Sarkar, A., Manjunath, B.S.: YASS: Yet another steganographic scheme that resists blind steganalysis. In: Furon, T., Cayre, F., Doërr, G., Bas, P. (eds.) IH 2007. LNCS, vol. 4567, pp. 16–31. Springer, Heidelberg (2008)
23. Roth, D., Zelenko, D.: Part of speech tagging using a network of linear separators. In: Proceedings of the 36th Annual Meeting of the Association for Computational Linguistics and 17th International Conference on Computational Linguistics, vol. 2, pp. 1136–1142. Association for Computational Linguistics (1998)
24. Li, X., Roth, D.: Exploring evidence for shallow parsing. In: Proceedings of the 2001 Workshop on Computational Natural Language Learning, vol. 7, p. 6. Association for Computational Linguistics (2001)
25. Li, X., Morie, P., Roth, D.: Robust reading: Identification and tracing of ambiguous names. Technical report, DTIC Document (2004)
26. Van Rijsbergen, C.: An algorithm for information structuring and retrieval. The Computer Journal 14(4), 407–412 (1971)

A Tool for Converting Different Data Representation Formats

Sanjay Chatterji, Subrangshu Sengupta,
Bagadhi Gopal Rao, and Debarghya Banerjee

Samsung R&D Institute India - Bangalore
{sanjay.chatt,subrangshu.s,
gopalrao.b,debarghya.b}@samsung.com

Abstract. Recently, data analysis and processing is one of the most interesting and demanding fields in both academics and industries. There are large numbers of tools openly available in web. But, different tools take inputs and return outputs in different data representation formats. To build the appropriate converter for a pair of data representation formats, we need both sufficient time and in depth knowledge of the formats. Here, we discuss CoNLL, SSF, XML and JSON data representation formats and develop a tool for conversion between them. Other conversions will be included in the extended version.

1 Introduction

Recently, data analysis and processing is one of the most interesting and demanding fields in both academics and industries. There are large numbers of tools like NLTK[1], CRF++ [2, 3], etc. which are openly available in web. The group, who wants to analyze a data, needs to download the tools and run them on their data. Different tools may be integrated in a Dashboard architecture like the Sampark architecture discussed in [4]. But, the main bottleneck of this process is as follows. Different tools take inputs and return outputs in different data representation formats. Even, some tools which are usable from programs like, Java programs, Python programs, etc. also require passing the data in proper format. If the format of the input of a tool does not match with the format of the data we wish to process, then we need to convert it to the appropriate format.

To build the appropriate converter for a pair of data representation formats, we need both sufficient time and in depth knowledge of the formats. The exact capabilities and functionalities of a format need to be understood and mapped to the other format functionalities. Again, the nature of the data may not be the same for each tagging. That is, the nature and fields of POS tagged data may not be the same as the nature and fields of chunk tagged data. A common tool, for converting all these formats between each other may be useful.

Conference on Natural Language Learning (CoNLL) format is used by some open source classifiers and data analysis tools. This format is able to represent all the words of a sentence, the features of the words and the relations between

R. Prasath et al. (Eds.): MIKE 2014, LNAI 8891, pp. 288–297, 2014.

the words. However, this format is not able to represent nested relations and relations between groups or chunks of words. Recently, the Shakti Standard Format (SSF) has become famous in representing Indian language sentences. This format is able to represent relations between chunks. Extensible Markup Language (XML) and JavaScript Object Notation (JSON) formats are famous in industries where a specific technology (like Java, C++, etc.) is able to create objects or structures from such instance representations.

In this paper, we discuss our tool for converting data in one format to the data in another format. As an initial task, we have considered four data formats which have similar capabilities in terms of the types of data they may represent. In Section 2 we discussed some work related to format conversion. In Section 3, we discussed different data representations formats with relevant examples. The methods of conversion are discussed in Section 4. The evaluation results and analyses are discussed in Section 5. Section 6 concludes the paper.

2   Related Work

Bharati et. al. [5] have discussed both the text level and sentence level SSF formats. [6] have defined the fields of each feature. Though these fields are fixed, there are scopes of extending these features for a particular task. The possible values of the features depend on the language of the text. Both in-memory representation and textual representation are possible in SSF format data. The reader is tool for converting textual to memory representation and the printer tool is used for converting memory to textual representation. They have also discussed the correspondence of metadata representations in SSF and CML formats.

Recently, a large amount of SSF format Indian language resources are created as part of the ILMT project[1]. Several Indian language analysts like [7–11] have converted these data from SSF format to CoNLL format for their individual tasks. These conversions may not be used for the tasks of other kinds. For example, the technique used for converting the SSF format Named Entity (NE) tagged data to the CoNLL format NE tagged data may not be useful for converting the SSF format Dependency RELation (DREL) tagged data to the CoNLL format DREL tagged data.

JavaScript Object Notation (JSON) is a lightweight data exchange text format evolved from JavaScript [12, 13]. JSON is now mostly used with all programming technologies due to its ease and very flexible nature of representing data types in both primitive (such as numbers, strings, boolean, null) and structured (such as arrays, objects) formats.

There are some work on conversion of formats. For example, the patents of [14, 15] have discussed about a possible conversion technique from source specific format to a self-describing format.

---

[1] Indian Language to Indian Language Machine Translation (ILMT) project is a consortium project carried out by more than 10 premier Indian institutes. This project is sponsored by Deity, MCIT, Govt. of India.

3   Introduction to Different Data Representation Formats

In this paper, we consider four mostly used data representation formats, namely, CoNLL, XML, JSON and SSF. Each of these data representation formats are explained below with relevant examples. The examples are English and Hindi sentences. Hindi words are presented in WX notation[16, 17].

In CoNLL data representation format[18], each token (word) of a sentence and their features are presented in a line. The fields of the features are same for each word. After each sentence there is a blank line. The CoNLL representation of the English sentence (Our home was in Bangalore.) is shown in Table 1. If a field is not appropriate for a word, like the tense field of noun, the corresponding value is represented by NA (Not Applicable).

Table 1. CoNLL format representation of the English sentence "Our home was in Bangalore." and the analyses of all its words.

Sl. No.	Word	Root	Lexical Category	Number	Per- son	Tense	Chunk Tag	Relation	Atta- chment
1	Our	I	pronoun	pl	1	NA	NP-B	possession	2
2	home	home	noun	sg	3	NA	NP-I	subject	3
3	was	is	verb	NA	NA	Past	VP-B	root	NA
4	in	in	preposition	NA	NA	NA	NP-B	preposition	5
5	Bangalore	Bangalore	noun	sg	NA	NA	NP-I	place	3
6	.	.	symbol	NA	NA	NA	NP-I	Symbol	5

In XML data representation format, after the initial tags like xml version, encoding, corpus lang, etc., the sentences are bounded by context and instance tags. Each word is followed by its POS tag in <p= "[tag_name]" /> format. The chunk of word is bounded by <P= "[chunk_tag_name] [head_word]"> and </P>. The XML representation of an English sentence (the company argued that its foreman needn't have told the worker not to move the plank to which his lifeline was tied because "that comes with common sense .") is shown below [19].

<P="S argued"> <P="NPB company"> the <p="DT"/> company <p="NN"/> </P> <P="VP argued">
argued <p="VBD"/> <P="SBAR-A that"> that <p="IN"/> <P="S-A have"> <P="NPB 't"> its <p="PRP"/>
foreman <p="NN"/> needn <p="NN"/> 't <p="NN"/> </P> <P="VP have"> have <p="VBP"/> <P="VP-
A told"> told <p="VBN"/> <P="NPB worker"> the <p="DT"/> worker <p="NN"/> </P> <P="SG-A to">
not <p="RB"/> <P="VP to"> to <p="TO"/> <P="VP-A move"> move <p="VB"/> <P="NP-A plank">
<P="NPB plank"> the <p="DT"/> plank <p="NN"/> </P> <P="SBAR to"> <P="WHPP to"> to <p="TO"/>
<P="WHNP which"> which <p="WDT"/> </P> </P> <P="S-A was"> <P="N-PB lifeline"> his <p="PRP"/>
lifeline <p="NN"/> </P> <P="VP was"> was <p="VBD"/> <P="VP-A tied"> tied <p="VBN"/> <P="SBAR
because"> because <p="IN"/> " <p="PUNC "'/> <P="S-A comes"> <P="NPB that"> that <p="DT"/>
</P> <P="VP comes"> comes <p="VBZ"/> <P="PP with"> with <p="IN"/> <P="NPB sense"> common
<p="JJ"/> sense <p="NN"/> . <p="PUNC."/> " <p="PUNC"/> </P> </P> </P> </P> </P> </P>
</P> </P> </P> </P> </P> </P> </P> </P> </P> </P> </P> </P>

In SSF data representation format, each sentence stars with <Sentence id= > and each chunk starts with the chunk number, "((", chunk tag and features of the chunk. Each word of a chunk contains the chunk number, the word number, the word, its POS tag and morphological features. The chunk and sentence ends are indicated by the ")))" and </Sentence>, respectively. The features of the words and chunks have two parts: mandatory part with 8 comma separated values (root, lexical category, gender, number, person, case, suffix for noun and suffix for verb) and the optional part. The SSF representation of the Hindi sentence (isalie isa JIla ko SeRanAga JIla kahA jAwA hE.) is shown below.

```
    <Sentence id="1">
1          ((        CCP       <fs af='isa,adv,,,,,lie,' name="CCP" drel="root:0">
1.1        isalie    CC        <fs af='isa,adv,,,,,lie,'>
           ))
2          ((        NP        <fs af='JIla,n,f,sg,3,o,,0' name="NP" drel="k2:VGF">
2.1        isa       DEM       <fs af='yaha,pn,any,sg,3,o,0,0'>
2.2        JIla      NN        <fs af='JIla,n,f,sg,3,o,0,0'>
2.3        ko        PSP       <fs af='ko,psp,,,,,,'>
           ))
3          ((        NP        <fs af='SeRanAga,n,m,sg,3,d,0,0' name="NP2" drel="k1:VGF">
3.1        SeRanAga  NNP       <fs af='SeRanAga,n,m,sg,3,d,0,0'>
           ))
4          ((        NP        <fs af='JIla,n,f,sg,3,d,0,0' name="NP3" drel="rs:NP2">
4.1        JIla      NNP       <fs af='JIla,n,f,sg,3,d,0,0'>
           ))
5          ((        VGF       <fs af='kaha,v,,sg,,,yA,yA' name="VGF" drel="ccof:CCP">
5.1        kahA      VM        <fs af='kaha,v,,sg,any,,yA,yA'>
5.2        jAwA      VAUX      <fs af='jA,vx,m,,2,,wA,wA'>
5.3        hE        VAUX      <fs af='hE,vx,,,any,,hE,hE'>
5.4        .         SYM       <fs af='&dot;,punc,,,,,,'>
           ))
</Sentence>
```

JSON represents data in text format as a logical sequence of tokens which supports structural tokens ([, {,], }, :, ,, ", ", etc.), literal tokens ("true", "false", "null", etc.) and whitespace characters (space, horizontal tab, new line, carriage return). Each JSON data representation consists of objects, arrays of objects or other allowed types within them. Here, data is represented as a "name:value" pair with the name and value written within double quotes. The JSON representation for the English phrase (mAz hE.) is shown below.

```
{ "sid":"1",
      "chunk":{ "cid":"1", "chunkTag":"NP", "relation":"k1", "attachment":"VGF"
            "values":{ "number":"sg", "root":"mAz", "lcal":"n", "gender":"f", "person":"3",
"case":"d", "noun_suffix":"0", "verb_suffix":"0" },
            "word":{ "name":"mAz", "wid":"1", "number":"sg", "root":"mAz", "lcal":"n",
"gender":"f", "person":"3", "case":"d", "postag":"NNP", "noun_suffix":"0", "verb_suffix":"0" },
      }
      "chunk":{ "cid":"2", "chunkTag":"VGF", "relation":"root", "attachment":"0",
            "values":{ "number":"sg", "root":"hE", "noun_suffix":"hE", "verb_suffix":"hE",
```

"lcat":"v", "case":"", "gender":"any", "person":"2" },

    "word":{ "name":"hE", "wid":"1", "number":"sg", "root":"hE", "noun_suffix":"hE",

"verb_suffix":"hE", "lcat":"v", "case":"", "gender":"any", "person":"2", "postag":"VM" },

    "word":{ "name":".", "wid":"2", "number":"", "root":"˙", "noun_suffix":"",

"verb_suffix":"", "gender":"", "case":"", "lcat":"punc", "person":"", "postag":"SYM" }

      }

 }

4   Techniques Used in Conversion of Different Data Representation Formats

We have considered 4 data representation formats. The basic definitions of these formats are discussed. In this section we discuss the techniques for converting each pairs of formats. We first discuss the techniques we have used for converting data from SSF to CoNLL and from CoNLL to SSF format. Then the techniques used for conversion between SSF and JSON and between JSON and XML are discussed.

4.1   SSF CoNLL Format Conversion

In sentences represented in SSF format, the words and the chunks have separate identity. The relations between the words and the relations between the chunks both are presented in SSF format. But, in CoNLL format sentences, there can be any one type of tokens and relations need to be defined between these tokens. We have considered, chunks as tokens in CoNLL format. Again, the SSF format may have some optional features which may be absent for some chunks. However, in CoNLL format representation, all fields are common for each token. The another main difference between these two representation formats is the representation of the dependency relations. Each SSF chunk of the sentence are given a name. The relations are defined for each chunk by name of the chunk to which it is related and the name of the relation. In CoNLL format representation generally the serial number of the chunk is used as name of the chunk.

If a chunk contains a single word then we consider that word as a head of the chunk. The chunk number, the chunk tag, the head word, the POS tag of the head word, and the features are considered as the fields of the chunk items. In the chunks containing multiple words, the head word is considered as the token and remaining words (RW) are considered as features of the token. The RW words may contain functional words like preposition, postposition and auxiliary or content words like adjective, noun and verb. When a functional word is attached with a noun or a verb to form a chunk, we can easily identify the head noun or verb. However, when a pair of nouns or a pair of verbs form a chunk, then we consider the first noun as the head noun and the second noun as a RW. We conbine the suffix of the head word with the RW (separated by _) and call it Bibhakti for head noun and Tanse-Aspect-Modality-Polarity-Emphasizer (TAMPE) for head verb. As the names suggest, the Bibhakti part of the noun chunk represents its case and the TAMPE part of the verb chunk represents the tense, aspect, modality, polarity and emphasizer features.

If the value of a field is unknown for a chunk, we fill it by a 'NA' value. For example, the outputs of simple parser may contain dependency relations for some chunks and the corresponding field for the remaining chunks are made blank. We fill it by the relation name 'NA' and attachment 'NA'.

4.2 SSF JSON Format Conversion

To convert a SFF format data representation to JSON format we have used a common JSON parsing API named Jackson. Jackson is a suite of data processing tools in Java which includes JSON parsing and generation library. The Java objects are converted into JSON and vice versa with the help of Jackson.

We have defined a Plain Old Java Object (POJO) for storing JSON data. Our POJO contains 2 nested inner classes. We have defined cid (chunk id), chunkTag, relation, attachment, values and word fields in upper class. Variables of values and words are root, lcat, gender, number, person, case, noun_suffix and verb_suffix. The word field has three extra fields namely, name, wid and poscat. The detailed procedure followed in conversion is given below.

1. Read the SSF format data and iterate through its contents line by line.
2. Define a POJO to map the SFF fields and the corresponding JSON fields.
3. Define some patterns in correspondence with the SSF mendatory fields.
4. If a line matches a pattern, store the values in the POJO fields.
5. Extract the optional part of the feature structure.
6. Split the optional part by space and each subpart by equal (=) symbol. The left part of the equal symbol is mapped with a field name in the POJO and and the right part is considered as the value of that field.
7. The 'drel' field is splitted again by collon (:) symbol. Each of these separated parts have predefined field names (relation and attachment) in POJO.
8. Now once the POJO is populated with the values from the SFF format data, use the Jackson API to convert the POJO to JSON format.

JSON to SFF conversion is the reverse process of SFF to JSON conversion described above. Here also we are using Jackson to parse JSON and create java object out of it. We create the java object known as a POJO and construct the SFF format data from the values of the POJO fields. The procedure followed in this conversion is discussed below.

1. Read the JSON from a file or String using Jackson and parse.
2. Define the POJO fields and mapping.
3. Create the POJO object with the values from the JSON.
4. From the POJO extract the values one by one and store in SFF format.

4.3 JSON XML Format Conversion

The process of XML to JSON conversion starts with the validation for the proper starting and ending tags in the XML data. Then conditions, sequence checking and validating various special characters used in an XML data are carried out. The appropriate error messages corresponding to the errors are analysed. Some very common errors are expecting a closing tag instead of "someValue", misshaped close tag, misshaped tag, and expected 'CDATA['. Finally, each XML tag of the input XML string is represented as an individual JSON object and if the XML tag has some values in itself then it is converted as the property of the corresponding attribute of the JSON object.

XML may not be in a plain format with the tag-value structure. It may have children for the tags and the values for the child tags. If the tag contains children having some tags and values then the JSON object will have childNodes property which is internally a collection of array of strings. This is repeated till the ending XML tags which will return a resultant JSON object.

The conversion of JSON to XML is the conversion of the input JSON object into a structural string following some rules which involve addition of tags (starting and ending) which makes an XML object. The first task in the conversion is to identify the first attribute in the JSON which behave as the parent node of the resultant XML object. This can be get by performing string operations (getString(0) in Java). The node is then appended with the special characters like "<" and ">" which gives it an initial structure of XML. Since every attribute of a JSON object resides in the key:value pair format our next task is to find the keys present in the JSON object which becomes the nodes of the XML object.

Once a list of keys and values in the object are obtained, the keys are considered as the XML nodes and the values for the keys as the values of the XML nodes. Since XML is not very good for supporting arrays, JSON arrays in the JSON object are broken into more simpler strings and (tag, value) pairs. They are used to follow the same methodology which is discussed above to find the tag names and the values. An implementation of the above JSON XML conversion technique is available as an open Java service.

5 Evaluation

We have tested the conversion tool for Bengali, Hindi and English sentences. There are POS, chunk, named entity and dependency annotated sentences for all these three langauges. The language of the data, the tag type and the number of sentences are presented in Table 1.

For evaluating the performance of the tool we have done the forward and then backward conversion and compared the initial text and the text we got after the two conversions. For example, to evaluate the SSF CoNLL conversion tool we convert the input SSF data to CoNLL and then the CoNLL data is again converted to output SSF. The input and the output SSF data is compared. We merge the data of all the three languages for evaluation. The performance of the tools, using this evaluation technique, are shown in Figure 1.

Table 2. Description of the tags and languages of the test data

Language	Tag type	No. of Sentences
Bengali	POS and Chunk	11252
	Named Entity	2440
	Dependency	4220
Hindi	POS and Chunk	511
	Named Entity	206
	Dependency	1881
English	POS and Chunk	14
	Named Entity	6
	Dependency	14

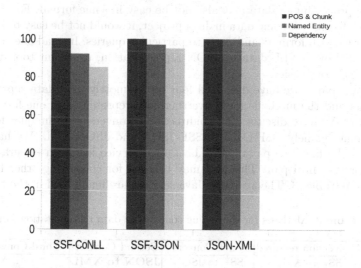

Fig. 1. Performance of the format conversion tools on converting POS & chunk, Named Entity and dependency annotated data

POS and Chunk annotated data are converted with 100% accuracy in all three convertions. However, for NE and dependency annotated data, convertors fail in certain cases. We analyse some issues in our tool. For SSF to CoNLL conversion we need to merge the root and suffix of the functional words and use it as feature of the corresponding head word. The features of the functional words are not stored. In the reverse process, the features are not recovered. For SSF to JSON conversion we need to create the POJO by mapping the exact fields of the SSF format. We sometimes write drel and sometimes deprel for denoting dependency relation. Similarly, we write NE, NEtag, etc. to denote named entities. So, sometimes, the field names in POJO are not matched with the SSF field names.

In POS and Chunk tagged data we conidered words as entity in CoNLL. Here we are able to store everything in CoNLL. But in NE and dependency annotated data, we considered chunks as entity in CoNLL and there the features of the functional words are missed. Again, the field names of some NE and dependency annotated data did not match. Therefore, the SSF to CoNLL and the SSF to JSON convertors have some failures in converting NE and dependency annotated data.

6 Conclusion

The data available for analysis may not be compatible with the format of the corresponding analyzer tool. Again the output format of an analyzer tool may not be compatible with the input format of the next analyzer tool in an integrated system. Here, our tool may be used as an intermediate convertor which will fill the gaps.

The processing of data may also not be easy in some format. For example, if we wish to analyze some data in Java project, it would not be easy to parse their SSF or CoNLL formats and reply to particular queries. Instead one can convert these SSF or CoNLL data to JSON representation, using our tool, which will make the work easier.

In this paper we have discussed four of the mostly used data representation formats and the methods for converting data represented in one format to the another. We have discussed the bidirectional conversion techniques for 3 pairs of formats namely, SSF-CoNLL, SSF-JSON and JSON-XML. We have implemented the first two pairs and a Java open service has been integrated in our tool for the third pair. This tool may be used for converting other formats as shown in Table 3. The available Java services are made bold in the table.

Table 3. Methods for converting some other data representation formats

Conversion required	First Convert	Second Convert	Third Convert
SSF to XML	SSF to JSON	**JSON to XML**	
XML to SSF	**XML to JSON**	JSON to SSF	
CoNLL to JSON	CoNLL to SSF	SSF to JSON	
JSON to CoNLL	JSON to SSF	SSF to CoNLL	
CoNLL to XML	CoNLL to SSF	SSF to JSON	**JSON to XML**
XML to CoNLL	**XML to JSON**	JSON to SSF	SSF to CoNLL

There are other JSON parsing APIs available like google's GSON, JSON data interchange format provided by JSON.org etc. which can also be used to parse and produce JSON strings. There are other data representation formats which will be taken care in the extended version of the paper.

References

1. Loper, E., Bird, S.: Nltk: The natural language toolkit. In: Proceedings of the ACL 2002 Workshop on Effective Tools and Methodologies for Teaching Natural Language Processing and Computational Linguistics, vol. 1, pp. 63–70. Association for Computational Linguistics (2002)

2. Lafferty, J.D., McCallum, A., Pereira, F.C.N.: Conditional random fields: Probabilistic models for segmenting and labeling sequence data. In: ICML, pp. 282–289 (2001)
3. Sha, F., Pereira, F.: Shallow parsing with conditional random fields. In: Proceedings of the 2003 Conference of the North American Chapter of the Association for Computational Linguistics on Human Language Technology, vol. 1, pp. 134–141. Association for Computational Linguistics (2003)
4. Kumar, P., Ahmad, R., Chaudhary, B., Sinha, M.: Enriched dashboard: An integration and visualization tool for distributed nlp systems on heterogeneous platforms. In: 2013 13th International Conference on Computational Science and Its Applications (ICCSA), pp. 105–114 (2013)
5. Bharati, A., Sangal, R., Sharma, D.M.: Ssf: Shakti standard format guide. Language Technologies Research Centre, International Institute of Information Technology, Hyderabad, India, pp. 1–25 (2007)
6. Bharati, A., Sangal, R., Sharma, D., Singh, A.K.: Ssf: A common representation scheme for language analysis for language technology infrastructure development. In: COLING 2014, p. 66 (2014)
7. Saxena, A., Madhyasta, P.S., Nivre, J.: Building the uppsala hindi-swedish-english parallel treebank
8. Agarwal, R.: Automatic Error Detection for Treebank Validation. PhD thesis, International Institute of Information Technology Hyderabad (2012)
9. Gade, R.P.: Dependency parsing approaches for Indian Languages: Hindi and Sanskrit. PhD thesis, International Institute of Information Technology Hyderabad (2014)
10. Tammewar, S.J.N.J.A., Sharma, R.A.B.D.M.: Exploring semantic information in hindi wordnet for hindi dependency parsing
11. Krishnarao, A.A., Gahlot, H., Srinet, A., Kushwaha, D.S.: A comparison of performance of sequential learning algorithms on the task of named entity recognition for indian languages. In: Allen, G., Nabrzyski, J., Seidel, E., van Albada, G.D., Dongarra, J., Sloot, P.M.A. (eds.) ICCS 2009, Part I. LNCS, vol. 5544, pp. 123–132. Springer, Heidelberg (2009)
12. Crockford, D.: Json: The fat-free alternative to xml. In: Proc. of XML, vol. (2006)
13. Ecma, E.: 262: Ecmascript language specification. ECMA (European Association for Standardizing Information and Communication Systems), pub-ECMA: adr (1999)
14. Tong, K.: Migrating data using an intermediate self-describing format. US Patent 7,290,003 (2007)
15. Clark, J., Tong, K., Wu, X., Vong, F.: Dynamically pipelined data migration. US Patent 7,299,237 (2007)
16. Gupta, R., Goyal, P., Diwakar, S.: Transliteration among indian languages using wx notation. g Semantic Approaches in Natural Language Processing, 147 (2010)
17. Sharma, S., Bora, N., Halder, M.: English-hindi transliteration using statistical machine translation in different notation. Training 20000(297380), 20000 (2012)
18. Buchholz, S., Marsi, E.: Conll-x shared task on multilingual dependency parsing. In: Proceedings of the Tenth Conference on Computational Natural Language Learning, pp. 149–164. Association for Computational Linguistics (2006)
19. Leacock, C., Towell, G., Voorhees, E.: Corpus-based statistical sense resolution. In: Proceedings of the ARPA Workshop on Human Language Technology, pp. 260–265 (1993)

Generating Object-Oriented Semantic Graph for Text Summarisation

Monika Joshi[1], Hui Wang[1], and Sally McClean[2]

[1] University of Ulster, Co. Antrim, BT37 0QB, UK
[2] University of Ulster, Co. Londonderry, BT52 1SA, UK
joshi-m@email.ulster.ac.uk, H.Wang@ulster.ac.uk,
sally@infc.ulst.ac.uk

Abstract. In this research paper we propose to extend the semantic graph representation of natural language text to object-oriented semantic graph representation and generate a summary of the original text from this graph. We have provided rules to construct the object-oriented semantic graph and rules to generate the text summary from it. This process has been elaborated through a case study on a news story. An evaluation of the generated summary shows the effectiveness of the proposed approach. This work is a new direction in single document text summarisation research area from semantic perspective and requires further analysis and exploration.

1 Introduction

Text summarisation is one of the major tasks of natural language processing and it has gained importance recently since the time it was first introduced in 1958 [1]. With the time internet users has loaded the web with volumes of text data and this has increased the requirement to shorten the lengthy text documents. Text summarisation methods researched till date are vast and vary from the type of input: single or multiple documents, to the kind of approaches followed to generate the summary. Examples of the well-known approaches in single document text summarisation are summarisers based on statistical features and simple linguistic features of text document such as frequency of term, sentence position, cue words [2][3], summarisers based on supervised and unsupervised machine learning methods of classification and clustering [4][5][6][7], and summarisers based on graphs[8][9][10][11]. Graph based text summarisation systems such as Text Rank [10], LexRank[8] and Semantic Rank [9] have shown good performance in the text summarisation tasks of Document Understanding Conferences. These systems prioritize sentences of the document by ranking graphical representation of its text content. Latest work on graph based text summarisation [12] utilizes connections from neighbouring documents to construct the graph and rank the sentences. Graphical representation of the document involves shallow or deep semantic parsing of its text to construct a connected graph of textual units extracted from it. Dense semantic graph based text summarisation [13] gives improved performance over triplet based text summarisation approach

R. Prasath et al. (Eds.): MIKE 2014, LNAI 8891, pp. 298–311, 2014.

[14]by exploiting the dependency links between nouns and adjectives of the text document and ranking them in a shortest distance dependency based semantic graph. However the nodes of the dense semantic graph are atomic entities, i.e. single words, and cannot be converted to summary directly. In other graphs where nodes are not words but sentence or clauses, we dont have ways to shorten them even further based on any ranking. In this research work we propose to overcome these gaps by enhancing the previously described dense semantic graph of atomic entities [13] to a graph of objects; an object-oriented semantic graph (O-O semantic graph). Enhancing the graphical representation of text to a more user understandable graph and then generating summary from it could be a good direction to text summarisation research. We plan to construct this graph by utilising the dependency relations of words within sentences. Novelty lies in the new rules to construct object-oriented semantic graph from unrestricted text and generating summary from this graph. Next section gives a review of earlier approaches for modelling the natural language text and their limitations. Section 3 gives details of the rules for object-oriented semantic graph generation from the syntactic and dependency parse of natural language text and rules for generating summary from the semantic graph. In section 4 we describe a case study for proposed approach on a simple news story.

2 Earlier Approaches and Their Limitations

Previous work to construct an object–oriented graph of a text document has been done for requirement analysis phase in software development lifecycle. Initially Abbott[15] gave a set of rules to identify objects/classes from nouns and relations from verbs. Later researches presented semiautomatic systems to model requirements written in a controlled natural language (CNL). In Elbendaks [16] work UML diagrams of the requirement specification documents are constructed by analysing part of speech tags and phrasal parse structures of the sentences and by combining user inputs. Vidhu balas work[17]was inspired by Elbendaks work [16], and is closely related to our proposed work. Their system works by analysing the dependency relations between words. But most of the research has been done for functional requirement specifications. We propose to generate O-O semantic graph for any unrestrained natural language text. Natural language text is an amalgamation of functional and non-functional text. Functional requirement poses a restriction that every sentence should describe some action. It doesnt include any sentence that describes the way system does the work in terms of quality or other measurable requirements. One example of functional requirement is sentence *This system adds value to binary tree and searches for value from binary tree.* Whereas the non-functional requirement sentence about same system is *System should be fast.* Here first we describe the general structure of a sentence and then we discuss the limitations of earlier approaches due to the structural difference of natural language text and functional text. From here onwards to differentiate the object of O-O semantic graph from the object of a verb, we will refer the later by v-object.

2.1   Sentence Structure

Sentences are made up of noun phrases (NP) and verb phrases (VP). Noun phrase consists of a head noun and optional pre or post modifier phrases. These modifier phrases are adjective phrases, participle phrases, prepositional phrases or subordinate clauses. Verb phrases consist of a verb group and its complement. Verb group consists of a lexical verb and optional auxiliary modifier verbs. Complement of a verb group can be noun phrases or subordinate clauses. These complements of the verb form the object of verbs; direct v-object, indirect v-object and prepositional v-object. This complementation decides the type of verbs. Verb phrase also includes optional prepositional phrase as modifier, which can be classified into adjunct adverbial, disjunct adverbial or conjunct adverbial [18]. Subordinate clauses have structure similar to sentence. They can be categorized into finite, non-finite and wh-clauses. Finite clauses contains proper forms of verb with tense information, whereas non-finite clauses contain verbs in four forms: (1) bare form (i.e. She made him darn her socks), (2) to-verb form (i.e. He is thought to be hiding in Brazil), (3) passive participle verb form (i.e. the palanquin loaded, we took a rest) and (4) verb-ing form (i.e. Getting up before dawn was not that good). Non-finite clauses has some missing syntactic constituent (mostly subject) and can be determined from main clause. Wh-clauses has connecting words from wh-words (what, where, which, who, when).

2.2   Limitations of Earlier Approaches

In semantic form, we describe the top level noun phrase of sentence as subject and top level verb phrase as predicate. Subject is the main element talked about in the sentence and predicate consists of event described about subject, and the v-objects of event. Modifier is another semantic unit which consists of descriptive words that restrict the described entity to identifiable instances. Previous approaches to model natural language text have focused on functional sentences where all semantic entities subject, v-object and modifiers can be drilled down to one syntactic unit (i.e. a compound noun/noun, verb, adjective) but did not consider cases where subordinate clauses can itself function as subject, v-objects modifiers or predicates. Here we discuss the limitations of earlier approaches in more detail.

Complex Sentence Structure. [18] As we can see from the sentence structure described above, natural language text sentences may contain sentences within them. These are called subordinate clauses and can have one or more syntactic element missing (mostly subject). It performs various functions; they could be subject of the verb, complement of verb or can function as modifiers of nouns or adjectives. In previous approaches dependency relations to clauses are not considered for subject/v-object extraction or extraction of properties for object/relation. Rules differ to identify subject from finite (that) and non-finite clauses complement. New rules to identify subject/v-object/predicate from these have been described in next section.

Modal Sentences. A modal sentence expresses the possibility of some event depending on certain conditions. This is expressed by additional auxiliary verbs along with the main lexical verb in the verb group[19]. In functional text, since every sentence describes some action, it either doesnt have modal verb or it is generally must or should.

Negation of Verbs. It is common to have sentences with negation of verbs. This information has been ignored in previous work to model requirement specifications.

All References Cannot Be Resolved to One Word or Entity Name. In natural language it is common to refer to the same entity by different references which are not pronominal but could be another name, type or adjectival clauses. In this case none of the reference can be replaced by other reference completely, as it will lead to information loss. But identifying co-reference relation between them can help in connecting the corresponding objects in the Object-oriented semantic graph.

Assumes Understanding of the Semantic Knowledge. Entities may have relation to each other which are not explicit in the text. These could be ontological relations which can be acquired from knowledge-base. Ex. Falcon and the bird can be two references to same entity.

Properties of Nouns from Post Modifying Phrases. In addition to adjectives and clauses, prepositional phrases can also modify nouns. Although we havent done work to resolve this issue in our work, it can be one good area to explore.

In the next section we describe the proposed rules to fulfil the gaps discussed in this section.

3   Proposed Rules for Graph Generation and Summary Generation

We propose the enhanced rules for O-O semantic graph design to bridge the gap left by earlier approaches to model natural language text. We propose to include related NLP tasks named entity recognition (NER), co-reference resolution and semantic mapping in this process. NER task can significantly improve the object identification process and can provide good scope for class identification of objects i.e. Person, Location. Here we discuss the rules used in this paper to construct the O-O semantic graph and then rules to generate summary from this graph. Rules taken from previous approaches have been cited with references.

3.1   Rules to Generate Object–Oriented Semantic Graphs

We have designed rules to construct the object oriented semantic graph (O-O semantic graph) from natural language text. O-O semantic graph consists of objects, relations between objects and properties of objects. We introduced a new feature, property of relation between objects. It is different from the previous approaches and is necessary for natural language text, because the relationships between objects may depend on certain conditions. Property of a relation accommodates the information about verbs (i.e. negation, modality). In earlier approaches, verbs are considered to be used in only affirmative sense, but modal sentences as discussed in section 2, adds different possibility of occurrence to the verb. In this paper new rules have been formed in addition to these rules, by analysing text and their manual summaries. These are described here:

Identification of Subjects, Relations and Properties from Clauses. In a sentence a verb has two prime arguments subject and v-object which are expressed by dependency relations nsubj, dobj or in passive sentence by dependency relations nsubjpass, agent. Dependency relations between two words describe the functional dependency of one word (dependent) on the other word (Governor) to make a complete meaning from the phrase i.e. nsubj(X,Y) represents Y is the subject of X and dobj(X,Z) represents Z is a direct v-object of X. Prepositional v-object to verb are denoted by dependency relation pobj. These dependency relations have been utilised in earlier research to generate an object oriented graph from functional requirements. But the differences between natural language text and functional requirement text, as explained in section 2.1, demands new rules to identify subject and additional information from clauses.

In open clausal complement of a verb phrase (Non-finite clause), subject is not present. We determine it to be same as the subject of that verb (V1) which is in xcomp relation with the verb (V2) in open clausal complement. It is mostly identified by parser with dependency relation xsubj. After identification of subject for V2, if V1 doesnt have separate v-object we combine V1 and V2 to form single relation. Ex. Tom likes to eat fish gives triple $Tom \rightarrow likes\_to\_eat \rightarrow$ fish. When in clausal complement, subject is internal we add V1 and its subject as property of relation V2. Ex. *I asked John to comment on this.* In this case *John* in the internal subject of clause *John to comment on this* and we add property '*I asked*' to relation '*comment*'. If *xcomp* relation connects verb V1 to non-verb word W2, then W2 becomes *v-object* of V1. It helps in preserving the logical cohesion information between clauses in the graph. Similar rules are applied for finite clauses ex. *I believed that they burnt the kebabs.* When there is a participle modifier of the noun phrase connected by *partmod* dependency relation, then we use it to derive $subject \rightarrow verb \rightarrow v-object$ triple. Example: *Truffles picked during the spring are tasty* generates triple: $Truffles \rightarrow picked\_during \rightarrow spring$. Clausal complement of verbs and adjectives, gives additional information about that verb or adjective. This information can be added to relations. Basically it will add value to property of that relation.

Rules to Identify Objects of the Graph. Rule 3.1.2.1. All named entities which are location, name of a person, name of an organization are made objects in the graph. This can be identified by Named entity recognizer tool. This is a new rule to identify objects.

Rule 3.1.2.2. All nouns which act as a subject of verb, in the triple *subject →
verb → v − object* are taken as object [17]. Pronouns are resolved to referring
nouns using co-reference resolver.

Rule 3.1.2.3. All nouns which are direct v-objects of verbs or are prepositional
v-objects of verbs may be objects in the graph if they act as subject of some
verb. [17] We modify this rule to include frequent v-objects of verbs even if they
are not subject of any verb. Frequency of v-object is determined by comparing
the count of its occurrence in all possible triples to a predefined threshold value.

Generally subject of verbs are considered worthy of being projected as an
object in the O-O semantic graph, but here we modify this rule to adjust the
story writing style of descriptive texts. In descriptive text a lot of information is
written about an entity (i.e. a person) that may never be subject of any verb (i.e.
performer of any action). But frequency of its occurrence indicates that entity
to be important enough to be projected as an object.

Rules to Identify Relation between Objects. Relation between objects
can be identified from these rules.

Rule 3.1.3.1. If two objects are in subject-¿verb-¿v-object triple from previous
identification of subjects and v-objects then this verb becomes a connecting
relation between them [15].

Rule 3.1.3.2. If two objects are in subject-¿verb-¿propositionally connected
word triple then this verb with the prepositional post modifier becomes a connecting relation [17].

Rule 3.1.3.2. As described in section 2.2 sometimes same entity may be referred
by two different words, which are not pronominal references. In this case if these
connections are identifiable from co-reference resolver then it forms a relation
between two indirect references to the same entity. We generally merge them
into one object and make one the property of other. It is a new rule added here.
An example for this has been shown in the case study done in this paper.

Rule 3.1.3.3. We propose to use ontologies (i.e. wordNet) to identify existence
of a hyponymy or synonymy relations between objects in the O-O semantic
graph. This is a new rule added here to fulfil the gaps stated in section 2.2.

Rules to Identify Properties of Objects. Object has certain behavioural
properties, which can be identified from connecting modifier words.

Rule 3.1.4.1. Adjectives represent additional information provided about the
object, which can differentiate this object from other objects of same class. All
adjectives are added as property of the corresponding objects [17].

Rule 3.1.4.2. Apostrophe gives additional information when it connects two
nouns i.e. possession.

Rule 3.1.4.3. Prepositional phrases which are connected to nouns works as post
modifiers as described in 2.1. We extract information about object from it.

Rule 3.1.4.4. If one noun acts as subject of other noun in dependency relation, then the other noun is considered as a property of the subject noun. Ex. in Mr. Clinton was the president of United States, president is the property of Clinton.

Rules to Identify Operations Performed on/by Object. Rule 3.1.5. All verbs connecting a object to a non-object nouns are considered as some event done by that object. They are sequenced according to their occurrence in the text, as an operation performed on/by this object[17].

Rules to Identify Properties of Relations. This is a novel attribute.

Rule 3.1.6.1. As stated earlier, natural language contains modal sentences, which expresses the possibility of event in different ways. Here we extract the modal information of verb by identifying word with dependency relation *aux* to main verb based on its part of speech tag *MD*, and attach it as a property to the relation.

Rule 3.1.6.2. If there exists a clausal dependency between a verb and other verb, then we add that as property to the relation derived from that sentence.

3.2   Rules to Construct Summaries from O-O Semantic Graph

A summary of a text document should present most important information of the document while coverage of the document text and diversification of summary content should be maximum. We have analysed a few news stories and their highlights provided by the editor in popular news site CNN. In this analysis we have observed that frequent named entities and most connected concepts/entities are included in the highlights. Most connected entity differs from frequent entity in the sense that although it may not be frequently mentioned, enough description about the entity is provided in the text by different references to it by other elements in the text. These named entities take the form of objects in object-oriented graph, along with other nouns which are subject of any action verb. So we can state that when generating summary from object-oriented graph, most connected objects should be ranked higher and given preference for inclusion in summary. Most connected objects can be identified by PageRank, one of the graph raking algorithms. The rank of an object Obj_i , $Rank(Obj_i)$, is calculated by combining its *strength* and *pageRank* score.

$$Strength(Obj_i) = Count(properties \in Obj_i) + Count(operations \in Obj_i) \quad (1)$$

$$Rank(Obj_i) = PageRank(Obj_i) + Strength(Obj_i)/(\max_i)Strength(Obj_i) \quad (2)$$

In eq.(2) strength is divided by maximum strength to bring both *PageRank* and *Strength* to scale of 0 to 1. Eq.(2) describes that an object with high degree of connections and with more properties and more operations is ranked high. After ordering the objects in descending order of rank, we identify their important properties, operations and relations to be included in the summary. We rank relations, operations and properties to generate summary at different abstraction levels.

Abstraction Level 1. Find the relation R_{ij} that lies in shortest distance path of length 1 between high ranking objects Obj_i and Obj_j. Construct a natural language sentence S_i by connecting objects Obj_i and Obj_j by this relation. Score of the sentences S_i is:

$$Score(S_i) = Rank(Obj_i) + Rank(Obj_j) + (Popularity(R_{ij}))/(\max_m)Popularity(R_m)$$

(3)

In Eq. (3), $Popularity(R_{ij})$ is the frequency count of relation R_{ij}. Include higher scoring sentences in the abstraction level 1 summary.

Abstraction Level 2. For all the objects present in abstraction level 1, calculate popularity of properties and operations by frequency count and conceptual similarity. Then start including properties and operations of objects to generate the summary at abstraction level 2.

Next section presents a case study based on the rules described in this section.

4 Case Studies

In this section we present a case study to construct object oriented semantic graph from a news story and to generate its summary from the graph.

4.1 Construction of the Object-Oriented Semantic Graph

Pre-processing of the Text (Co-reference Resolved and POS Tagged). First we pre-process the original text to extract a list of all named entities and identify co-references to replace pronominal references with the named entities or nouns. We maintain a list of all mentions and their corresponding head mentions to be used later for merging objects. We use Stanford CoreNLP library[20] for preprocessing. Pre-processed text is shown below with replaced co-references in bold-italics; part of speech tag added after every word and identified named entities(only Location,Person or Organization) are underlined.

*Serbian/NNP   construction/NN   workers/NNS   were/VBD   digging/NN when/WRB **Serbian/NNP construction/NN workers/NNS** found/VBD something/NN shocking/JJ ./. **Bomb/NN** was/VBD an/DT unexploded/JJ bomb/NN from/IN World/NNP   War/NNP   II/NNP   ./. Experts/NNS say/VBP   that/IN   the/DT   one-ton/JJ   bomb/NN   was/VBD   made/VBN in/IN Germany/NNP ./. **The/DT one-ton/JJ bomb/NN** contains/VBZ 620/JJ kilograms/NNS of/IN explosives/NNS ./. Local/JJ residents/NNS*

Table 1. Dependency relations between the words of pre-processes text

Head	Dependency relation	Dependent	Head	Dependency relation	Dependent
Digging	Subject	Serbian Construction Workers	Made	Prep_in	Germany
Found	Subject	Serbian Construction Workers	Contains	Subject	Bomb
Shocking	Subject	Something	Contains	Object	Kilograms
Found	Xcomp	Shocking	Kilograms	Prep_of	Explosives
Bomb	Subject	Bomb	Residents	Amod	Local
Bomb	Advmod	Unexploded	Evacuated	Nsubjpass	Residents
Bomb	Prep_from	World_war II	Evacuated	Prep_from	Area
Say	Subject	Experts	Transferred	Nsubjpass	Bomb
Bomb	Amod	One-ton	Evacuated	Conj_and	Transferred
Made	Nsubjpass	Bomb	Base	Amod	Military
Say	Ccomp	Made	Transferred	Prep_to	Base
			Bomb	Passive sub-ject	Will be de-stroyed

were/VBD evacuated/VBN from/IN the/DT area/NN and/CC the/DT bomb/NN was/VBD transferred/VBN to/TO a/DT military/JJ base/NN where/WRB **the/DT bomb/NN** *will/MD be/VB destroyed /VBN safely/RB ./.*

Dependency Relation between Words to Be Used in Graph Generation. We parse the pre-processed text using Stanford dependency parser to identify the dependency relations between words which are shown in table 1. After dependency parsing, we identify objects, relations, properties and operations from these dependency relations given in table 1, by applying rules of section 3.1.

Identified Objects, Relations, Properties, Operations and Properties of Relations. Following objects are identified by applications of rules given in brackets, on the dependency relations shown in table 1 and on the pre-processed data shown in 4.1.1.

(i) Serbian construction workers [Rule 3.1.2.2] , (ii) Bomb [Rule 3.1.2.2] , (iii), Germany [Rule 3.1.2.1] and (iv) Residents [Rule 3.1.2.2]

Following relations are identified by applications of rules shown in bracket.

(i) *Bomb → Made_in → Germany* [Subject→verb→Prepositional v-object] [Rule 3.1.3.2] (ii) *Serbian construction workers → found → something* [Subject→verb→ xcomp's head(when xcomp connects verb to no-verb word)] [Rule 3.1.1]

Following properties are identified by applications of rules shown in bracket.

(i) *Bomb: Something Shocking* [Rule 3.1.3.3] (ii) *Bomb: Unexploded* [Rule 3.1.4.1] (iii) *Bomb: From world_war II* [Rule 3.1.4.3]

Following operations are identified by applications of rule described in 3.1.5.

(i) *Bomb: Contains (620 Kilograms of Explosives)* [Subject → verb → v-object] (ii) *Bomb: will be destroyed( )* [Passive subject which has no agent → verb]

(iii) *Bomb:transferred_to(Military Base)*[Passive subject (which has no agent) → verb → Prepositional v-object] (iv) *Residents: Evacuated_from(Area)* [Passive subject(no agent)→ verb → Prepositional v-object]

We have the triple identified as *Experts → say → Made_in (Subject → Verb → Verb's clausal complement, which is again a verb)*. From rule 3.1.6.2 *Expert → say*, will be added as property of already identified relation *Made_in* .

Post-processing. In post-processing step we identify more possible connection between objects. Initially through co-reference resolver we have created a list of all words and their possible corresponding head mentions.We see that clause *Digging when they found something shocking* in sentence 1 refers to same entity which is also referenced by *Unexploded bomb from World War II* in sentence 2 and thus share same head mention number. While going through all 5 objects identified in 4.1.3 we see that *something* and *bomb* refers to same mention. Object *Bomb* having more number of properties and operations becomes the primary object and the secondary object *something* is merged into the primary object. It generates the following relation: *Serbian construction workers → found → bomb* [Rule 3.1.3.2] The constructed O-O semantic graph has been shown in Fig. 1. The automated O-O semantic graph generated from our implementation of proposed methodology is shown in fig. 2. We have shown another automated graph generated along with the text in fig. 3.

Summary Generation. After the construction of the graph, we have applied the proposed approach discussed in section 3.2 to generate a summary from this graph. First we will identify the important object by calculating the rank of objects by eq.(2). According to eq.(2) Bomb is most important Object, with 2 connections to other Objects and most number of properties and operations. After that Objects *Serbian construction workers* and *Germany* has same number of connections, with *Serbian construction workers* having 1 property. So the identified important objects in descending order are: 1. *Bomb*, 2.*Serbian construction worker*, 3. *Germany*

Identifying Important Operations and Properties of Important Objects. For most important object *Bomb* we cannot decide ranking based on popularity of operation as all operations are mentioned only once for this object, but we can decide conceptual similarity of operations *explosion* and *destruction* to *Bomb* and can determine that *contains (explosives)* and *will be destroyed in (military base)* are conceptually related to object *Bomb*, thus ranks high.

Ranking of properties is similar to ranking of operations, thus conceptually related property *unexploded* will be included in important properties. We generate the summary at abstraction level1 from O-O graph by connecting important objects by the relations that lies in the shortest distance traversal path of length 1. Here we generate *Serbian construction workers found bomb. Bomb was made in Germany, expert say*. After adding important properties identified previously

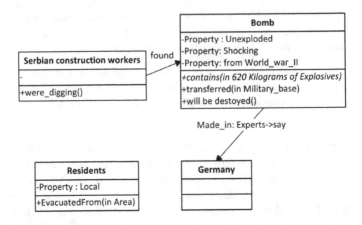

Fig. 1. Object-oriented semantic graph generated for text 1

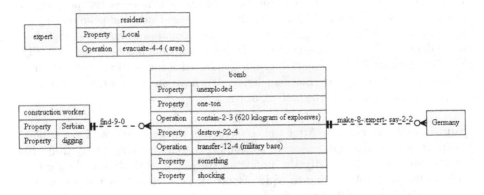

Fig. 2. Automated Object-oriented semantic graph generated for text 1

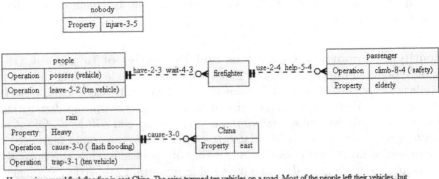

Heavy rains caused flash flooding in east China. The rains trapped ten vehicles on a road. Most of the people left their vehicles, but passengers on a bus were elderly and children. They had to wait for firefighters. Firefighters used ladders to help the passengers climb to safety. Nobody was injured.

Fig. 3. Automated Object-oriented semantic graph generated for text 2

with the first appearance of object in text and its important operations, the summary at abstraction level 2 is: *Serbian construction workers found unexploded bomb. Bomb was made in Germany, expert say. Bomb contains 620 Kilograms of explosives; Bomb will be destroyed.* Different abstraction levels can be achieved for summary depending on the required summary length.

5   Evaluation

We have automated the graph generation process. By manually comparing the automated graph with the proposed graph we can see that it is a realizable methodology.

To evaluate the effectiveness of the proposed summarisation system, we have compared the summary generated at abstraction level 2 from O-O semantic graph with the summary generated from dense semantic graph [2]. The three high ranking summary sentences taken from dense semantic graph based text summarisation . In [21]Jensen-Shannon divergence has resulted as an effective measure for summary evaluation. We got similarity score 0.926 for O-O semantic graph based summary and 0.923 for the dense semantic graph based summary. Approximately equal similarity scores over probabilistic topic distribution indicate both summaries cover the topics equally. Next we evaluated diversity of the summary which is inversely proportional to the redundancy of the information. To calculate redundancy we used wordNet based semantic similarity measure to calculate the similarity of sentences in each system generated summary. We calculated normalised scores of the sum of the maximum semantic similarity of a sentence with any other sentence in the same summary using SEMILAR library[22]. Redundancy score for the summary generated from O-O semantic graph is 2.34 and from dense semantic graph is 2.8. It indicates that summary generated from O-O semantic graph is more diverse than generated from dense semantic graph. This simple evaluation shows that O-O semantic graph benefits text summarisation in terms of diversification of information.

6   Conclusion and Future Works

The object-oriented semantic graph proposed in this paper provides an efficient structure to represent information present in a text document. It can be pruned to generate abstractions of the original information present in the document. This semantic graph can also be used as a basic structure in other natural language applications i.e. Question/answering. We have automated the graph generation process and it has been developed by utilising the available resources for natural language processing. Future research will be more focussed on automation of ranking of objects and properties to generate automatic summary from the object-oriented semantic graph. Overall this approach provides a new direction to text summarisation and can provide good results with complete automation of the process.

References

1. Luhn, H.P.: The automatic creation of literature abstracts. IBM J. Res. Dev. 2, 159–165 (1958)
2. Edmundson, H.P.: New methods in automatic extracting. J. ACM 16, 264–285 (1969)
3. Barzilay, R., Elhadad, M.: Using lexical chains for text summarization. In: Proceedings of the ACL Workshop on Intelligent Scalable Text Summarization, pp. 10–17 (1997)
4. Nomoto, T., Matsumoto, Y.: A new approach to unsupervised text summarization. In: Proceedings of the 24th Annual International ACM SIGIR Conference on Research and Development in Information Retrieval, SIGIR 2001, pp. 26–34. ACM, New York (2001)
5. Svore, K.M.: Enhancing single-document summarization by combining ranknet and third-party sources. In: Proceedings of the 2007 Joint Conference on Empirical Methods in Natural Language Processing and Computational Natural Language Learning (EMNLP-CoNLL). Association for Computational Linguistics (2007)
6. Dunlavy, D.M., O'Leary, D.P., Conroy, J.M., Schlesinger, J.D.: Qcs: A system for querying, clustering and summarizing documents. Inf. Process. Manage. 43, 1588–1605 (2007)
7. Wong, K.-F., Wu, M., Li, W.: Extractive summarization using supervised and semi-supervised learning. In: Proceedings of the 22nd International Conference on Computational Linguistics, COLING 2008, pp. 985–992. Association for Computational Linguistics, Stroudsburg (2008)
8. Erkan, G., Radev, D.R.: Lexrank: Graph-based lexical centrality as salience in text summarization. J. Artif. Int. Res. 22, 457–479 (2004)
9. Tsatsaronis, G., Varlamis, I., Nørvåg, K.: Semanticrank: ranking keywords and sentences using semantic graphs. In: Proceedings of the 23rd International Conference on Computational Linguistics, pp. 1074–1082. Association for Computational Linguistics (2010)
10. Mihalcea, R., Tarau, P.: Textrank: Bringing order into texts. Association for Computational Linguistics (2004)
11. Yeh, J.-Y., Ke, H.-R., Yang, W.-P.: ispreadrank: Ranking sentences for extraction-based summarization using feature weight propagation in the sentence similarity network. Expert Systems with Applications 35(3), 1451–1462 (2008)
12. Wan, X., Xiao, J.: Exploiting neighborhood knowledge for single document summarization and keyphrase extraction. ACM Transactions on Information Systems (TOIS) 28(2), 8 (2010)
13. Joshi, M., Wang, H., McClean, S.I.: Dense semantic graph and its application in single document summarisation. In: Proceedings of the 7th International Workshop on Information Filtering and Retrieval co-located with the 13th Conference of the Italian Association for Artificial Intelligence (AI*IA 2013), Turin, Italy, December 6, pp. 25–36 (2013)
14. Leskovec, J., Milic-Frayling, N., Grobelnik, M., Leskovec, J.: Extracting summary sentences based on the document semantic graph. Microsoft Research, Microsoft Corporation (2005)
15. Abbott, R.J.: Program design by informal english descriptions. Communications of the ACM 26(11), 882–894 (1983)
16. Elbendak, M., Vickers, P., Rossiter, N.: Parsed use case descriptions as a basis for object-oriented class model generation. Journal of Systems and Software 84(7), 1209–1223 (2011)

17. Vidya Sagar, V.B.R., Abirami, S.: Conceptual modeling of natural language functional requirements. Journal of Systems and Software 88, 25–41 (2014)
18. Burton, R.N.: Analysing sentences: An introduction to english syntax (1986)
19. Klinge, A.: The english modal auxiliaries: from lexical semantics to utterance interpretation. Journal of Linguistics 29, 315–357 (1993)
20. Manning, C.D., Surdeanu, M., Bauer, J., Finkel, J., Bethard, S.J., McClosky, D.: The Stanford CoreNLP natural language processing toolkit. In: Proceedings of 52nd Annual Meeting of the Association for Computational Linguistics: System Demonstrations, pp. 55–60 (2014)
21. Louis, A., Nenkova, A.: Automatic summary evaluation without human models. In: Notebook Papers and Results, Text Analysis Conference (TAC 2008), Gaithersburg, Maryland (USA) (2008)
22. Rus, V., Lintean, M.C., Banjade, R., Niraula, N.B., Stefanescu, D.: Semilar: The semantic similarity toolkit. In: ACL (Conference System Demonstrations), pp. 163–168. Citeseer (2013)

Ontology-Based Information Extraction from the Configuration Command Line of Network Routers

Anny Martínez[1], Marcelo Yannuzzi[1], René Serral-Gracià[1],
and Wilson Ramírez[2]

[1] Networking and Information Technology Lab (NetIT Lab)
[2] Advanced Network Architectures Lab (CRAAX)
Technical University of Catalonia (UPC), Barcelona, Spain
{annym,yannuzzi,rserral,wramirez}@ac.upc.edu

Abstract. Knowledge extraction is increasingly attracting the attention of researchers from different disciplines, as a means to automate complex tasks that rely on bulk textual resources. However, the configuration of many devices in the networking field continues to be a labor intensive task, based on the human interpretation and manual entry of commands through a text-based user interface. Typically, these Command-Line Interfaces (CLIs) are both device and vendor-specific, and thus, commands differ syntactically and semantically for each configuration space. Because of this heterogeneity, CLIs always provide a *"help"* feature—i.e., short command descriptions encoded in natural language—aimed to unveil the semantics of configuration commands for network administrators. In this paper, we exploit this feature with the aim of automating the abstraction of device configurations in heterogeneous settings. In particular, we introduce an Ontology-Based Information Extraction (OBIE) system from the Command-Line Interface of network routers. We also present ORCONF, a domain Ontology for the Router CONFiguration domain, and introduce a semantic relatedness measure that quantifies the degree of interrelation among candidate concepts. The results obtained over the configuration spaces of two widely used network routers demonstrate that this is a promising line of research, with overall percentages of precision and recall of 93%, and 91%, respectively.

Keywords: Ontology-Based Information Extraction, Knowledge Discovery, Routers, Configuration Management, Data Mining, Semantic Relatedness.

1 Introduction

One of the most complex and essential tasks in network management is router configuration. Overall, router configuration encompasses a large number of functions, which include, setting routing protocols, security filters, interface parameters, forwarding rules, QoS policies, etc. Currently, network administrators rely on Command-Line Interfaces (CLI), as the means for configuring their

R. Prasath et al. (Eds.): MIKE 2014, LNAI 8891, pp. 312–322, 2014.

routers [1]. This is typically the case of administrators operating in Internet Service Providers (ISPs), data centers, corporate networks, public administrations, etc. Due to the lack of a better alternative, administrators must deal with the complexities associated with this practice, since CLIs are generally device and vendor-specific. This is further exacerbated by the heterogeneity of today's network infrastructures, as networks are typically composed of routers from different vendors. Accordingly, network administrators are forced to gain specialized expertise for a wide range of devices, and to continuously keep knowledge and skills up to date as new features, operating system upgrades, or technologies appear in the routing market. In light of all this, router misconfiguration is a common event, which often leads to serious service disruptions [2].

From reviewing the literature, it becomes evident that, over the last few years, academic and industrial communities have devoted considerable efforts to overcome the inherent complexities of the so-mentioned CLIs or the cumbersome and rarely used configuration means provided by the Simple Network Management Protocol (SNMP) [3]. Clearly, the need for standards has always been among the best interests of the Internet community, since they are essential to achieve interoperability at a global level. In light of this, the NETCONF protocol [4] emerged as an attempt for standardizing router configurations. However, the lack of a common data model, along with its slow and scarce adoption, have made it fail as the preferred mechanism for router configuration [5]. Some industrial initiatives have also attempted to provide uniform configuration means, by developing and maintaining dedicated software agents. Unfortunately, this approach demands serious development efforts, which are neither scalable nor easy to maintain under the dynamics of current networking environments.

In light of this, there is a need to explore other fields that can help pave the way toward seamless network device configuration. The use of ontologies and other semantic techniques has been considered in the past for achieving interoperability in the network management domain [6–11]. However, many of these works target interoperability at the network-level, but they do not approach the underlying interoperability issue at the element-level, i.e., device configuration management. The work in [9] is the most closely aligned to our research goals. Their main contribution is a semantic similarity function that enables mappings between ontologies of different router configuration domains. In that work, ontologies for each available device are assumed to be given in advance, but the task of building an ontology is quite challenging and sufficiently complex by itself. Unless vendors start providing the ontologies—which formally define the knowledge of their own configuration space—and the corresponding plugins, scalability is a major concern for a solution of this nature. Our approach is considerably different since it aims to be independent of device-specific knowledge models (i.e., ontologies) and instead, we automatically build these models from the information natively provided by network vendor's in their CLIs.

Despite numerous efforts, there is still a long research path to follow in order to significantly benefit from the field of ontologies and semantic technologies. It is worth highlighting that, the exploitation of knowledge from natural language resources has been the focus of several research initiatives in numerous domains, but—to the best of our knowledge—researchers in the networking field have not explored this path yet. In this paper, we follow this path, by targeting a semantic-based approach that can automatically build specialized forms of knowledge from the information already available in routers' CLI.

The remainder of this paper is organized as follows. Section 2 introduces our semantic approach toward the abstraction of device configuration, and describes the architecture and fundamentals of our OBIE System. Section 3 presents the experimental results carried out over the configuration spaces of two different routers that are widely used in the networking community. Finally, Section 4 concludes the paper.

2   Approach

In this section, we describe an Ontology-Based Information Extraction system for the router configuration domain. Our central hypothesis is that configuration knowledge can be acquired from CLIs, by exploiting two distinctive features, namely, *(i)* the information provided in *"help"* descriptors, and *(ii)* the knowledge inherent to the hierarchical arrangement of commands. In light of this, we developed both a router domain ontology, and an ontology-based learning approach for information extraction (IE) from CLIs. Note that, herein, we focus on the router configuration domain, but the fundamentals of our approach can be extended to other domains wherein configurations rely on the utilization of CLIs (e.g., medical equipment, printer stations, etc).

The general architecture of our OBIE system is depicted in Fig. 1. Notice that, there are two inputs, namely, *(i)* the routers' CLI—as natively provided by

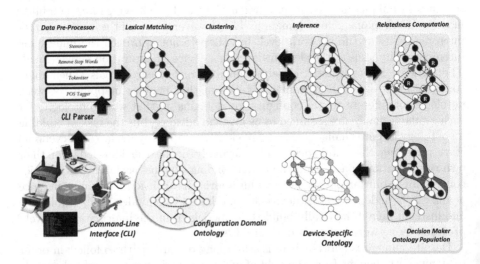

Fig. 1. General architecture of our OBIE system for device command instantiation from Command-Line Interfaces (CLIs)

device vendors—and *(ii)* the router configuration domain ontology. The output of our system is a device-specific ontology which is the result of populating the domain-ontology with instances of commands and variables for a particular device. In other words, a device-specific ontology provides the configuration knowledge (commands and variables) for a concrete network element. These device-specific ontologies are further stored in a database to enable future potential applications (e.g., outsource of configuration tasks to third-party systems). We shall first describe the ontology modeling phase, and then provide details on our OBIE process, which comprises a multi-stage algorithm for the semantic-based instantiation of commands.

2.1 ORCONF: The Ontology for the Router Configuration Domain

We developed and implemented ORCONF, the **O**ntology for the **R**outer **CONF**iguration domain, motivated by the fact that the knowledge expressed in routers' CLIs—disregarding proprietary technologies—mostly corresponds to protocols and technologies that are broadly accepted by practitioners in the field. On this basis, it is not *"what"* CLIs provide what mostly concerns network administrators, but instead, *"how"* they provide it (i.e., the terminologies and corresponding semantics), a fact which, after all, determines the usability of the CLI.

The ontology modeling stage relies on the knowledge provided by networking experts in addition to that extracted from configuration manuals and textbooks. The design of an ontology is closely related to the ultimate use of the model. In our approach, ORCONF constitutes a valuable resource as we make use of this knowledge to guide the extraction of configuration information from the CLI. In light of this, we have defined two distinct layers (cf., Fig. 2), namely, the router *resource layer* and the router *operation layer*. The former defines all physical and virtual resources of the routing space (e.g., port, interface, domain-name, etc.), and integrates a domain lexicon that will aid in the IE. The latter defines the set of atomic operations (e.g., delete traffic filter, set bandwidth, etc.) that can be performed over resource(s) of the previous layer. With this in mind, concepts within the *operation layer* are semantically associated to concepts in the *resource layer* through the non-hierarchical property "to configure" (cf., Fig. 2). In short, a router *resource* represents a component that can be supplied or consumed in the *configuration*. Both layers of the domain ontology include numerous taxonomies, which are used to indicate all hierarchical relations. The intuition of building the ontology in two separate layers lies on the fundamentals of our learning approach, wherein we first, identify resources and verb phrases from the CLI help—i.e., we identify routing concepts from the *resource layer*—and second, we derive the semantics of the complete sequence of commands from the set of identified resources—i.e., we determine the corresponding configuration action from the *operation layer*. ORCONF was formally defined using the Ontology Web Language (OWL) and built with Protégé. Moreover, we used the Protégé API to access, create and manipulate ontology resources from the OBIE process.

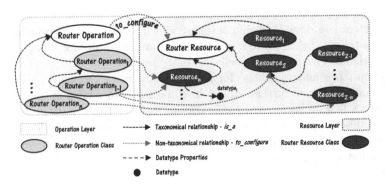

Fig. 2. Layered structure of the Domain-Ontology (ORCONF)—further details including full access to the ontology can be provided after the peer-review process

2.2   OBIE Process from the Command-Line Interface

In this section, we describe our ontology-based approach for IE from CLIs. The final goal is to semantically instantiate CLI commands, and automatically generate device-specific versions of the domain ontology. Next, we will provide details on each stage of our semantic approach (cf., Fig. 1).

CLI Parser: The CLI Parser has a two-fold functionality, *(i)* to identify the structural components of the CLI—i.e., distinguish between commands, variables and help descriptors—and *(ii)* to scan the hierarchy from top to bottom in order to build all executable configuration statements—i.e, valid sequences of commands and variables that combined represent a configuration operation. For example, the sequence: $set \rightarrow system \rightarrow host\_name \rightarrow \langle name \rangle$, represents a valid statement for the configuration of a router's name. Due to the simplicity of CLIs structure, these tasks are somewhat straight-forward, but still key to our solution. Consider the following—in the context of our approach—the information included in "help" descriptors is not target of instantiation. However, they represent valuable sources of information for determining the semantics of *commands* or *variables*—which are ultimately our goal of instantiation. In practice, the syntax and semantics of *commands* and *variables* are prone to customization, as vendors struggle to distinguish from others. For this reason, *help descriptors* are fundamental for setting common foundations, which guide network administrators on the interpretation and use of the CLI. Note that, a command's name per se is neither sufficient nor determinant for extracting the semantics. Even if the help descriptors do not explicitly provide a common reference for disambiguation, the context can aid in the inference of a term's sense—this feature will be further exploited in our learning approach.

Data Pre-processing: This stage combines shallow Natural Language Processing (NLP) tools for basic data preprocessing. The first NLP resource applied to our input stream is the Part-Of-Speech (POS) Tagger, aimed to identify verbs. The motivation for this stems from the following: (i) information given in CLIs is concise, hence, there is no need for in-depth POS analysis; (ii) CLIs lack of verbosity and proper grammar, i.e., descriptions are barely restricted to verbs and resources; (iii) the number of actions (verbs) are finite and well-known in the domain, e.g., "delete", "add", "set", "show", etc. For this reason, verb identification in an early stage can limit the scope of the semantic search. In our implementation, we used the Stanford Log-linear POS Tagger. Next, we separate data into tokens, remove stop words, and reduce inflectional forms of a word to its common base form for further processing. Notice that, in our approach—as done by many search engines—words with the same stem are treated as synonyms to increase the probability of hits when performing lexical matching.

Lexical Matching: We rely on lexical matching for initial identification of routing entities from the CLI. To this end, we automatically build a gazetteer (an entity dictionary) from the knowledge provided in ORCONF. From now on we will consider the ontology as a semantic graph—with nodes representing concepts and edges relations between concepts—the outcome of this stage is a set of "activated" (i.e., highlighted) nodes as for every identified routing entity in the CLI. The notion of "activated" nodes is depicted in Fig. 1. Our strategy admits both partial and exact matching. In the case for which the matching keywords are the same for several gazetteer entries, our approach keeps the most general concepts, i.e., we generalize in the absence of information. Moreover, if a concept is not identified in the stage of lexical matching (not even a generalized form of the same), mainly because of the terminology used by vendors, further stages can infer this knowledge based on context information. For example, if we identify in contiguous levels the concepts $\langle ip-address\rangle$ and $\langle bandwidth\rangle$, we can decide upon activating the $\langle interface\rangle$ concept, based on the notion that, subsequent levels specify properties that derive from this general concept.

Clustering and Inference: In this stage directly related concepts for adjacent levels of the CLI (i.e., contiguous "activated" nodes in the semantic graph) are grouped into clusters (cf., Fig. 3). Notice that, the notion of concepts being part of semantic clusters is based on the premise that, commands are arranged in the hierarchy by association—i.e., commands become specific down in the tree structure. Accordingly, concepts in adjacent levels are expected to be semantically related to a certain extent. Notice that, the degree to which concepts are related depends on the granularity of the CLI—which differs for every configuration space. For this reason, clustering is not sufficient and we require a means to measure the degree of semantic relatedness.

Furthermore, we perform basic semantic inference with a two-fold purpose. First, to reason over equivalent axioms of the ontology, i.e., activate an entity if a set of conditions is fulfilled for a *defined* class and further allocate it within a cluster. For example, if the entity $\langle route\rangle$ was not activated by lexical matching, but still we identify $\langle hop\rangle$, $\langle source\rangle$ and $\langle destination\rangle$ entities, we can infer from the equivalent axioms of our ontology that we are referring to the $\langle route\rangle$ concept. Second, to generalize or specify already "activated" concepts based on context information. For example, if the domain entity $\langle interface\rangle$ was activated for a certain level, and further IE identifies "exclusive" properties of a child concept, we infer that, the most specific concept is most likely to be the asserted entity (e.g., a specific interface type).

Relatedness Computation: The semantic relatedness of a set of clusters $\{\mathcal{C}\}$ is as a means to determine the degree to which candidate entities are associated by meaning. As mentioned earlier, due to the hierarchical nature of CLIs, we assume that maximum interrelated concepts are most likely representative of an executable sequence of commands. To exemplify this, let us consider the example shown in Fig. 3, where the entities "OSPF_Area" $\in \mathcal{C}_B$ and "Local_Area_Network" $\in \mathcal{C}_C$ are two possible candidates for the CLI term "area". Observe that both entities and the clusters to which they belong to are disjoint, as only one ontological class can represent the semantics of the term. From a lexical perspective, the succinctness of the CLI is what generates ambiguity between both concepts. However, based on the contextual background, the entity "OSPF_Area" seems to be a better candidate, as it is semantically related to concepts identified in adjacent levels (i.e., concepts $\in \mathcal{C}_A$). Thus, our relatedness measure must contribute to the problem of word sense disambiguation, i.e., picking the most suitable sense of the word and constraining the potential interpretation of terms in our system. To this end, the relatedness measure $\mathcal{R}$ that we introduce next quantifies the degree to which a set of clusters $\{\mathcal{C}\}$ are semantically related.

Let $G(\mathcal{C}, R)$ be a directed graph, where the vertex $\mathcal{C}$ represents a cluster $\in$ G, and the edge R represents a relationship among two adjacent clusters. Let $G_k \subseteq G$ represent a connected subgraph of G, and $\mathcal{C}_k^i$ be the ith cluster $\in G_k$. As depicted in Fig. 4, the ontological class l within $\mathcal{C}_k^i$ shall be denoted as c_k^{il}. Equation (1) shows the relatedness measure $\mathcal{R}$ used in our model, which consists of two components: *(i)* the connection density $d(G_k)$, and *(ii)* the maximum information content coverage $\mathcal{I}(G_k)$. The density component $d(G_k)$ is shown in (2), and it is basically a measure of the semantic connectivity of graph G_k. It is computed as the relation between the number of "activated" entities along the shortest path between any pair of clusters in graph G_k, and the total number of connections (i.e., the number of relations between entities) in those shortest paths. More specifically, let $\mathcal{C}_k^i$ and $\mathcal{C}_k^j$ be a pair of clusters in G_k, and let $\mathcal{P}(c_k^{il}, c_k^{jp})$ denote a path between a pair of entities $c_k^{il} \in \mathcal{C}_k^i$, and $c_k^{jp} \in \mathcal{C}_k^j$. The

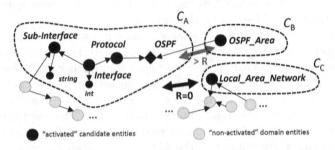

Fig. 3. The rationale behind the quantification of the Semantic Relatedness

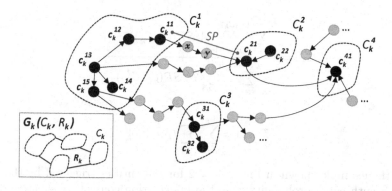

Fig. 4. An example of Semantic Relatedness

shortest path between two clusters is defined as $\mathcal{SP}(\mathcal{C}_k^i, \mathcal{C}_k^j) = \min \mathcal{P}(c_k^{il}, c_k^{jp})$, $\forall$ c_k^{il}, c_k^{jp} in clusters $\mathcal{C}_k^i$, and $\mathcal{C}_k^j$, respectively. To illustrate this, consider the paths between the clusters $\mathcal{C}_k^1$ and $\mathcal{C}_k^2$ as shown in Fig. 4. In this case, the shortest path between any pair of entities (c_k^{1l}, c_k^{2p}), i.e., paths with source in cluster $\mathcal{C}_k^1$ and termination in $\mathcal{C}_k^2$, or vice-versa, is $\mathcal{SP}(\mathcal{C}_k^1, \mathcal{C}_k^2) = [(c_k^{11}, x), (x, y), (y, c_k^{21})]$. Now, let the function $\mathcal{A}(\mathcal{P})$ return the total number of "activated" entities (i.e., the ontological classes) in path $\mathcal{P}$. In our example, $\mathcal{A}(\mathcal{SP}(\mathcal{C}_k^1, \mathcal{C}_k^2)) = 2$, which are c_k^{11} and c_k^{21}. Observe that the source of a path $\mathcal{P}$ is always an "activated" entity—recall that the clusters are composed of activated entities only—hence the number of "active connections" along a path $\mathcal{P}$ is $(\mathcal{A}(\mathcal{P}) - 1)$ (cf. (2)). Similarly, the function $\mathcal{H}(\mathcal{P})$ in the denominator of (2) returns the total number of hops in path $\mathcal{P}$. For instance, in the example shown in Fig. 4, $\mathcal{H}(\mathcal{SP}(\mathcal{C}_k^1, \mathcal{C}_k^2)) = 3$. Observe that when the clusters $\mathcal{C}_k^i$ and $\mathcal{C}_k^j$ are not adjacent, the shortest path will traverse other clusters. Hence, in a connected graph, the number of activated entities always satisfies $\mathcal{A}(\mathcal{SP}(\mathcal{C}_k^i, \mathcal{C}_k^j)) \geq 2$.

The second term of the relatedness measure $\mathcal{R}$ is the Information Content $\mathcal{I}(G_k)$, which is shown in (3). This term represents a measure of the knowledge covered by the entities (i.e., the ontological classes) in their corresponding clusters. Let t_k^{il} denote the number of CLI terms that "triggered" the activation of an entity $c_k^{il} \in \mathcal{C}_k^i$ in the semantic graph G_k. Observe that in (3), the contribution of an entity c_k^{il} to the domain knowledge is weighted by two factors, m_k^{il}, and o_k^{il}. Let, m_k^{il} be the "matching" factor of the lth entity, which is 1 for entities identified by perfect match; otherwise, its value is chosen as $\frac{1}{(e+1)}$, with e the total number of entities identified for the same CLI term. In other words, in case of partial match, the weighting factor m_k^{il} represents the probability of being any of the e entities identified for the same CLI term—including none of them $(+1)$.

$$\max_k \quad \mathcal{R}(G_k) = d(G_k) \cdot \mathcal{I}(G_k) \tag{1}$$

$$d(G_k) = \frac{\sum\limits_{i=1}^{|\mathcal{C}_k|-1} \sum\limits_{j=i+1}^{|\mathcal{C}_k|} [\mathcal{A}(\mathcal{SP}(\mathcal{C}_k^i, \mathcal{C}_k^j)) - 1]}{\sum\limits_{i=1}^{|\mathcal{C}_k|-1} \sum\limits_{j=i+1}^{|\mathcal{C}_k|} \mathcal{H}(\mathcal{SP}(\mathcal{C}_k^i, \mathcal{C}_k^j))} \leq 1 \qquad (2)$$

$$\mathcal{I}(G_k) = \sum\limits_{i=1}^{|\mathcal{C}_k|} \sum\limits_{l=1}^{|c_k^{il}|} t_k^{il}.m_k^{il}.o_k^{il} \qquad (3)$$

In the example shown in Fig. 3, $e = 2$ for the entities triggered by the term "area", with equal probability from the information content perspective of being *"OSPF_Area"*, *"Local_Area_Network"*, or none of them. Moreover, the other weighting factor, i.e., o_k^{il}, is a measure of the "occurrence" of a candidate entity in the CLI, over the total number of occurrences of its exclusive disjoint entities. This measure is computed during the IE process at the lexical matching stage.

Finally, observe that we compute the semantic relatedness $\forall G_k \subseteq G$, that is, over the total number of connected cluster subgraphs of G. As indicated in (1), the relatedness measure that we chose is the maximum obtained $\forall G_k$. It is worth mentioning that, even though at first sight our model might look a bit intricate, its computation is actually quite straightforward. The nature and hierarchical structure of CLIs typically yields a small number of interrelated clusters, and more importantly, the system outlined in Fig. 1 operates in offline mode, so the only and fundamental goal is the accuracy of the OBIE process. Indeed, the results that we present in the next Section confirm the strengths of our model and the approach proposed in this paper. Also observe that, although the ontology and some of the descriptions made in this Section are application-specific—i.e., they consider particular features of CLI environments for routers—the essence of our model can be generalized and applied to other contexts, especially, those that rely on hierarchical CLIs for device configuration.

3   Evaluation

In this section, we present the experimental results of our semantic approach, when carried out over the configuration spaces of two widely used routers. In order to ensure heterogeneity, we have selected both a commercial and an open-source router environment, namely, *Juniper* and *Quagga*. Observe that the number of commands available in a router can be significantly large, but the ones that are commonly used in practice represent a relatively small set. For this reason, our evaluations are centered on the set of features that are widely used in practice. In order to define a relevant set of configuration features, we thoroughly selected dissimilar branches of the CLI hierarchy, in an effort to encompass a broad set of functionalities. We used a set of approximately 150 commands (over 70 configuration statements) including not only protocol or technology-dependent settings but also router-related functions—e.g., administrative configurations. We consider that this a significant set as typical

Table 1. Performance Results of our OBIE Process

	Bag						Per-Level					
	Traditional			Augmented			Traditional			Augmented		
	P	R	F_1	AP	AR	AF_1	P	R	F_1	AP	AR	AF_1
Juniper	89%	88%	88%	92%	92%	**92%**	78%	87%	82%	81%	91%	**86%**
Quagga	91%	88%	89%	94%	91%	**92%**	81%	81%	81%	85%	85%	**85%**
Overall	90%	88%	89%	**93%**	**91%**	92%	80%	84%	82%	83%	87%	**85%**

routers' configurations contain around 100 configuration statements [12], taking into account that many settings are recurrent, e.g., the configuration of interfaces. Moreover, notice that there is no semantic approach following our same line of research, so without comparison possibilities, we can only evaluate the overall performance of our instantiation process.

Typical measures to assess the performance of IE systems are inadequate when using ontologies [13]. For this reason, we rely on the Balanced Distance Metric (BDM) [14]—a cost-based component that measures the degree of correctness according to the ontological structure. Notice that, the BDM per se does not provide a means for evaluation. To this end, we measure the Augmented Precision (AP), Augmented Recall (AR), and Augmented F_1-measure (AF_1), as introduced in [14]. The evaluation results are depicted in Table 1. We report percentage values of traditional and augmented Precision, Recall and F_1-Measure for both configuration spaces separately. Observe that, we have considered two criteria for reporting our results, namely, per-level and "Bag". These criteria are just for evaluation purposes. The former, strictly considers an entity valid if it was identified in the level that it actually corresponds to in the configuration statement, while the latter handles all identified entities regardless of the level to which the entity is attributed. Overall, the results for the "Bag" criteria are better than those reported per-level. This is mainly because the information within a level is not always sufficient to accurately determine its semantics. Instead, the inference stage further derives knowledge from the information obtained from subsequent levels. Moreover, evaluation results are also affected by assertions made by vendor's, which are not strictly aligned to the domain knowledge. The apparent miss-classification of entities due to inconsistencies of this nature are actually a degree of correctness of our learning approach.

In a nutshell, our system achieved an overall augmented F_1-Measure of **92%**. Notice that, augmented measures are in all cases greater than traditional ones, basically because our ontology-based approach has the ability to generalize or specify a concept according to contextual knowledge, and the BDM metric adds a cost-based component to this matched entity.

4   Conclusions

Overall, our experimental results show that the semantic approach proposed in this paper opens a promising line of research in the router configuration domain, and that the knowledge provided by CLIs is a valuable source for Information

Extraction (IE). The results indicate that we achieve remarkable precision and recall for both case studies, but still, there is significant room for improvement.

Acknowledgment. This work was supported in part by the Spanish Ministry of Science and Innovation under contract TEC2012-34682.

References

1. Schonwalder, J., Björklund, M., Shafer, P.: Network configuration management using NETCONF and YANG. IEEE Communications Magazine 48(9), 166–173 (2010)
2. Le, F., Lee, S., Wong, T., Kim, H.S., Newcomb, D.: Detecting Network-Wide and Router-Specific Misconfigurations Through Data Mining. IEEE/ACM Transactions on Networking 17(1), 66–79 (2009)
3. Pras, A., Schonwalder, J., Burgess, M., Festor, O., Perez, G.M., Stadler, R., Stiller, B.: Key Research Challenges in Network Management. IEEE Communications Magazine 45(10), 104–110 (2007)
4. Enns, R., Bjorklund, M., Schoenwaelder, J., Bierman, A.: Network Configuration Protocol (NETCONF). RFC 6241, IETF (June 2011)
5. Martinez, A., Yannuzzi, M., López, V., López, D., Ramírez, W., Serral-Gracia, R., Masip-Bruin, X., Maciejewski, M., Altmann, J.: Network Management Challenges and Trends in Multi-Layer and Multi-Vendor Settings for Carrier-Grade Networks. IEEE Communications Surveys Tutorials (99), 1 (2014)
6. López, J., Villagrá, V., Berrocal, J.: Applying the web ontology language to management information definitions. IEEE Communications Magazine 42(7), 68–74 (2004)
7. Xu, H., Xiao, D.: A Common Ontology-Based Intelligent Configuration Management Model for IP Network Devices. In: First International Conference on Innovative Computing, Information and Control, ICICIC 2006, vol. 1, pp. 385–388 (August 2006)
8. Xu, H., Xiao, D.: Applying semantic web services to automate network management. In: 2nd IEEE Conference on Industrial Electronics and Applications, ICIEA 2007, pp. 461–466 (May 2007)
9. Wong, A., Ray, P., Parameswaran, N., Strassner, J.: Ontology mapping for the interoperability problem in network management. IEEE Journal on Selected Areas in Communications 23(10), 2058–2068 (2005)
10. López de Vergara, J., Guerrero, A., Villagrá, V., Berrocal, J.: Ontology-Based Network Management: Study Cases and Lessons Learned. Journal of Network and Systems Management 17(3), 234–254 (2009)
11. Colace, F., De Santo, M.: A Network Management System Based on Ontology and Slow Intelligence System. International Journal of Smart Home 5(3), 25 (2011)
12. CISCO Sample Configuration, http://www.cisco.com/c/en/us/td/docs/routers/access/1800/1801/software/configuration/guide/scg/sampconf.html (Online; accessed October 2014)
13. Maynard, D., Peters, W., Li, Y.: Metrics for Evaluation of Ontology-based Information. In: WWW 2006 Workshop on Evaluation of Ontologies for the Web (May 2006)
14. Maynard, D., Peters, W., Li, Y.: Evaluating evaluation metrics for ontology-based applications: Infinite reflection. In: LREC. European Language Resources Association (2008)

Using Association Rule Mining
to Find the Effect of Course Selection
on Academic Performance in Computer Science I

Lebogang Mashiloane

School of Computer Science,University of the Witwatersrand,
Johannesburg, South Africa
lebogang.mashiloane@wits.ac.za

Abstract. It is important for first year students in higher educational institutions to get the best advice and information with regards to course selection and registration. During registration students select the courses and number of courses they would like to enroll into. The decisions made during registration are done with the assistance of academics and course coordinators. This study focuses on the first year Computer Science students and their overall academic performance in first year. Computer Science I has Mathematics as a compulsory co-requisite, therefore after selecting Computer Science I, the students have to enroll into Mathematics and then select two additional courses. Can data mining techniques assist in identifying the additional courses that will yield towards the best academic performance? Using a modified version of the CRISP-DM methodology this work applies an Association Rule Mining algorithm to first year Computer Science data from 2006 to 2012. The Apriori algorithm from the WEKA toolkit was used. This algorithm was used to select the best course combinations with Computer Science I and Mathematics I. The results showed a good relationship between Computer Science I and Biology on its own, Biology with Chemistry and Psychology with Economics. Most of the rules that were produced had good accuracy results as well. These results are consistent in related literature with areas such as Bio-informatics combining Biology and Computer Science.

Keywords: Educational Data Mining, Association Rule Mining, Apriori algorithm.

1 Introduction

Registration is one of the most important days in a first year students' academic career. On this day, the student makes decisions about which courses and how many courses they would like to enroll into. These decisions could potentially influence that students' academic performance. At Wits, registration is done at the beginning of the year. Each group of students (by faculty and year of study) get a day or more devoted to them for registration. During registration

R. Prasath et al. (Eds.): MIKE 2014, LNAI 8891, pp. 323–332, 2014.
© Springer International Publishing Switzerland 2014

course coordinators and academics advise and assist the students. Some students are prepared and have done adequate research on the available courses and the courses' prerequisites. While other students have little knowledge of the available courses and the details around registration.

The advice that is given by the course coordinators and academics, present at registration, are based on co-requisites and/or past experience. This makes the advice subjective. Is there a way that the experience of these academics could be supported by results from an investigation of the historical data? Can data mining techniques be used to extract knowledge that could assist with this course selection and influence the best possible outcome? Focusing more specifically at the first year Computer Science students, how do the additional courses they select influence overall performance in first year? Data mining techniques will be used to attempt to identify the courses that, when combined with Computer Science, yield the best possible academic performance for first year students. This study will try to answer the following question:

How strong is the association between the courses selected during registration by Computer Science I students and the overall academic performance?

2   Background

2.1   Computer Science I Class and Registration

All first year students in the Faculty of Science at Wits are required to select atleast four courses to register into. Most courses have recommended and/or compulsory co-requisite courses. Computer Science I has Mathematics I as a compulsory co-requisite and Computational and Applied Mathematics (CAM) as a recommended co-requisite. Therefore all Computer Science I(CS-I) students have to register into Mathematics I. Additionally, most CS-I students register into CAM. There are many courses available for the CS-I students to select from. Science students are allowed to select courses from most of the schools in the university including schools from other faculties. The registration period is during the first month of the academic year and all the students in each year of study are given dates to come register. During registration lecturers and administrators from the different schools are gathered in a hall to advise and register the students into their selected courses. Some courses have pre-requisites and/or requirements of minimum marks for specific high school subjects. This information is available in the Wits prospectus book and provided during the registration process.

Apart from the courses that the CS-I students select, within the Computer Science course itself, these students are required to complete four modules. These modules are namely: Basic Computer Organization (BCO), Data and Data Structures (DDS), Fundamental Algorithmic Concepts (FAC) and Limits of Computation (LOC).

2.2   Association Rule Mining

Association Rule Mining is a data mining technique also known as Relationship Mining. The purpose of this technique is to extract frequent patterns or associations in a database or data set [1]. These patterns come in the form of a rule. The rules consist of an antecedent implying a consequent [1]. Association Rule Mining can be used to find frequent or least frequent item sets [7], this study will look for frequent item sets.

The Apriori algorithm is the most commonly used of the available Association Rule Mining algorithms. A high level description of the steps is shown in Figure 2.2. This study uses the predictive Apriori algorithm from the WEKA toolkit.

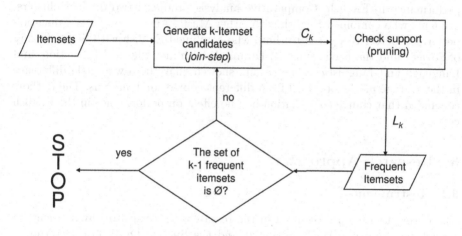

Fig. 1. Apriori Algorithm [5]

2.3   Related Work

Related work has shown ways in which data mining can be used to assist in recommending courses to students. A study by [4] aimed to create a tool that recommends courses to students. The four Apriori algorithms from the WEKA toolkit were used on a dataset which consisted of records from the Department of Information Technology, Computer Science and Engineering. The study found that students interested in Switching Theory & Logic Design and Operating Systems are more inclined towards Data Structures. The recommendation system suggested in this research focuses on suggesting courses based on profiling, therefore using the idea that students with similar profiles will prefer the same courses.

Another study by [6] analyzed data from 2002 to 2008 from the School of System Engineering at Universidad de Lima. This university uses an online registration system, therefore the students can enroll in some or all of the courses in their study plan or curriculum without consultation with any academics. [6] uses classification to create decision trees that would produce rules based on the selection of courses of previous students and their academic performance. The approach taken was using the data to predict the best academic outcome by looking at the course selection using the C4.5 algorithm. Four iterations of the algorithm were done. Conclusions in the paper showed that satisfactory results can be found when trying to use data mining to recommend the number of courses and which courses to take for students. Another benefit that was found is that it also assisted with improving academic performance of the students.

[8] conducted a study at Griffith University, involving 251 students enrolled in courses such as drama, music, etc. Excluded from the study were students whose course combinations could not be identified, including those who did not submit all assignments. As part of of the study the aim was to look into how different course combinations influenced academic performance, more specifically performance in English. Comparative analysis produced two distinct clusters. Students who combined English with either Music, Art or Computer Education performed better compared to those who combined English with SOSE (Study of Society and the Environment), Drama, Health and Physical Education, and Language and Linguistics. The results showed that there was little difference in the average marks obtained from different course combinations. The authors concluded that course combination had no effect on performance in the English course.

3   Research Approach

3.1   Instruments

There were two main tools used in the research: the Waikato Environment for Knowledge Analysis (WEKA) toolkit and the Success Or Failure Determiner (SOFD) tool, which was created specifically for this research. The predictive Apriori algorithm from WEKA was used to analyze the data set and this model was then integrated into the SOFD tool for further analysis and investigation.

3.2   Research Design

The research approach which was selected was the CRISP-DM methodology [3]. A modified version of this methodology is described in the phases presented below:

Data Understanding. The data used in this investigation was from the School of Computer Science at Wits. The data was extracted from the university database for the years 2006 to 2012. The data set consisted of a record for each student with the courses they enrolled into and their final academic outcome. The options for the the academic/progression outcome are shown in Table 1.

Data Processing. The data that was selected for the investigation consisted of all the courses enrolled into by that student and their final progression outcome. There were 564 student records. The data set required no cleansing. During the

Table 1. Overall Result Decision Codes

Decision Code	Meaning
PCD	Proceed
Q	Qualified
RET	return to year of study
CAN	canceled
MBP	Minimum requirements not met. Renewal of registration permitted by Examinations Committee (Proceed)
MBR	Minimum requirements not met. Renewal of registration permitted by Examinations Committee (Return)
MRNM	Minimum requirements not met. Needs permission to re-register
XXXX	Result depends on outcome of deferred examination(s)
****	Result not available
FTC	Failed to qualify

investigation the full data set was initially used then a reduced data set was used. The first was the full data set as extracted from the database. The second was a reduced data set where all the courses with less than 10 Computer Science students enrolled were removed from the full data set. Additionally all the second year courses were removed from the data set. This reduced the records from 564 to 528. Both these files were converted into .arff files which is the preferred WEKA format.

Modelling. The Predictive Apriori algorithm from the WEKA toolkit was applied on both the full data set and the reduced data set. The results from this are presented in Section 4. The models created from applying the Predictive Apriori algorithm to the data sets was then integrated into the SOFD toolkit. This is a GUI which has the WEKA General API embedded into it. This will allow for further investigation.

4   Results and Discussion

4.1   Full Data Set

Data Profiling. The initial data analysis was the profiling of the data set. This revealed the courses that were most preferred by Computer Science I students. These are shown in Figure 2. These courses exclude Mathematics I, which is a compulsory co-requisite of Computer Science I (COMS1000) and Computational and Applied Mathematics, which is a recommended co-requisite of COMS1000. Majority of the COMS1000 students were enrolled into the latter two courses. It is clear from Figure 2 that Physics and Economics are also very popular selections.

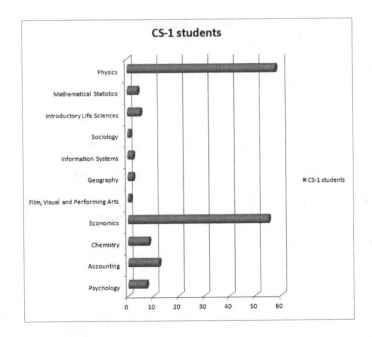

Fig. 2. Common Course Combinations

Analysis of Rules. Table 3 presents the rules produced after applying the Predictive Apriori algorithm on the full data set. All of the top ten rules have a good level of accuracy. From the ten rules, two progression outcomes are presented: 'RET'(which is failing the year) and 'PCD' (which is passing the year). Both these academic outcomes are described in Table 1. In this investigation the 'RET' would be seen as a negative rule and the 'PCD' was seen as the positive rule. 'RET' would result in a student returning to the same year of study. 'PCD' would result in the Computer Science student being allowed to proceed to the second year of study. From the 564 students in the data set: 240 of them obtained the 'PCD' outcome, 128 of them obtained the 'RET' outcome and the rest had the other outcomes mentioned in Table 1. Six of the top ten rules lead to an unsuccessful academic outcome for the Computer Science I students. The course codes presented in Table 3 are explained in Table 2.

Table 2. Course codes and titles

Course code	Course title
COMS1000	Computer Science
ELEN1002	Concepts of Design
INFO1003	Information Systems I
MATH1034	Algebra I
MATH1036	Calculus I
APPM1021	Applied Mathematics for Applied Computing I
PHYS1023	Physics for Applied Computing I

Table 3. Predictive Apriori Phase 1 Results

Attribute	Best Rules Generated	Accuracy
APPM1021	1. {APPM1021=yes, INFO1003=yes} $\Longrightarrow$ RET	0.83330
PSYC1002	2. {PHYS1023=yes, INFO1003=yes} $\Longrightarrow$ RET	0.83330
CHEM1013	3. {INFO1003=yes, ELEN1002=yes} $\Longrightarrow$ RET	0.83330
APPM1006	4. {BIOL1000=yes, CHEM1013=yes} $\Longrightarrow$ PCD	0.83330
ELEN1002	5. {BIOL1000=yes} $\Longrightarrow$ PCD	0.82353
ECON1000	6. {ECON100=yes, PSYC1002=yes} $\Longrightarrow$ PCD	0.79997
ECON2001	7. {ELEN1002=yes} $\Longrightarrow$ RET	0.75004
INFO1003	8. {APPM1006=yes, STAT1003=yes} $\Longrightarrow$ RET	0.75004
BIOL1000	9. {APPM1006=yes, ECON2001=yes} $\Longrightarrow$ PCD	0.74998
STAT1003	10. {APPM1006=yes, INFO1003=yes} $\Longrightarrow$ PCD	0.74998
PHYS1023		

Rule 4 and rule 5 both yield towards a positive academic outcome. Additionally, they both also have BIOL1000 as an antecedent in the rule. This is consistent with related literature which recognizes the link between Computer Science and Biology [2]. The area of Bio-informatics is an integration of Computer Science and Biology and one of the examples that show the association between the two courses. Another noteworthy finding in this result is that the INFO1003 course in three of the 'negative' rules. This is the Information Systems course offered by the Faculty of Commerce, Law and Management at Wits. It is clear from the results that COMS1000 students who select this course as one of their additional courses have a higher probability of failing their first year of study. This is an interesting result because Information Systems and Computer Science are usually branched off together. This result is therefore consistent with industry norms and related literature.

4.2 Reduced Data Set

The full data was then reduced by removing all the courses with less than ten Computer Science student enrollments and all courses which formed part of the second year of study. The new reduced data set is presented in Figure 3. From the reduced data set it is clear that most COMS1000 students enroll into Computational and Applied Mathematics, which is expected. Economics, Physics (Major) and Physics (Auxiliary) also enroll a lot of the COMS1000 students.

The top ten rules extracted from applying the Predictive Apriori algorithm to the reduced data set are shown in Table 4. Most of the resulting rules have good accuracy, however the last rule (tenth rule) has a low accuracy rate. It is also visible that most of the rules are 'positive', with only one rule being 'negative'. The first two rules with the highest accuracy are 'positive' rules with BIOL100/ Biological Sciences as a selection. This reaffirms the findings in the full data set which state that the selection of the Biological Science course by first year Computer Science students will increase the probability of them successfully completing their first year of study. The only other rule which had an

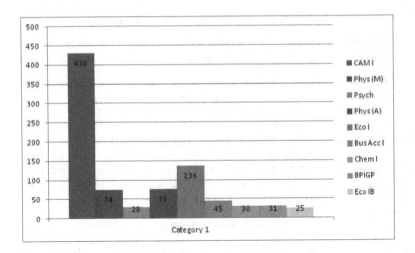

Fig. 3. Courses in Reduced Dataset

Table 4. Predictive Apriori Phase 2 Results

Attribute	Best Rules Generated	Accuracy
APPM1021	1. {BIOL1000=yes, CHEM1013=yes} $\Longrightarrow$ PCD	0.83333
PSYC1002	2. {BIOL1000=yes} $\Longrightarrow$ PCD	0.82353
CHEM1013	3. {ECON1000=yes, PSYC1002=yes} $\Longrightarrow$ PCD	0.80000
APPM1006	4. {ELEN1002=yes} $\Longrightarrow$ RET	0.76471
ELEN1002	5. {APPM1006=yes, STAT1003=yes} $\Longrightarrow$ PCD	0.75005
ECON1000	6. {APPM1021=yes} $\Longrightarrow$ PCD	0.73684
ECON2001	7. {PHYS1023=yes} $\Longrightarrow$ PCD	0.73684
INFO1003	8. {CHEM1013=yes} $\Longrightarrow$ PCD	0.69231
BIOL1000	9. {STAT1003=yes} $\Longrightarrow$ PCD	0.69231
STAT1003	10. {INFO1003=yes, APPM1021=yes, PHYS1023=yes, ELEN1002=yes} $\Longrightarrow$ PCD	0.25000
PHYS1023		

accuracy of 0.8 and higher shows Psychology and Economics as resulting in a successful academic progression outcome. From Figure 3, it is shown that only a few Computer Science I students enroll into Psychology I. This should raise awareness to the course administrators and academics to advise more students to include Psychology I as one of their additional courses with Computer Science I. It is also significant that the only 'negative rule' was the selection of Concepts of Design as an additional course. This will require further investigation.

5   Conclusion and Future Work

During the registration process students select the courses they would like to enroll into. This decision is mainly made by the student but can be influenced by advice from course coordinators and academics present during registration. Assistance from the academics could be based on research findings rather than previous experience and personal opinions. The aim of this investigation was to use data mining and historical data to find the association between course selection and academic performance. Results of applying the Apriori algorithm to historical data showed a relationship between course selection and academic performance. These results could be used to assist academics and course coordinators in preparation for registration.

Courses such as Biology, Economics and Psychology seem to have a better relationship with Computer Science I. Students who select these courses have a higher probability to be successful in their overall first year academic performance. These results are consistent with related literature.

This research showed two main things. Firstly, data mining can be used to find associations between courses and secondly, there are specific courses which when paired with Computer Science I yield successful academic performance. From this result, the SOFD tool can be considered as a recommendation tool for academics in the School of Computer Science. Course recommendation systems are becoming popular in the Educational Data Mining community. Most of the systems do focus on profiling the likes of students from historical data and using that to recommend courses to new students. This work uses academic performance as the key factor in assisting with the recommendation of courses to students. Future work can look at all the courses in Faculty of Science and considering which courses are best taken together to increase the first year pass rate. Although Computer Science I may be best matched with Economics, you may find another course which Economics I students enroll into, which is best matched with Economics. This can be taken further and investigated in other faculties and higher education institutions.

A limitation of this research is in the fact that the data set included students who are either repeating the first year of study or repeating some courses in the first year while enrolled in some second year courses. Although they are in the minority, they still influence the rules found. It is also noteworthy that the aim of this research was to find courses that would be good recommendations for first year students enrolling into Computer Science I, not necessarily to blame certain courses for failure in Computer Science I or first year.

Acknowledgments. I would like to acknowledge my supervisor Mr. Mike Mchunu for his assistance and guidance throughout the research. I would also like to thank Ms Bindu Cherian for assistance in the editing of this paper. And finally my husband, Landi Mashiloane, and family for their prayers and continuous support.

References

1. Kotsiantis, S., Kanellopoulos, D.: Association rules mining: A recent overview. GESTS International Transactions on Computer Science and Engineering 32(1), 71–82 (2006)

2. Nilges, M., Linge, J.: Bioinformatics (2013), http://www.pasteur.fr/recherche/unites/Binfs/definition/bioinformatics_definition.html
3. Shearer, C.: The crisp-dm model: The new blueprint for data mining. Journal of Data Warehousing 5(4), 13–22 (2000)
4. Sunita, B., Lobo, L.: Article: Best combination of machine learning algorithms for course recommendation system in e-learning. International Journal of Computer Applications 41(6), 1–10 (2012); Published by Foundation of Computer Science, New York, USA
5. Vannozzi, G., Della Croce, U., Starita, A., Benvenuti, F., Cappozzo, A.: Journal of neuroengineering and rehabilitation. Journal of Neuroengineering and Rehabilitation 1, 7 (2004)
6. Vialardi, C., Bravo, J., Shafti, L., Ortigosa, A.: Recommendation in higher education using data mining techniques. International Working Group on Educational Data Mining (2009)
7. Zailani, A., Tutut, H., Noraziah, A., Mustafa, M.D.: Mining significant association rules from educational data using critical relative support approach. Procedia - Social and Behavioral Sciences 28, 97 (2011), http://www.sciencedirect.com/science/article/pii/S1877042811024591; World Conference on Educational Technology Researches - 2011
8. Penn-Edwards, S.: They do better than us: A case study of course combinations and their impact on English assessment results. Educating: Weaving Research into Practice (2004)

Information Extraction from Hungarian, English and German CVs for a Career Portal

Richárd Farkas[1], András Dobó[3], Zoltán Kurai[3], István Miklós[1], Ágoston Nagy[1], Veronika Vincze[2], and János Zsibrita[3]

[1] University of Szeged, Institute of Informatics, Szeged, Hungary
{rfarkas,nagyagoston,mikist}@inf.u-szeged.hu
[2] MTA-SZTE Research Group on Artificial Intelligence, Szeged, Hungary
vinczev@inf.u-szeged.hu
[3] Nexum Magyarország kft., Szeged, Hungary
{dobo,zsibrita}@inf.u-szeged.hu, zoltan.kurai@nexum.hu

Abstract. Recruiting employees is a serious issue for many enterprises. We propose here a procedure to automatically analyse uploaded CVs then prefill the application form which can save a considerable amount of time for applicants thus it increases user satisfaction. For this purpose, we shall introduce a high-recall CV parsing system for Hungarian, English and German. We comparatively evaluate two approaches for providing training data to our machine learning machinery and discuss other experiences gained.

Keywords: CV parsing, recruitment process, text mining.

1 Introduction

For large companies it imposes a serious problem to manage the recruitment of new employees. Due to insufficient human resource, it is common practice to select a small proportion of the applicants randomly, and the human resource managers only consider these applicants for the position. Here we propose an application procedure where first the applicant uploads his/her curriculum vitae (CV) which will be automatically analysed and the multi-page application form will be prefilled. Then the user can edit the form and submit it. This approach saves a considerable amount of time for applicants compared to filling out the form from scratch. Hence, it can increase user satisfaction. This procedure also enables the gathering of more detailed (but still manually verified) information about the applicant as applicants are usually reluctant to give detailed information in an empty form while they can upload their well-edited CV in just a second.

To reach this goal, we constructed a method that is able to extract data from the applicants' CVs into a uniform data structure, namely the popular HR-XML format. The extracted data includes the personal data, contact details, education and work history, language skills and many others. This method was first implemented for Hungarian CVs, and was later adapted for the English and German languages too.

R. Prasath et al. (Eds.): MIKE 2014, LNAI 8891, pp. 333–341, 2014.
© Springer International Publishing Switzerland 2014

We had access to a huge amount of manually filled application forms and the uploaded CVs in case of Hungarian and English applications, which raises the opportunity of exploiting this database as a training dataset for our machine learning-based machinery. Besides this training data we also manually annotated several hundreds of CVs to obtain a classical training dataset for information extraction. We comparatively evaluate the two approaches starting with these two kinds of training data.

The chief contributions of this paper are:

- we introduce a machine learning-based CVparser system consisting of several stages,
- we compare two training data gathering approaches,
- we discuss our experiences on language adaptation.

2   Related Work

There have been several commercial applications developed for automatically extracting data from CVs and for ranking CVs based on the extracted data in case of English and some other languages. For instance, [5] ranks CVs on the basis of criteria specified by the end user with the help of SVMrank. [7] transforms job descriptions into queries which are then searched in a database of Dutch CVs and the best-ranked candidates are then selected automatically. [6] applies a Structured Relevance Model to select CVs for a given job or to select the best jobs for a given applicant based on their CV. Besides information gained directly from CVs, some authors also apply external information while ranking the candidates with regard to the specific position: for instance, social media information is also exploited in [8]. Moreover, data collected from the LinkedIn profile of the candidate is also taken into consideration and personality traits are also calculated from the personal blogs of the candidates [9].

Although these previous studies also parse the textual content of CVs, they aim for high precision as they focus on CV retrieval/ranking. On the other hand, for the application form-prefilling use case, high recall is also a must as the main goal is to force users to input as many information about themselves as possible. Because of this issue we consider the task to be a standard information extraction problem, and we will employ the standard evaluation metrics of information extraction. Hence our results are hardly comparable to those obtained by previous studies.

3   The CVparser

The extraction of the relevant data of the applicants for a given position is done by our CVparser method. As a first step, the CVs have to be converted to a uniform file format, and be further preprocessed. After this, a machine learning method can automatically learn how to extract the relevant data fields from the CV texts. Below, the detailed description of these modules follows.

3.1   Preprocessing the CVs

Converting the CVs to a Common File Format (positioned simple text). One of the most salient problems for data extraction was that the applicants usually submit their CVs in many different file formats to the enterprises, including DOC, DOCX, RTF, PDF, TXT, XLS, XLSX and HTML among others. We convert them to the positioned simple text format (TXT), in which the text is in simple text format, but the original positioning of the text in the CVs is roughly preserved. For this conversion, the easiest method we found is to first convert the different files into PDF format, and then to convert these to positioned simple text files using the Poppler PDF rendering library[1]. We decided to drop the document formatting information (like "presented in DOC") because the same layout can be achieved by many techniques (e.g. enumeration, multiple spaces, tabs, tables etc).

Normalisation of CV Texts. As a next step, the CV texts are normalised, character encoding issues are resolved, unnecessary special characters are omitted or replaced, important signs (like enumeration) are unified. After all of the character issues are resolved, the structure of the CV texts is refined. There are CVs with very diverse structure, most of which might be easily processed by humans but cannot be directly processed by computers (e.g. columns, multi-level enumerations). We use a tree to represent the structure of the document, where the subordinate and coordinate relations between textual nodes are encoded in the tree. The construction of the tree is carried out by hand-crafted rules which employ statistics from the document rather than fix magic numbers.

3.2   Extracting the Relevant Data from CVs

Recognition of Some Important Parts of CVs. The first stage of our multi-level CVparser is to recognise bigger parts of the CVs reporting the education, employer, hobby and other competency descriptions. After the big parts are identified, their subparts, i.e. individual records are separated (e.g. a record of a particular previous employer). To recognise these parts and subparts of the CV, the location and the content of the text nodes were analysed employing hand-crafted rules.

The Machine Learning Technique. After the structured CV text files are ready, the data extraction can take place. For this, we employed supervised machine learning algorithms. We experimented with two sequential labelling techniques, namely the maximum-entropy Markov model [1] (MEMM) and the conditional random fields model [2] (CRF). We finally chose MEMM over CRF, as their results were very similar, with MEMM having significantly shorter runtime.

The Two-Level Annotation Scheme. For the data extraction from the CVs, we used a two-level annotation scheme: meaning that several second-level annotations can form a first-level annotation. The reason for applying this annotation structure was that there are complex data structures to be annotated: e.g. a first-level Education

---

[1] http://poppler.freedesktop.org/

annotation can comprise an EduOrgName (denoting the name of the institution), a DegreeName (denoting the degree obtained), an EduStartDate and an EduEndDate annotation (denoting the time period of the education), among others. Similarly, the first-level PersonName, EmployerHistory, Language and Address annotations also have their corresponding second-level annotations, while all the others like Email or TelephoneNumber are simple second-level annotations without a parent.

Prediction. The MEMM we used had to classify each token into one of the possible second-level annotation categories or leave it without any annotation, after which neighbouring second-level annotations of the same type were merged to form a single second-level annotation phrase. We trained special MEMMs for the different types of document parts, i.e. two different models were employed for the education and the employer parts.

The record boundaries of the document in question were also used for restricting MEMM to extract only one mention for the second-level classes. For example, at most one employer name can occur in a particular *previous employer* record. To achieve this we defined the probability of a tagged phrase as the average of posteriori probabilities of its tokens and keep only the phrase in the record with the highest probability.

Beside this MEMM prediction, a rule-based prediction proved to work better for some simple classes. These classes were DrivingLicence (like *A* or *D+E*), Special-Competency (like *Microsoft Word*), OtherCompetency (mostly personal traits, like *ability to work in groups*) and Hobby; in the case of all four, a regular expression-based prediction was employed. At the end, all the extracted data were converted to the commonly used HR-XML format, to be used by the CV ranking method in the future.

Feature Set. In our MEMM, numerous different types of features were used. A detailed explanation of several of these can be found in Chapter 2.2.2 of [3]. These included dictionary features (lexicons), e.g. a list of common given names, family names, position titles, degree names and company types among many others. Furthermore, we also used many orthographical features, for instance word form, capitalization, word length, sentence position, word suffix, frequency information, numerous regular expressions and a number of other features.

An important attribute of most the features was that they were compact, meaning that the actual and surrounding word forms were not added to them, thus reducing the number of features by much. Moreover, if a feature was activated for a word, then this feature was also added to the surrounding words, together with the distance from the word on which the feature fired. As most features were language-independent, they worked with all three languages without modification. Those features that also contained some language-specific information, for example most of the lexicons, had to be translated for the algorithm to work with all the languages correctly.

3.3   Creating the Training Data

In order to apply a supervised machine learning technique like the MEMM used here, pre-annotated training data is required. Therefore we had to create such training data, which we have done in two different ways.

Manually Annotated CVs. First, we asked linguists to manually annotate a small part of the available CVs for each language. This was done by employing an annotator tool developed directly for such purposes. Personal data like the applicant's name, date of birth and address were distinctively marked, together with contact data such as telephone numbers and e-mail addresses. For the complex first-level annotations that also had children, both the first-level and the second-level classes were tagged.

Some additional manual annotation work was also carried out on those CV files that were used for testing. In this test corpus, driving licences, IT-related skills, other competencies and hobbies were also marked. Currently, there are approximately 1000 Hungarian, 700 English and 500 German manually annotated CVs.

Automatically Annotated CVs. Although annotating CVs manually results in very high-quality training data, it is very costly and time-consuming. It could especially pose a problem if the algorithm needs to be adapted to new languages, as in this case a new manually annotated corpus needs to be created and it would require much human resource to create a manually annotated corpus comprising several thousands of CVs. Therefore, we chose to experiment with creating training data automatically.

For this task we created an automatic annotator tool, which required the most relevant data of applicants in a uniform data structure together with their CVs as input. Luckily, in case of the Hungarian and English CVs, both were available from some career portals for which we created our CVparser method. Unfortunately, for the German CVs such data was not available for us at the time, therefore we could not create automatically annotated CVs for German. Having both the CV and the structured data as input, this annotation task could be performed. For this, we used the same preprocessing steps as in the case of the CVparser, then the structured data were mapped to the CV text in a pre-defined order.

Most simple second-level annotations were mapped using special regular expression rules. In the case of complex data structures, which also had second-level annotations as children, special rules were employed. For example, sub-data of the same education or work record had to be matched inside the same education or work entity. To provide the best mapping possible, a reliability measure for matching was created, and for each education and employer record, the most reliable entity was chosen.

Although at first this automatic annotation seemed to be an easy task, we faced several problems. First, as already indicated before, every CV is individually structured, and there are many CVs that are structured badly or not at all. Although this was tried to be resolved during the preprocessing phase, extreme cases could not be converted to well-formed tree structure. Second, although it seemed to be safe to assume that the data detailed in the CV texts and filled out in the career portal forms match each other, this was hardly the case. In many cases the same data were present both in the CV and the form, but in very different format, and it was also common that some of the data was present in just one place. Furthermore, there were many spelling and other grammatical mistakes that also made our work harder. To manage these, we used different normalisation methods, employed different patterns, and created separate annotator functions for the different annotation classes. Despite the problems we faced, we think that we could develop a method that is able to automatically annotate CVs with success. Altogether, we have annotated approximately 42,000 Hungarian and 11,000 English CVs automatically.

4   Results

For determining how our methods succeeded in extracting the relevant data from the CVs, we performed several tests using different settings, training data and testing data. First, we had to define disjoint sets of CVs for training and testing. To get reliable results, for testing only manually annotated CVs were used. Therefore, all of the Hungarian and English CVs not annotated manually could be used for training (due to unavailable data, for German we could not annotate CVs automatically), and we divided our available Hungarian, English and German manually annotated CVs into a training and a testing set, with roughly 90% used for training and the rest for testing. Results for all the different settings can be found in Table 1.

Table 1. Phrase and information tuple-level $F_{\beta=1}$ evaluation scores for different settings. N/A: not available yet.

Training data	Feature set	Test data		
		Manual Hu	**Manual En**	**Manual De**
Manual Hu	Hu	0.695	0.488	0.422
Manual En	En	0.485	0.641	0.461
Manual De	De	0.405	0.495	0.511
Automatic Hu	Hu	0.570	N/A	N/A
Automatic En	En	N/A	0.410	N/A

We note that while the scores seem to be not high enough, they are the result of a very strict perfect matching evaluation setup. According to the subjective opinions of a few recruitment experts the output is about 0.7 for the most difficult type of information (like working experiences) and over 0.9 for the simpler ones (like contact data).

Table 1 shows that training on the manually annotated CVs yields better performance. This could be due to the fact that although there was much more automatic training data, the manual training data is of much better quality. This suggests that it is enough to annotate around a thousand CVs manually to be able to obtain good prediction results. When comparing between the different settings, it seems that much better results can be achieved when testing on the same language as the training was done, which is not surprising. Nevertheless, the results on the other languages are also fair, which show that a significant part of our methods are language-independent.

Table 2. Detailed results on the Hun. manual testing data (most important annotation classes). Auto: Automatically evaluated results on the whole manually annotated Hungarian test CV set. Man: Manually evaluated results on 10 manually annotated Hungarian test CVs.

Annotation class	Recall		Precision		$F_{\beta=1}$	
	Auto	Man	Auto	Man	Auto	Man
PrimaryEmail	0.830	1.000	0.907	1.000	0.867	1.000
PersonName	0.980	1.000	0.980	1.000	0.980	1.000
GivenName	0.980	1.000	0.942	1.000	0.961	1.000
FamilyName	0.980	1.000	0.980	1.000	0.980	1.000
BirthDate	0.975	1.000	0.975	1.000	0.975	1.000
Address	0.870	0.800	0.909	0.889	0.889	0.842
PostalCode	0.854	0.900	0.875	1.000	0.864	0.947
CityName	0.956	0.900	0.977	1.000	0.966	0.947
AddressLine	0.909	0.800	0.952	0.889	0.930	0.842
PrimaryTelephone	0.959	1.000	0.959	1.000	0.959	1.000
Education	0.633	0.806	0.413	0.806	0.500	0.806
DegreeName	0.743	0.875	0.545	0.913	0.629	0.894
EduOrgName	0.796	1.000	0.578	1.000	0.670	1.000
EduStartDate	0.918	0.929	0.736	1.000	0.817	0.963
EduEndDate	0.988	0.935	0.649	0.935	0.783	0.935
Language	0.878	0.857	0.935	0.857	0.905	0.857
LanguageName	0.918	1.000	0.978	1.000	0.947	1.000
CEF-Level	0.914	1.000	0.941	1.000	0.928	1.000
HasCertificate	0.200	0.000	0.500	0.000	0.286	0.000
Employer	0.489	0.714	0.584	0.781	0.532	0.746
PositionTitle	0.706	0.800	0.837	1.000	0.766	0.889
EmpOrgName	0.593	0.853	0.723	1.000	0.652	0.921
EmpStartDate	0.829	0.806	0.871	1.000	0.849	0.893
EmpEndDate	0.697	0.893	0.931	0.962	0.797	0.926
SpecificCompetency	0.989	0.522	0.695	0.923	0.817	0.667
OtherCompetency	0.500	0.692	0.246	0.500	0.330	0.581
DrivingLicence	0.935	1.000	0.853	1.000	0.892	1.000
GenderCode	1.000	1.000	1.000	1.000	1.000	1.000
Average (first level)	**0.836**	**0.866**	**0.788**	**0.896**	**0.804**	**0.875**

When comparing the results for the Hungarian, English and German CVs, it is revealed that, although the results for English and German are also good, the results are considerably better for Hungarian (especially compared to German). This is no surprise, as our CVparser method was originally developed for Hungarian, and there were also much more training data for Hungarian than for English and German. And as much less adaptation work was done for German than for English so far, it does not surprise us that the results for English are much better than for German.

So, the results for both English and German are very promising (especially in case of English), despite the fact that there was not much of adaptation work done: only some features were translated and some training and test data were created. This means that our algorithm can be fairly easily adapted to other languages, obtaining relatively good results without much additional work (which of course could be improved with further optimization for the given language).

All in all, the best results were achieved by training on the Hungarian manual training data with the feature set for the Hungarian language and evaluated on the Hungarian manual testing data, for which the detailed results can be found in Table 2. This table contains the Recall, Precision and $F_{\beta=1}$ scores for the most important annotation classes. As it is impossible to create a completely accurate automatic evaluation tool, and our automatic evaluation of the annotations is very strict, we also evaluated the results on a part of the Hungarian manual testing data (namely on 10 CVs) manually. This enabled a more accurate and fine-grained evaluation than the automatic version. These results can also be found in the table, and they show that in almost all cases the results are actually better than they seem using the automatic evaluation.

5   Conclusions and Future Work

In this paper, we have developed an algorithm that is able to extract the most relevant data of an applicant from his/her CV, which can be used to help filling the forms on the career portal of large enterprises.

From our results it can be seen that the method performs well on Hungarian CVs, especially in the case of simpler data structures. Based on these, we think that our method can be used in real-life systems with success, and it can truly help both job applicants and human resource managers. Furthermore, our algorithms can be easily adapted to other languages, such as we adapted it to English and German: it only requires some feature translation and some manual CV annotation. This also increases the usability of our method significantly.

In the future, beside testing our current versions further, we would like to extend our method for additional languages too. Later on, we would like to further improve our results for all three of the currently supported languages. One of the promising ideas is to extend some of our feature lists automatically, so that these features have better recall. We think that this could be achieved by methods calculating the semantic similarity of words automatically, for instance using the method of [4]. Further, it would also be interesting to test how the results would evolve if the combination of all the feature sets for the different languages were used in the data extraction. We believe that these and other feature engineering will further improve our results.

References

1. McCallum, A., Freitag, D., Pereira, F.: Maximum Entropy Markov Models for Information Extraction and Segmentation. In: Proceedings of the 17th International Conference on Machine Learning, pp. 591–598. Morgan Kaufmann Publishers Inc. (2000)
2. Sutton, C., McCallum, A.: An Introduction to Conditional Random Fields. ArXiv e-prints (2010)
3. Szarvas, Gy.: Feature Engineering for Domain Independent Named Entity Recognition and Biomedical Text Mining Applications. University of Szeged, Szeged (2008)
4. Dobó, A., Csirik, J.: Computing Semantic Similarity Using Large Static Corpora. In: van Emde Boas, P., Groen, F.C.A., Italiano, G.F., Nawrocki, J., Sack, H. (eds.) SOFSEM 2013. LNCS, vol. 7741, pp. 491–502. Springer, Heidelberg (2013)
5. Patil, S., Palshikar, G.K., Srivastava, R., Das, I.: Learning to Rank Resumes. In: Proceedings of FIRE 2012, ISI Kolkata, India (2012)
6. Yi, X., Allan, J., Croft, W.B.: Matching Resumes and Jobs Based on Relevance Models. In: Proceedings of SIGIR 2007, Amsterdam, The Netherlands, pp. 809–810 (2007)
7. Rode, H., Colen, R., Zavrel, J.: Semantic CV Search using Vacancies as Queries. In: Proceedings of the 12th Dutch-Belgian Information Retrieval Workshop, Ghent, Belgium, pp. 87–88 (2012)
8. Bollinger, J., Hardtke, D., Martin, B.: Using social data for resume job matching. In: Proceedings of DUBMMSM 2012, Maui, Hawaii, pp. 27–30 (2012)
9. Faliagka, E., Ramantas, K., Tsakalidis, A., Tzimas, G.: Application of Machine Learning Algorithms to an online Recruitment System. In: Proceedings of ICIW 2012, Stuttgart, Germany, pp. 215–220 (2012)

Fuzzy Cognitive Map of Research Team Activity

Evgenii Evseev and Ivan Kovalev

Saint-Petersburg State University
Universitetskaya nab. 7-9, 199034 Saint-Petersburg, Russia
{e.evseev,i.kovalev}@spbu.ru
http://www.spbu.ru

Abstract. A cognitive model of activity of research team is considered. Object of research is a R&D department (223 persons) in a large scientific and industrial enterprise for sea prospecting works. The model (fuzzy cognitive map) that represents the activity of this department is based on the results of applied sociological research. The fuzzy cognitive map containing 14 concepts, divided into 3 groups (Personal, Group and Organizational). The list of concepts, their initial values and weight matrix are based on an assessment of several experts from the studied organization. The behavior of target concepts at various values of parameters of model is studied results testify to the favorable tendencies in department activity.

Keywords: fuzzy cognitive maps, research team management, socio-economic systems.

1 Introduction

Managing the activities of research team and increasing the productivity of its performance is the actual practical task. Unfortunately, a variety of the points of view, a lack of rigorous theories and large amount of interdependent, often poorly formalized, psychological factors makes the solution of this task very difficult. Use of the models based on fuzzy cognitive maps (FCMs) for elaborating the solution of similar problems seems to be very promising. FCMs have a wide range of applications in social sciences [12,11,10,4,3,8,2], however they could be used more.

In study [5], based on the results of applied sociological investigation of research department, an attempt to construct and analyze a cognitive map of its activity was made. The main goal of this research was to develop administrative decisions, which would lead to improvement of department activity. The model represents a linear FCM [8] including 14 concepts and 67 links. The list of factors, their initial values and weight matrix based on interviews of experts from the studied organization. The analysis of a cognitive map was carried out according to logic of the structural and target analysis [1] for the formulated strategy of management. The final values of concepts were found, coherence of target and initial values were established.

R. Prasath et al. (Eds.): MIKE 2014, LNAI 8891, pp. 342–350, 2014.

The initial weight matrix of a cognitive map was unstable — not all its eigenvalues less modulo than 1 — for its stabilization the columns of an initial matrix were normalized. However, further analysis, according to steps of the structural and target analysis is executed not for initial, but for the stabilized matrix, that represents some excessive idealization of actual processes in the studied object. Furthermore, the constructed model did not consider enough fuzzy nature of expert knowledge of the studied object. The fuzzy cognitive maps are flexible and powerful instrument of modelling and analyzing the dynamic systems. In the investigation of research department mentioned above the quantitative assessment of both initial values of concepts, and their interferences fuzzy techniques were used. Therefore, the choice of FCM as a mathematical basis of model considered to be quite reasonable.

2   Structure of Model

The model of activity of research department represents the fuzzy cognitive map based on information received during sociological survey of one of research departments of the large state scientific and industrial enterprise "Sevmorgeo". Department consist of 223 persons and 4 laboratories. Activity of department consists in development, improvement of techniques, technologies and programs for the analysis and data processing of field researches, drawing up methodical recommendations and other leading documents and also processing, studying, synthesis of geologic-geophysical data and creation of models of a deep structure.

Some heads from the studied organization were used as experts. On the first stage of our research we defined the list of concepts characterizing the studied department and goal parameters of department research activities according to theoretical concepts of research team activities. In order to assess the initial values of concepts their substantial interpretations were offered to the experts during the interviews – as a result the sense and structure of the corresponding concepts was revealed, the system of indicators on which it is possible to judge the considered concept and options of an assessment of intensity of their manifestation were offered. Based on these analytical designs experts estimated degree of expressiveness of each concept in linguistic scale ("poorly"–"moderately"– "so-so"–"obviously"–"strongly"). Then the received linguistic estimates of each expert were transferred to a numerical scale (0–0,25–0,5–0,75–1) and estimates of all experts were average with identical weights — thus, values of concepts belong to an interval $[0, 1]$ — the value, the close to 1 means the strong expressiveness of the characteristics corresponding to the considered concept in activity of the department.

As goal concepts which behavior characterizes studying object the following characteristics were considered: *Timeliness* (c_1) — timeliness of realization of department tasks and *Quality* (c_2) — quality of realization of department tasks.

At an assessment of values of this concepts the quantity, quality and deadlines of the performed tasks corresponded to department working schedule. The concept *Timeliness* was estimated on the relation of quantity of the tasks performed in time to total amount of tasks of department for the considered period — value of this concept the close to 1 corresponds to a high share of the tasks performed in due time. As the department performs tasks of various research type, value of a concept *Quality* represents judgment of experts of compliance of the performed tasks to the requirements established for them — therefore it is possible to say that large values of this concept (the close to 1) mean that the larger share of tasks is executed on high professionally level and also meets all requirements.

As the concepts describing a status of department, the following parameters united in three groups are considered:

1) Personal factors — *Motivation* (c_3), *Level of Professionalism* (c_4), *Physical Condition of The Employee* (c_6). The motivation of employees of department and level of their professionalism are complex characteristics and, in turn, depend on many factors (including psychological reasons) therefore values of these concepts were estimated by experts proceeding from their subjective ideas of the ideal research department. Level of physical condition of staff of the department was estimated on the basis of an average on department of percent of the days passed due to illness — in this case value, the close to 1 corresponds about top physical condition of employees.

2) Group factors — *Level of Technical Equipment*(c_8), *Social and Psychological Climate* (c_5), *Turnover of Staff* (c_{14}). Level of technical equipment was estimated by experts on the basis as objective index — a share of security with the computer equipment in comparison with standards, and a subjective index — opinion of employees on sufficiency of providing department with the computer equipment — the great value of this concept testifies to a high level of technical equipment. The assessment of social and psychological climate in department was carried out by experts subjectively on the basis of an assessment of both personal, and business relationship in collective, and the maximal assessment corresponded to good shape of social and psychological climate. For an assessment of turnover of staff in department the calculated coefficients of turnover of staff were compared with empirically recommended values (3-5%) thus, values of this concept the close to 1 testify to the considerable turnover of staff in collective and values, the close to 0 — to practical lack of turnover of staff.

3) Organizational factors — *Physical Office Conditions* (c_7), *Organizational Culture* (c_9), *Administrative Methods of Management* (c_{10}), *Social and Psychological Methods of Management* (c_{11}), *Economic Methods of Management* (c_{12}), *Level of Communications* (c_{13}). Physical working office conditions were estimated by comparison of observed parameters (temperature, humidity, noise, etc.) to existing sanitary standards — the great value of this concept corresponds to good physical working conditions in department. Value of a concept *Organizational Culture* was defined by experts on the basis of results of research on a technique of OCAI and represents an assessment of force and unity of the prevailing type of organizational culture in department. The assessment of concepts, reflecting methods of management represents subjective judgment by experts of degree of expressiveness of these methods in department management. Level of

communications was estimated by experts also subjectively — for formation of an assessment descriptions and features of various types of communication in the organization (the sense of concept communication revealed, various forms and its possible intensity of collective communication were given) were offered them before estimation. Thus, the large value of this concept testifies from a high level of communications (as the formal and informal) in collective.

At the next step, the matrix of interference of concepts was constructed. Experts gave estimates of intensity of influences in a linguistic form (in a classical 9-mark scale from "the strong decrease" to "the strong increase"). In the estimates of influences of concepts experts were based on the following semantic interpretation of influence: if $w_{ij} \approx 1$, it means that "large value of a concept c_i leads to fast increase in a concept c_j" [2]. Then by routine procedure these estimates were transferred to a numerical interval [-1, 1] on the basis of the corresponding membership functions. The weight matrix W is presented in Table 1.

Table 1. Weight matrix W

	c_1	c_2	c_3	c_4	c_5	c_6	c_7	c_8	c_9	c_{10}	c_{11}	c_{12}	c_{13}	c_{14}
c_1	0	0	0	0	0,25	0	0	0	0	0	0	0	0	0
c_2	0	0	0,25	0	0,25	0	0	0	0	0	0	0	0	0
c_3	0,5	0,5	0	0,25	0,25	0	0	0	0,25	0	0	0	0	-0,25
c_4	0,5	0,5	0	0	0	0	0	0	0	0	0	0	0	0
c_5	0,5	0,5	0,25	0	0	0	0	0	0	0	0	0	0	-0.25
c_6	0,5	0,75	0	0	0,25	0	0	0	0	0	0	0	0	0
c_7	0,25	0	0	0	0,25	0,5	0	0	0	0	0	0	0	0
c_8	0,5	0,75	0,25	0,25	0	0	0	0	0	0	0	0	0,5	0
c_9	0,25	0	0,5	0,25	0,5	0	0	0,5	0	0,75	0,5	0,25	0,5	-0,25
c_{10}	0	0,5	0,25	0,25	0,25	0	0	0	0	0	-0,25	-0,5	0,25	0,25
c_{11}	0,25	0,25	0,75	0	0,5	0	0	0	0	-0,25	0	-0,25	0,25	0
c_{12}	0,5	0.5	0	0	0	0	0,25	0	0	-0,25	-0,25	0	0,5	-0,25
c_{13}	0,75	0,25	0	0	0,25	0	0	0	0	0	0	0	0	-0,25
c_{14}	0,25	0,25	0	-0,25	0	0	0	0	0	0	0	0	0	0

As a mathematical form of FCM the next one is used [7,13]:

$$c_j(t) = f\left(k_1 c_j(t-1) + k_2 \sum_{\substack{i=1 \\ i \neq j}}^{n} w_{ij} c_i(t-1)\right),$$

where n is the number of concepts, $c_j(t)$ is the value of concept j at time t, w_{ij} is the weight between concepts i and j, and f is a nonlinear threshold function that restricts the possible concept values to interval [0, 1]: $f(x) = 1/(1 + e^{-x})$.

It is possible to say that this model is generalization of classical FCM of Kosko [9] and coefficients k_1 and k_2 gives larger flexibility in reflection of features of dynamics of the studied systems. As it is emphasized in [2], in this case the necessary semantic interpretation of influences of concepts is provided.

The coefficient k_1 reflects effect of influence of a concept on itself ("memory effect") — value of self-influence at all concepts is identical, as $w_{ii} = 1$ for all i. The coefficient k_2 reflects effect of influence on a concept c_j of all other concepts, bound to the considered concept according to a digraph of influences. It is possible to say that k_2 represents "gain ratio" of forces of influence on each concept of all other related concepts. In many studies only the case $k_1 = k_2 = 1$ is considered.

It is important that coefficients k_1 and k_2 should not be considered only from the formal point of view as dynamics of concepts of FCM may significantly depends on their values. In [6] for a special case it is studied influences of these coefficients on dynamics of system and some interpretations of these coefficients are offered. For example, it is noted that value of coefficient of k_1 influences dynamics of model, doing it to more "smooth". It is apparent also that value and influence of these coefficients on dynamics is intimately bound to type of the used function of transformation and weights matrix [13]. These parameters have to be defined at a step of creation of FCM and be estimated in the same way (by the same technique) as well as values of influence of concepts. In [6] it is supposed that these coefficients are fuzzy and must be estimated by the fuzzy technique. If we want to keep substantial interpretations of results of simulation than the assessment by experts of values of coefficients k_1 and k_2 also has to be based on some scheme of their interpretation. It is possible to offer, for example, the following technique of an assessment and interpretation of values of these coefficients.

1. First of all, value of the sum of coefficients $k = k_1 + k_2$ is estimated, only for $k_1, k_2 \geq 0$. This sum can be interpreted as some common "gain ratio" of influence strength in model. In the absence of any reasonable assumptions of value of the sum of coefficients it is reasonable to put $k_1 + k_2 = 2$.

2. Next, it is necessary to determine the relative values of coefficients by one of two schemes, in each the estimated value represents some measure of "sensitivity" of concepts to influence of other concepts of model:

   (a) Estimate the relative effect of influence of all other concepts in comparison with self-influence of a concept — i.e. to estimate the relation $\alpha = \frac{k_2}{k_1}$. Value α in this case can be interpreted as the relative gain ratio of "external" influence to concepts: the more α than more changes of a concept depend on values of other concepts. In case of lack of a reasonable assessment let $\alpha = 1$, i.e. $k_1 = k_2$.

   (b) Estimate a share of effect of influence of all other concepts in cumulative effect influence on a concept — i.e. to estimate the relation $\beta = \frac{k_2}{(k_1 + k_2)}$, $0 < \beta < 1$. Value β in this case is reasonable to express as a percentage and to interpret as degree of "responsibility" of concepts to influence of "an environmental context" or as a measure of their "stability". Also perhaps value interpretation β as measures of "importance" of influence of all other concepts in comparison with self-influence. In case of lack of a reasonable assessment let $\beta = \frac{1}{2}$, i.e. $k_1 = k_2$.

3. After estimation α or β (it is trivial $\beta = \frac{\alpha}{1+\alpha}$) absolute values of coefficients k_1 and k_2 are uniquely determinate on the basis of value k.

From the substantial point of view, such two-step indirect determination of coefficients seems to be adequate to the common technique of an assessment of the FCM parameters.

3   Simulation Results and Discussion

During research, our experts didn't state reasonable assumptions concerning coefficients k_1 and k_2, therefore let $k_1 = k_2 = 1$. The analysis of the considered FCM in case $k_1 = k_2 = 1$ shows that value of goal concepts *Timeliness* and *Quality* are stabilized for the first 2-3 steps at the level, the close to 1 (see Fig. 1 — values of these concepts practically coincide). That is why it is possible to draw a conclusion on positive tendencies in research department activity: value of a concept *Timeliness*, the close to 1, testifies that department performs practically all tasks in time. Value of a concept *Quality* almost equal 1 means that all these works are performed at a high professional level and meet all established requirements. Besides, value of a concept *Motivation* quickly increases (from initial value 0,3) and is stabilized at the level of 0,93 — it, according to experts, testifies to the strong positive relation of employees to work. The increased value of a concept *Level of Professionalism* to value 0,81 is also positively interpreted — it testifies to rather high level of competence of workers. Value of a concept *Turnover of Staff*, established at the level of 0,39 it is possible to interpret as some decrease of coefficients of staff turnover in comparison with empirical recommended coefficients values (though increased in comparison with an initial level 0,2). According to experts the reason of it — non-optimum personnel structure of department: average age of employees is 47 years, the share of employees of a retirement age is about 25% and a share of employees till 29 years — only 15%, the share of employees till 39 years is only 27%.

It may be noted some more tendencies which are not presented on diagrams because of limitation of a place. Value of a concept *Administrative Methods of Management* decreases while values of concepts *Social and Psychological Methods of Management* and *Economic Methods of Management* increase and become approximately equal, thus value of a concept *Social and Psychological Climate* increases almost to 1 — it can be interpreted as formation of more balanced and efficient strategy of management. It should be noted also the increase of values of concepts the *Level of Communications* and *Organizational Culture*.

Influence of values of coefficients k_1 and k_2 on FCM dynamics can be estimated according to charts on Fig. 2–3. So, at $k_1 = 0$ $k_2 = 1$ that corresponds to classical FCM of Kosko [9] in the absence of effect of self-influence of concepts, values of target concepts *Timeliness* and *Quality* were quickly stabilized: their values increase and aspire to 0,95 (see Fig. 2). In case of $k_1 = 0$, $k_2 = 2$, that similar to Kosko model without self-influence of concepts [9] on condition of $k_1 + k_2 = 2$, dynamics of goal concepts practically coincides with the previous case — values of concepts are quickly stabilized at the level of 0,99.

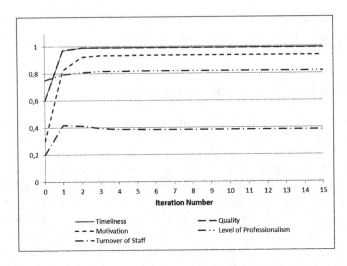

Fig. 1. Dynamic of FCM concepts in case of $k_1 = k_2 = 1$

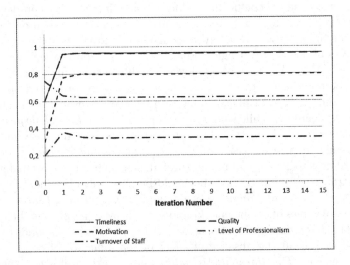

Fig. 2. Dynamic of FCM concepts in case of $k_1 = 0$, $k_2 = 1$

Dynamic in case of equal "importance" of all concepts, i.e. case of $k_1 = \frac{1}{14} = 0{,}14$, $k_2 = \frac{13}{14} = 1{,}86$, $\beta = 0{,}93$ is presented on Fig. 3.

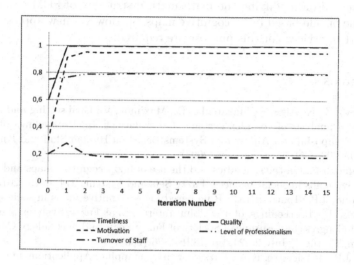

Fig. 3. Dynamic of FCM concepts in case of $k_1 = 0{,}14$, $k_2 = 1{,}86$

Thus, it is possible to claim that in considered model relative value of coefficients of k_1 and k_2 has no essential impact on behavior of goal concepts *Timeliness* (c_1) and *Quality* (c_2). Behavior of goal concepts *Timeliness* and *Quality* remains almost invariable for the cases considered above, changes concern, generally their final values and behavior of concepts *Level of Professionalism* and *Turnover of Staff* in the considered cases.

4 Conclusion

The paper presented a model of research team activity based on FCMs. The FCM analysis was useful for understanding the dynamic of concepts and allowed us to use fuzzy expert knowledge. The model is based on a system of concepts and their interrelations offered by experts during the research procedure. The choice of goal concepts was based on experts opinion and theoretical models of research activity. The practical approach to the assessment of some parameters of model is offered — their values can theoretically have a strong impact on model dynamics.

The analysis of the FCM showed positive tendencies in activity of department, their character does not strongly depend on the considered FCM parameters. The reason seems to be found not only in weight matrix, but also in the type of the non-linear threshold function used in model. It is possible to state that our goal concepts have enough behavior stability.

Future research will be focused on conducting further real life experiments in order to test and promote the usability of the model and also to identify the adequacy of model of actual organization behavior. Furthermore, future research will be focused on modifying the mathematical structure of FCM according to the scheme of rule-based fuzzy cognitive maps [3]. However new applied surveys and expert interviews for this purpose are required.

References

1. Avdeeva, Z., Kovriga, S., Makarenko, D., Maximov, V.: Goal setting and structure and goal analysis of complex systems and situations. In: Proceedings of the 8th IFAC Symposium on Automated Systems Based on Human Skill and Knowledge, pp. 899–903. Göteborg, Sweden (2003)
2. Carvalho, J.P.: On the semantics and the use of fuzzy cognitive maps and dynamic cognitive maps in social sciences. Fuzzy Sets and Systems 214, 6–19 (2013)
3. Carvalho, J.P., Tome, J.A.B.: Rule based fuzzy cognitive maps in socio-economic systems. In: Proceedings of the Joint International Fuzzy Systems Association World Congress and European Society of Fuzzy Logic and Technology Conference, Lisbon, Portugal, July 20-24, pp. 1821–1826 (2009)
4. Cole, J.R., Persichitte, K.A.: Fuzzy cognitive mapping: Applications in education. International Journal of Intelligent Systems 15, 1–25 (2000)
5. Evseev, E.A., Stankevich, A.M.: Cognitive model of research team management. Vestnik St. Petersburg University 1, 206–215 (2012)
6. Glykas, M.: Fuzzy cognitive strategic maps in business process performance measurement. Expert Systems with Applications 40, 1–14 (2013)
7. Groumpos, P.P.: Fuzzy cognitive maps: Basic theories and their application to complex systems. In: Glykas, M. (ed.) Fuzzy Cognitive Maps. Advances in Theory, Methodologies, Tools and Applications. Studies in Fuzziness and Soft Computing, vol. 247, pp. 1–23. Springer (2010)
8. Knight, C.J.K., Lloyd, D.J.B., Penn, A.S.: Linear and sigmoidal fuzzy cognitive maps: An analysis of fxed points. Applied Soft Computing 15, 193–202 (2014)
9. Kosko, B.: Fuzzy cognitive maps. Internetional Journal of Man-Machine Studies 1, 65–75 (1986)
10. Mago, V.K., Morden, H.K., Fritz, C., Wu, T., Namazi, S., Geranmayeh, P., Chattopadhyay, R., Dabbaghian, V.: Analyzing the impact of social factors on homelessness: a fuzzy cognitive map approach. BMC Medical Informatics and Decision Making 13(94), 1–19 (2011)
11. Özesmi, U., Özesmi, S.: A participatory approach to ecosystem conservation: Fuzzy cognitive maps and stakeholder group analysis in Uluabat Lake, Turkey. Environmental Management 31(4), 518–531 (2003)
12. Papageorgiou, E.I.: Review study on fuzzy cognitive maps and their applications during the last decade. In: EEE International Conference on Fuzzy Systems, Taipei, Taiwan, June 27-30, pp. 828–835 (2011)
13. Stylios, C.D., Groumpos, P.P.: Mathematical formulation of fuzzy cognitive maps. In: Proceedings of the 7th Mediterranean Conference on Control and Automation (MED 1999), Haifa, Israel, June 28-30, pp. 2251–2261 (1999)

Knowledge Acquisition for Automation in IT Infrastructure Support

Sandeep Chougule, Trupti Dhat, Veena Deshmukh, and Rahul Kelkar

Tata Research Development and Design Centre,
54 B Hadapsar Industrial Estate, Pune 411013, India

Abstract. In todays IT-driven world, the IT Infrastructure Support (ITIS) unit aims for effective and efficient management of IT infrastructure of large and modern organizations. Automatic issue resolution is crucial for operational efficiency and agility of ITIS. For manually creating such automatic issue resolution processes, a Subject Matter Expert (SME) is required. Our focus is on acquiring SME knowledge for automation. Additionally, the number of distinct issues is large and resolution of issue instances requires repetitive application of resolver knowledge. Operational logs generated from the resolution process of issues, is resolver knowledge available in tangible form.

We identify functional blocks from the operational logs, as potential standard operators, which the SME will validate and approve. We algorithmically consolidate all the steps the resolvers have performed historically during the resolution process for a particular issue, and present to the SME a graphical view of the consolidation for his assessment and approval. We transform the graphical view into a set of rules along with the associated standard operators and finally ensemble them into a parametrized service operation in tool agnostic language. For an ITIS automation system, it is transformed into a configuration file of a targeted orchestrator tool. Bash and powershell script transformations of service operations are executed by resolvers manually or via an automation web portal.

1 Introduction

In a Service Desk (SD) of ITIS, a small set of common problems are raised again and again, hence SD resolvers spend most of their time dealing with problems that have been previously addressed, enriching their experience of handling similar problems, and tend to get more efficient and more systematic in handling such problems over the time. Though sharing of these experiences among the resolvers takes place few times, in general, the skills for solving a commonly encountered problem are unevenly distributed among the resolvers.

There is a large industry focus on increasing the IT infrastructure production support efficiency and optimizing cost through automation. This requires that most of the issues be resolved automatically without human dependence or with minimal human intervention. Use of existing knowledge is a key to this exercise.

R. Prasath et al. (Eds.): MIKE 2014, LNAI 8891, pp. 351–360, 2014.

This knowledge exists in various forms today, one of such forms is operational logs. The operational logs are steps performed by the resolvers during the issue resolution are logged into system in form of text file, CSV file or database entries. The operational logs are associated with sufficient context information and can be used effectively for automation.

Traditional method of production support results in operational logs on individual infrastructure elements with limited context information, resulting into a knowledge source of little value. However, using some advanced next generation production support tool, we get operational logs that are rich in context i.e. the logs contain information about operations performed on various resources (for example a server) related to resolver and ticket. A ticket is a service request raised by a user facing an issue while using some resource. As the support activity continues, the log gets enriched. The log acts as a rich knowledge store which is used for various purposes, one of them being automation.

Identification of automation candidates and generation of parametrized service operations as the automation library, through analysis of the captured operational logs (Table 3) is a challenging industrial problem, which we have tackled here. Automated service operation is an executable entity capturing all states of infrastructure elements and associated decisions taken to accomplish issue resolution.

In this paper, major contribution of our work is a variant of apriori algorithm [3] to find ordered and consecutive sequence patterns from logs having noise to some extent.

Our another contribution is a combinatorial approach to find an isomorphic directed sub graph from directed acyclic graphs. We combine the graphs using hyper-nodes and hyper-links technique. For the information visualization, we show the control flow (Figure 1) to a Subject Matter Expert (SME). After assessment and approval from the SME, we create rules out of the control flow and assign standard operators to the consolidated steps. Then the SME assigns name, summary and description for service operation to be created.

We describe motivation and related work in section 2. Problem formulation is described in section 3. In section 4 and 5, we go through the solution approach and the algorithms explained. In section 6, to describe a use case scenario for knowledge repository, we show a full fledged ITIS automation system. We cover conclusion and future work in section 7.

2   Motivation and Related Work

ITIS Automation systems (Figure 2) use manually created rules for orchestrator, a centralized control system governing the workflow for the requested service based upon the pre-defined rules/configuration. Our motivation behind this paper is to target automatic rules generation as part of knowledge acquisition. We identify frequent logical patterns to list down potential standard operators. In frequent itemset mining [3] on transactional databases, and for sequential pattern mining [6], [4,5] in noisy environment, good amount of research has been accomplished. Yang has used boarder collapsing approach with considering replacements (mutation) of the pattern (for example a gene expression) symbols as the noise. Yang decides noisy match using string match scoring and *min_match* threshold. In our domain the noise is not replacement but intrusion of extra symbols. We don't use any string match scoring but use exact match on hashmaps to allow noise in the pattern. We store pattern instances using two hash maps used in variant of apriori algorithm (Table 2) for patterns identification.

3   Problem Formulation

Operational log sequences are a major source of existing procedural knowledge, which we process while creating a knowledge repository. Every operational log sequence is an ordered set of actions performed by the resolver. For each action, the user role, date stamp, command executed and output for that command are the main attributes which are captured by the system. For creation of a knowledge repository of service operations, we have two objectives:

1. Identification of standard operators
2. Creation of service operation rules

For both objectives, we perform analysis on a homogeneous set of operational log sequences with few classification attributes.

Operational log sequences contain functional blocks which are nothing but frequently occurring sets of commands. Once identified, these command sets are considered as automatable units and are offered as standard operators. Some major challenges involved in doing this are as follows:

– Logical sections of logs are identified and their frequency is calculated. The logs contain some noise, which has to be identified and dealt with.
– It is least probable to identify exact matches so there should be a provision to allow some extent of inexactness.

For service operation rule generation, first a control flow is generated for a particular issue by unifying all relevant command sequences. One command sequence captures all ordered actions performed by a resolver for resolution of a ticket. Some major challenges involved in doing this are as follows:

– A method to generalize command options and parameters is needed before actual comparison of commands.
– The order of merging the sequences, affects the accuracy of the final result. To enforce the sequence merging order, we need to cluster the sequences into groups. For clustering appropriate sequences, a similarity scoring mechanism needs to be devised.

4   Identification of Standard Operators

We give a homogeneous set of sequences to our engine which uses a variant of apriori algorithm [3] for mining frequent subsequences. Based on the user inputs

Table 1. Standard Operator Template Example

---

1. local getFileSystem__Return=$1; shift
2. local getFileSystem__Output=$1; shift
3. local DomainName="$1"; shift
4. local LoginName="$1"; shift
5. local HostName="$1"; shift
6. local FileSystem="$1"
7. local TS_AIX_getFileSystem_4_int_Output =
    $(ssh -T -q ${LoginName}@${HostName}
    bash -s "$DomainName" "$FileSystem" <<
    "EOF"
        DomainName="$1"; shift
        FileSystem="$1"
        <command>
    EOF)
8. local TS_AIX_getFileSystem_4_int_Return='echo $?'
9. eval $getFileSystem__Return='$TS_AIX_getFileSystem_4_int_Return'
10. eval $getFileSystem__Output='$TS_AIX_getFileSystem_4_int_Output'

---

about the pattern frequency threshold and pattern size threshold, the sequences are filtered out and further the user validates the sequences eligible for forming the standard operators. The last step is the assisted script creation where the user will select valid set of commands for each user selected pattern and even parameterize them.

For understanding the standard operator examples we use a template (Table 1) with <command> as template parameter which is an executable step. Few examples of standard operators (pertaining to mountFileSystem Issue) are as follows:

1. TS_AIX_getFileSystem, in which we use *"lsfs ${FileSystem}"* in place of <command>
2. TS_AIX_getMountFileSystem_4(), in which we use *"mount | grep ${FileSystem}"* in place of <command>
3. TS_AIX_mountFileSystem_4(), in which we use *"mount ${FileSystem}"* in place of <command>

4.1   Apriori Based Sequence Mining Algorithm

In the standard operator finding problem, we aim to find the block of consecutive and ordered commands as a repeating pattern [3] over same or across all operational log sequences from the input homogeneous set.

If we have to solve this problem with very basic and intuitive technique, then we create all possible blocks in $O(l^2)$ time complexity for a given sequence of the length l, i.e. by selecting block start index l_s and block end index l_e in $\binom{l}{2}$ different ways. If there are such n sequences, then the time complexity will be

Table 2. Frequent Sequence Generation Algorithm

FUNCTION: frequentSequenceGeneration

Input : hash maps h_r & h_o, noise c, support min_{sup}
Output: Complete set of frequent sequences, L
1: $h_r' = \phi$ and $h_o' = \phi$
2: $\forall$ key α_k such that $(\alpha_k, T_r) \in h_r, (\alpha_k, T_o) \in h_o$ do
3: $\forall (\beta_r, \gamma_o)$ pairs such that $\beta_r \in T_r$ and $\gamma_o \in T_o$ do
4: $\forall (o, s_o, e_o) \in$ set $\overline{\gamma_o}$ and $(r, s_r, e_r) \in$ set $\overline{\beta_r}$ do
5: if $r = o, s_o \leq s_r \leq e_o, e_r - s_o \leq k + c$ do
6: $h_r' += (\gamma_o, (r, s_o, e_r))$
7: $h_o' += (\beta_r, (r, s_o, e_r))$
8: endif
9: prune h_r' and h_o' by $h = h \backslash \{(\alpha_{k+1}, \sigma_{k+1}) | min_{sup} > |\overline{\sigma_{k+1}}|\}$
10: $L = L \cup \{\sigma_{k+1} | \sigma_{k+1} \subset$ value sets of h_r' or $h_o'\}$
11: if $\left(h_r' \neq \phi\right)$ and $\left(h_o' \neq \phi\right)$ do
12: frequentSequenceGeneration$\left(h_r', h_o'\right)$
13: endif

$O(nl^2)$. We use a hash map to store the sequences and frequency and finally just sort the minimum support satisfying sequences in the decreasing order of their length and frequency. The apriori version will be an iterative generation of patterns by pruning the candidates violating the minimum support criteria, which will reduce the time complexity substantially.

Our problem has a generalization over the pattern by incorporating some acceptable level of noise [6], provided by the user as a parameter. As every resolver from ITIS domain may not be an expert, the resolvers are used to run few unnecessary commands in between. This unintentional noise may prevent an occurrence of a pattern from being recognized and in turn reduces the support of a pattern.

Let us denote the set of all command sequences (operational logs) by set $P = \{P^i\}$ and the jth command from the ith sequence by P_j^i. Every occurrence of a pattern is represented by an integer triplet $\{(i, s, e)\}$, i.e. the contiguous subsequence starting at index s and ending at index e from the ith sequence. Let such a contiguous subsequence $\left(P_s^i, P_{s+1}^i, \ldots, P_e^i\right)$, be denoted by a notation sequence (i, s, e). Initially the sequence data is parsed and the F_2 set is created by generating all distinct $\left(P_j^i, P_{j+1}^i\right)$ pairs of consecutive commands along with, set of all triplets corresponding to the occurrence of pattern $\left(P_j^i, P_{j+1}^i\right)$ simply denoted by $\overline{\left(P_j^i, P_{j+1}^i\right)}$. A post hash map h_o is created to store a set of frequent patterns $\left(P_s^i, P_{s+1}^i\right)$, with the command P_{s+1}^i as the hash key. Similarly a pre hash map h_r is created to store frequent patterns $\left(P_s^i, P_{s+1}^i\right)$, with the command P_s^i as the hash key. For 2-length patterns, the sequence end index $e = s + 1$. That means initially h_r will have a hash map entry with key command a, and hash map value as a set of all patterns starting with command a, i.e. ab, ac, ad

etc. Similarly h_o will have a hash map entry with key command z, and hash map value as set of all patterns ending with command z i.e. wz, xz, yz etc.

Lets define the frequent patterns of size k by $\{\alpha_k\}$. Let the entries of hash map h_r, be denoted by $(\alpha_k, \{T_r\})$ where $T_r = \{\overline{\beta_r}\} = \{\{(r, s_r, e_r)\}\}$. The set of triplets $\{(r, s_r, e_r)\}$ for a pattern β_r is denoted by $\overline{\beta_r}$. Similarly the h_o hash map entries are denoted by (α_k, T_o) where $T_o = \{\overline{\gamma_o}\} = \{\{(r, s_o, e_o)\}\}$. The set of triplets $\{(r, s_o, e_o)\}$ for a pattern γ_o is denoted by $\overline{\gamma_o}$.

Initially data structure entries only for patterns of size 2 are generated. For both hash maps, the patterns not satisfying the frequency support criteria are excluded. That is patterns β such that $min_{sup} > |\beta|$, where min_{sup} is the frequency support threshold value, are excluded. Using frequentSequenceGeneration (Table 2) recursive algorithm, data structure entries for patterns of other sizes are generated.

5   Creation of Service Operation Rules

First we will create a graph called a control flow using operational log data, which will be transformed into set of rules for a new service operation.

Here we consider 8 operational log sequences (Table 3) of mount point issue from a technology service, 'OS - Linux'. For every operational log sequence,

Table 3. Set of Operational Log Sequences

ID	Sequence
1	ping -C4 $Hostname, echo "$HostName host does not exist"
2	ping -c4 ${HostName}, ssh ${HostName} mount | grep ${FileSystem}, echo "FileSystem $FileSystem is mounted already on $HostName"
3	ping -c4 ${HostName}, ssh ${HostName} mount | grep ${FileSystem}, ssh ${HostName} lsfs ${FileSystem}, ssh ${HostName} mount ${FileSystem}, echo "FileSystem $FileSystem has not been mounted on $HostName"
4	ping -c4 ${HostName}, ssh ${HostName} mount | grep ${FileSystem}, ssh ${HostName} lsfs ${FileSystem}, ssh ${HostName} mkdir dirpath, echo "FileSystem $FileSystem has not been mounted on $HostName"
5	ping -c4 ${HostName}, ssh ${HostName} mount | grep ${FileSystem}, ssh ${HostName} lsfs ${FileSystem}, ssh ${HostName} mkdir dirpath, ssh ${HostName} mount ${FileSystem}, echo "FileSystem $FileSystem has not been mounted on $HostName"
6	ping -c4 ${HostName}, ssh ${HostName} mount | grep ${FileSystem}, ssh ${HostName} lsfs ${FileSystem}, ssh ${HostName} mkdir dirpath, ssh ${HostName} mount ${FileSystem}, echo "FileSystem $FileSystem has not been mounted on $HostName"
7	ping -c4 ${HostName}, ssh ${HostName} mount | grep ${FileSystem}, ssh ${HostName} lsfs ${FileSystem}, ssh ${HostName} mkdir dirpath, ssh ${HostName} mount ${FileSystem}, echo "FileSystem $FileSystem has been mounted successfully on $HostName"
8	ping -c4 ${HostName}, echo "$HostName host does not exist"

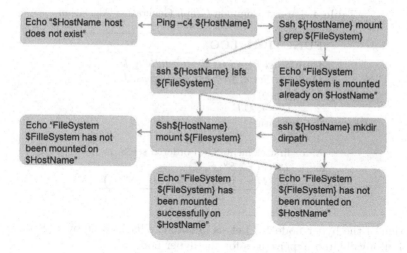

Fig. 1. Generated Control Flow

initially the graph is simply a uni-directional path. We keep on merging the sequences and finally we get a directed acyclic graph (Figure 1). Finally we create rules out of the control flow and assign standard operators to the consolidated steps. Then the SME will assign name, summary and description for service operation to be created.

5.1 Graph Mining Algorithm

In the control flow generation problem, our primary goal is to take union of all operational log sequences from the input homogeneous set (solution set for a particular problem/issue). For union of sequences, there are 2 fundamental steps which are run iteratively on the input graphs: 1. finding largest common graph from the inputs [1], [4,5], 2. union of inputs keeping single instance of the elements from largest common graph and ensuring the embedding remains intact for all inputs. As the initial log sequences are simple path graphs, we will store those as digraphs with number of directed edges' count, exactly less by 1 than the commands' count i.e. length of the sequence. Here we will examine the algorithm to generate the largest common graph. Let us denote the set of all command graphs by set $P = \{P^i\}$ and the jth command from the ith graph by P^i_j. These graphs are stored with DAG data structure as those are simply directed acyclic graphs. For that we initially construct the set of all valid hyper nodes of dimension d, if the number of input graphs for union is d i.e. $P = \{P^i | 0 < i \leq d\}$. Let us denote the set of valid hyper nodes by $S = \{S^i | P^k_{S^i_k} = \alpha^i\}$ i.e. all the nodes corresponding to the hyper node S^i, correspond to the same command α^i. This is an element of the cross product of the all graphs from ordered set P.

Similarly we will create *InvalidMap I*, i.e. a hash map of hyper node S^i as key and set of hyper edges for set S, which will get invalidated because of the

Table 4. Largest Common Graph Generation Algorithm

FUNCTION: LCG
Input: *Set S, subset E, InvalidMap I*
Output: *Set of frequent sub graphs, U*
1: if $(S = \phi)$
2:     $U = U \cup E$
3:     return
4: endif
5: for first element S^0 in ordered set S do
6:     $LCG\left(S \setminus S^0, E, U\right)$
7:     $LCG\left(S \setminus \left(I\left(S^0\right) \cup S^0\right), E \cup S^0, U\right)$

selection of the hyper node S^i. Let us denote the hash value by $I\left(S^i\right)$, i.e. the set of all invalidated hyper edges for the hyper node S^i.

Our LCG function (Table 4) is analogous to the constrained combinatorial function. We will give the initial call of this function with complete valid hyper nodes set S as first parameter and remaining two parameters as empty sets, i.e. the call will now be framed as $LCG\left(S, \phi, \phi\right)$.

5.2   Service Operation Rules

After assessment and approval of the control flow from the SME, the control flow is translated into predicate-action pairs. These predicate action pairs are tool agnostic and can be further translated to other executable forms like bash and powershell or as configurations for other orchestrator tools like Arago, Service Now, Scorch, etc. Let's see an example of a rule created in our system language, from the control flow of mount point issue (Figure 1). This rule corresponds to the edge emphasizing the fact that "*mount | grep ${FileSystem}*" command is the next logical step after the successful execution of "*ping -c4 ${HostName}*" command

```
<Rule>
    <Name> MountLocalFileSystem_getFileSystemExecution </Name>
    <Predicate> (TS_AIX_checkHostReachability_1_Return = 0) </Predicate>
    <Action> (Return1, Output1) = TS_AIX_getMountFileSystem_4(
            $DomainName, $LoginName, $HostName, $FileSystem)
    </Action>
</Rule>
```

We show here the above rule's transformation into bash script format.

```
Log Rule : MountLocalFileSystem_getFileSystemExecution Starts :'date'
if (([[ -n "$TS_AIX_checkHostReachability_1_Return+x" ]]
&& [[ $TS_AIX_checkHostReachability_1_Return -eq 0 ]])); then
```

TS_AIX_getFileSystem_4 TS_AIX_getFileSystem_4_Return
TS_AIX_getFileSystem_4_Output
"$DomainName" "$LoginName" "$HostName" "$FileSystem"
fi

6 Use Case: ITIS Automation System

The standard service catalog of operations consists of a set of routine technical operations and troubleshooting procedures. In spite of having a large scope of standardization and automation, managing applications largely remains a people intensive task delivered using dedicated teams because of the lack of a good enough knowledge repository that contains knowledge for automated execution of different service operations. We create this kind of repository using a semi automated knowledge acquisition system explained in our paper above.

Repository knowledge is used to generate bash and powershell scripts, which are stored and executed manually or via some web portal for automation. This knowledge is applied to configure certain orchestrator tools such as Arago[1], ServiceNOW[2], Scorch[3]. This knowledge repository is also used in larger systems with IT Service Management (ITSM) Monitoring tools.

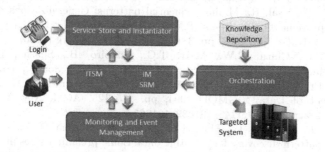

Fig. 2. Example of a system based upon Knowledge Repository

Figure 2 shows a use case scenario of a full fledged ITIS automation system using a knowledge repository. The Service Store and Instantiator, is a web portal for user to login and select a service and initiate the execution. The IT Service Management module consists of Incident Management (IM), Service Request Management (SRM), which represents the ticketing system in ITIS environment. BMC Remedy[4] and Service Now Incident Management[5] are examples of well-known ITSM tool. The issue/request instantiated via ITSM is handled by an orchestrator.

---

[1] https://www.arago.de
[2] http://www.servicenow.com
[3] http://scorch.codeplex.com
[4] http://www.bmc.com/it-solutions/it-service-management.html
[5] http://www.servicenow.com/products/it-service-automation-applications/
incident-management.html

7   Conclusions and Future Work

In this paper we described knowledge acquisition using automatic rules generation and identification of standard operators in presence of noisy commands. We devised the constrained combinatorial approach of subgraphs finding, and iterative graph merging to lead to a control flow of an issue resolution process.

As future work, we found that for the graph merge iterations of control flow generation, the sequence order of input graphs should be devised smartly, so that the output control flow will be more accurate. For accomplishing that we plan to employ clustering on graphs using graph seriation [7] and graph edit distance [2] methods.

References

1. Inokuchi, A., Washio, T., Motoda, H.: An apriori-based algorithm for mining frequent substructures from graph data. In: Zighed, D.A., Komorowski, J., Żytkow, J.M. (eds.) PKDD 2000. LNCS (LNAI), vol. 1910, pp. 13–23. Springer, Heidelberg (2000)
2. Robles-Kelly, A., Hancock, E.R.: Graph edit distance from spectral seriation. IEEE Transactions on Pattern Analysis and Machine Intelligence 27(3), 365–378 (2005)
3. Srikant, R., Agrawal, R.: Mining sequential patterns: Generalizations and performance improvements. In: Apers, P.M.G., Bouzeghoub, M., Gardarin, G. (eds.) EDBT 1996. LNCS, vol. 1057, pp. 3–17. Springer, Heidelberg (1996)
4. Wang, J.T.L., Chirn, G.W., Marr, T.G., Shapiro, B., Shasha, D., Zhang, K.: Combinatorial pattern discovery for scientific data: Some preliminary results. In: Proceedings of the 1994 ACM SIGMOD International Conference on Management of Data, SIGMOD 1994, pp. 115–125. ACM, New York (1994), http://doi.acm.org/10.1145/191839.191863
5. Wang, J.T.-L., Chirn, G.-W., Marr, T.G., Shapiro, B., Shasha, D., Zhang, K.: Combinatorial pattern discovery for scientific data: Some preliminary results. SIGMOD Rec. 23(2), 115–125 (1994), http://doi.acm.org/10.1145/191843.191863
6. Yang, J., Wang, W., Yu, P.S., Han, J.: Mining long sequential patterns in a noisy environment. In: Proceedings of the 2002 ACM SIGMOD International Conference on Management of Data, SIGMOD 2002, pp. 406–417. ACM, New York (2002), http://doi.acm.org/10.1145/564691.564738
7. Yu, H., Hancock, E.: String kernels for matching seriated graphs. In: Proceedings of the 18th International Conference on Pattern Recognition, vol. 1, pp. 224–228 (2006)

Developing eXtensible mHealth Solutions for Low Resource Settings

Yvonne O' Connor[1], Timothy O' Sullivan[1], Joe Gallagher[2],
Ciara Heavin[1], and John, O' Donoghue[3]

[1] Health Information System Research Centre, University College Cork, Ireland
{y.oconnor,t.osullivan,c.heavin}@ucc.ie
[2] gHealth Research Group, University College Dublin, Ireland
joejgallagher@gmail.com
[3] Global eHealth Unit, Imperial College London, United Kingdom
j.odonoghue@imperial.ac.uk

Abstract. Over the last ten years there has been a proliferation of mHealth solutions to support patient diagnosis and treatment. Coupled with this, increased attention and resources have been attributed to the development of technologies to improve patient health care outcomes in low resource settings. Most significantly, it is the development of highly extensible, portable and scalable technologies which have received the most attention. As part of an mHealth intervention in Malawi Africa, an agnostic clinical guideline decision-support rule engine has been developed which uses classification and treatment rules for assessing a sick child defined in XML; namely, Integrated Management of Childhood Illness (IMCI) and Community Case Management (CCM). Using a two-phased approach, 1) the rules underpinning the cloud-based mobile eCCM application were devised based on the widely accepted WHO/UNICEF paper based guidelines and 2) subsequently validated and extended through a user workshop conducted in Malawi, Africa.

Keywords: mHealth, Low Resource Settings, XML, Cloud-based technology, Community Case Management (CCM), electronic-CCM (eCCM).

1 Introduction

The level of research attention within the area of adoption and use of Information and Communication Technology (ICT) in developing regions is currently on the rise [1, 2]. Developing regions, in the context of this paper, are countries operating within ICT resource-constrained conditions and where the population has limited knowledge in utilising ICT solutions [3]. Distinguishing between developed and developing countries, from the perspective of ICT implementation, is often referred to in literature as the "digital divide". Bridging the gulf between developed and developing regions can be reduced by leveraging ICT [4] and in recent years, progress insofar as reducing the digital divide from the perspective of broad access to IT is declining [5]. This may be due to the recent hype surrounding "Information and Communication Technology for Development (ICT4D or ICT4Dev)".

R. Prasath et al. (Eds.): MIKE 2014, LNAI 8891, pp. 361–371, 2014.
© Springer International Publishing Switzerland 2014

In the past few years, ICT4D has become a buzz term for developing technological solutions to support rural development for low resource settings. Since the term first emerged in the literature, circa mid-1950s, a wide array or opinions have emerged as to what ICT4D constitutes (cf. [6]) there continues to be a lack of consensus on an agreed definition. For example, Sutinen and Tedre [7] define ICT4D purely from a computer science perspective stating it refers to the opportunities of ICT as an agent of development. Unwin [8] defines ICT4D as the use of ICT to enable progress and growth towards an economic, social and political "better good". Dhir et al., [9] convey ICT4D to encompass the development of ICT to solve global societal challenges, which is "how technology can solve challenges faced in global development (p.70)". Despite the various disagreements between authors surrounding their approach to theorising ICT4D, there is a consensus that ICT4D is an attempt to bridge the digital divide between resource-rich and resource-poor settings, subsuming ICT with the intentions of assisting disadvantaged populations, primarily those living in developing countries.

ICT4D initiatives have been applied to various industries in developing countries such as education and retail with the healthcare industry receiving a plethora of attention. The recent rise in mHealth ICT4D initiatives in developing countries may be the result of attempts to achieve Millennium Development Goals (MDG) established by United Nations Millennium Declaration. Eight MDGs were proposed ranging from reducing child mortality to minimising the spread of HIV/AIDS and improving maternal health, all by the target date of September 2015. It is reported by Blaya et al., [10] that the adoption of ehealth technologies is rapidly growing in resource-poor environments, primarily through the use of mobile technologies [11, 12]). While such ICT4D initiatives have made initial attempts in assisting disadvantaged populations, the vast majority of these projects, especially in low resource settings, have been designed as pilot projects to explore a development concept [13, 14]. A dearth of research exists on how such ICT4D initiatives can be scaled beyond the pilot study phase and sustained in the long term [15, 16, 17].

In order to bridge this gap in extant literature extensible mHealth solutions must be developed as part of ICT4D initiatives. It is widely recognised than the development of scalable [18], transparent [19] and customisable [20] mHealth solutions are paramount for ICT4D to succeed in the long term. In order to support the ICT4D vision this paper presents the design of extensible, transparent and customisable clinical guideline decision-support rule engine that is agnostic with respect to the instance of guidelines being applied (e.g. IMCI, CCM, etc.). These guidelines are defined in XML, underpinning a mHealth artifact aimed at improving health care workers decision making whilst attempting to reduce child mortality in remote, rural low resource settings. The development of a comprehensive XML schema which is interpreted by the custom built rule engine, presented in this paper, may be extended and used beyond the scope of this project.

This paper is structured as follows: Section 2 describes the research methodology underpinning the development of the clinical guideline algorithms to be used on mobile technology. Section 3 discusses the implications of using XML for this project and wider ICT4D initiatives.

2   Theoretical Grounding

Integrated Management of Childhood Illness (IMCI) is an approach as part of the management of child health which focuses on the well-being of the whole child in low and middle income countries. IMCI is an initiative introduced by the World Health Organisation (WHO) and United Nations Children's Fund (UNICEF) with the aim to reduce death, illness and disability while promoting improved growth and development from 2 months to five years of age. To help ensure that the aims outlined by IMCI are achieved, it is imperative that the following three strategies are implemented in developing regions: (1) Improving case management skills of health-care staff, (2) Improving overall health systems, and (3) Improving family and community health practices.

IMCI case management training equips health workers at first-level health clinics (i.e. nurses and clinicians) with skills to manage children for a combination of illnesses, identify those requiring urgent referral, administer appropriate treatments, and provide relevant information to child carers. IMCI case management consists of six core steps including (1) Assessing the sick infant/child; (2) Classifying the Illness; (3) Identifying treatment; (4) Treating the infant/child; (5) Counselling the caregiver and (6) Providing follow-up care advice. In total, healthcare providers implementing IMCI guidelines are required to follow numerous complex algorithmic processes to determine the assessment, classification and treatment for sick children.

To extend the delivery of healthcare services beyond first-level healthcare facilities to the community level WHO and UNICEF, in association with other partners, established a shorter, more refined and easier set of protocol for less experienced health workers (i.e. Community Health Workers) to follow. This strategy is commonly referred to as Community Case Management (CCM). In Marsh et al., [21] it was reported that CCM is a "feasible, effective strategy to complement facility-based management for areas that lack access to facilities". CCM is defined by George et al., [22] as "a strategy that enables trained community health workers or volunteers to assess, classify, treat and refer sick children who reside beyond the reach of fixed health facilities". CCM initially addressed a single disease; namely, malaria. Building from the initial guidelines CCM was subsequently extended to focus on four infectious diseases which include pneumonia, diarrhoea, neo-natal sepsis, and malaria. Similar to IMCI this strategy also requires the healthcare worker to sequentially follow six steps (refer to the previous paragraphs for these steps). For the purpose of this paper, the researchers focus on CCM solely. CCM is a subsection of the wider IMCI guidelines and the data presented herein can be applied to IMCI guidelines.

2.1   mHealth

Mobile markets and telecommunications in developing worlds have grown immensely in the past decade [23], leading to a proliferation of mobile technology development initiatives in low resource settings. One core sector which has witnessed a surge of mobile development projects is that of the healthcare sector [11, 12]. The application of mobile technology to healthcare, referred herein as mHealth, is reported to offer health workers the potential for flexible and mobile access to patient information quickly, efficiently and securely independent of location at any point in time [24].

Various mHealth initiatives developed for low resource settings include treatment and disease management, data collection and disease surveillance, health support systems, health promotion and disease prevention, communication between patients and health care providers, and medical education [25]. In recent years attempts have been made to digitise clinical guidelines such as IMCI and CCM for use by health workers in low resource settings [26, 27]. The next section considers in more detail the development of eCCM based on existing clinical guidelines.

2.2   Building Clinical Algorithms for mHealth Using XML

Digitising clinical guidelines (i.e. eCCM) could potentially overcome the reported negative impacts of non-adherence to paper-based guidelines [27, 28, 29]. However, this may not be achieved on a large, wide scale if the mobile application (eCCM) cannot be extended beyond the pilot study. Noteworthy, CCM guidelines can vary according to contextual factors. For example, some countries primarily focus on pneumonia, diarrhoea and malaria as these infectious diseases are commonly associated with high mortality rates. If applications are hard-coded with the guidelines then this would restrict the ability of scaling the application to other countries [30]. Further to this, the World Health Organisation has in the past modified the original guidelines and may do so again in the future [31] due to the changing nature of global health. Without the ability to customise existing guidelines, eCCM solutions could become redundant thus, endangering the longevity of the application. To overcome this, a guideline agnostic decision-support rule engine was constructed which operates on XML-based definitions of specific guideline instances. XML involves creating a language to describe data vis-à-vis tags, similar to HTML [32]. Using XML facilitates for the creation of user defined markup languages for storing structured data in text files [33, 34]. The implementation of XML is completely transparent to the end user [35]. This is especially advantageous as it presents the capability for medical professionals to define and script the guideline rules which will be interpreted by the rule engine. As part of this study XML is used to define the CCM guidelines, including illness classification and treatment rules, this paper illustrates how XML may be used to enable the development of extensible mHealth solutions.

3   Research Methodology

3.1   Developing Clinical Guideline Algorithms for mHealth Solution in Malawi

The development of this cloud-based mHealth solution was undertaken as part of the Supporting LIFE project. This European funded project aims to apply a novel combination of smartphone technology and expert decision support systems (DSS) to equip the Malawian Ministry of Health's front-line healthcare workers known as Health Surveillance Assistants (HSA) with mobile devices and applications to assist in their assessment, diagnosis and treatment of seriously ill children. Integral to this is the development of the clinical guidelines algorithm which acts to underpin the mobile application which is outlined as part of this paper.

3.2 Mobile and Server Development

The use of mobile technologies in developing countries is on the rise, as previously highlighted. As a result, this project focused on developing eCCM on mobile devices to assist healthcare providers at the point of care. The target device was an ASUS / Google Nexus 7 16GB 3G-enabled tablet. The device was chosen for its relatively large screen, compact size, 3G capabilities and the Open Source nature of Android. The application (eCCM) was developed for Android 3.0 (Honeycomb) or above using Java and the Android Software Development Kit. The Eclipse Integrated Developer's Environment (IDE) and the Android Developer Tools (ADT) Plugin were used to facilitate this development. Eclipse was widely considered the de facto standard development tool for Android at the time of writing. An alternative such as Google's IDE, called Android Studio, was rejected because it was still in early Beta release when this project commenced.

The source code is stored and shared through a GitHub repository. GitHub is an online repository for Open Source and proprietary code storage. It was chosen over similar providers (e.g. BitBucket) for the ease with which it integrated with Eclipse. The project required several software libraries to be installed on each developer's machine. Maven creates a list of these software dependencies in eXtensible Markup Language (XML). When other developers download the source code with the attached XML file, Maven installs all libraries and frameworks required to support that source code.

A Java Enterprise Edition (J2EE) infrastructure with a Spring-based Model View Controller framework was selected for the back-end architecture. Data communication between the Android device and the cloud-based server is facilitated through RESTful requests using JSON data structures. The system architecture underpinning eCCM is depicted in Figure 1. A continuous integration environment has been constructed to support the development process and a suite of automated tests have been created to ensure the stability, integrity and robustness of the decision-support rule engine.

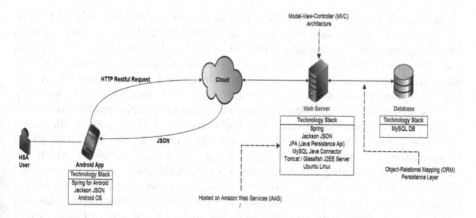

Fig. 1. System Architecture for eCCM on Mobile Device

3.2.2   Requirements Gathering

As part of every ehealth development process it is imperative to understand the requirements underpinning the project. For this project, requirements gathering were initially captured by reviewing the existing paper-based guidelines developed by WHO, UNICEF, and Save the Children Foundation. Using this as a baseline, the developers replicated these guidelines, using exact wording, converting them into electronic format Figure 2 illustrates the paper-based CCM guidelines and electronic CCM (eCCM) guidelines. These rules act to underpin the development of IMCI (as considered in Section 2) and subsequently the CCM classification and treatment XML documents.

Figure 2 (a) illustrates the paper-based CCM guidelines used by HSAs in their diagnosis and treatment of children in Malawi. This is only a small snapshot of the extensive guidelines that must be completed as part of a patient assessment. Figure 2 (b) depicts just one of the many eCCM screens underpinned by the CCM guidelines algorithm.

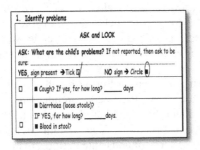 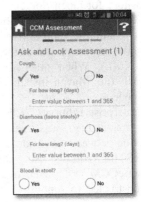

Fig. 2. CCM Paper-Based (a) and electronic CCM (eCCM) (b)

As illustrated in Figure 3 the XML documents are readable and help to provide transparency into the decision-making process of the IMCI/CCM rule engine. The rules can at times be complex (especially in the case of IMCI) due to the number of symptoms and subsequently the number of permutations of the symptoms that result in patient diagnosis however the XML schema is able to support symptom multiplicity and granular symptom specification, whilst also having the capability to dynamically examine the presence of other classifications when determining exact patient illness.

There are currently 20 distinct CCM classifications which can be diagnosed following a patient assessment. Diagnosis is determined by the DSS through its interpretation of multiple combinations of the 34 distinct symptoms which are recorded by a HSA during a medical assessment. Each classification has an associated grouping of recommended treatments which are themselves composed of granular treatment rules. These treatment rules investigate the presence of individual symptoms, combinations of specific symptoms, and/or the presence of other severe classifications when determining the treatments to recommend.

The DSS is time efficient with benchmark tests indicating that CCM classification support requires on average 30ms to complete. This time expenditure covers the reading of the xml-based classification rules from the file system into memory and then cross-referencing the rules against the specific patient symptoms recorded to accurately diagnose patient classifications. A comparable benchmark test for the CCM treatment component of the DSS indicated average time completion in 20ms.

Adopting a rule-based infrastructure facilitates clean separation of the graphical component of the tool and the decision logic. This introduces a key benefit whereby the logical soundness of the decision-making capability of the rule engine can be robustly interrogated and tested for correctness by an automated suite of regression tests. Additionally, the rule-based infrastructure can be easily scaled to support other clinical guidelines (e.g. IMCI) or variants of the CCM guidelines to be deployed in different geographical settings. The veracity of this adaptability and scalability was witnessed by the development team who required only two days development effort to draft the xml-based classification and treatment rules for IMCI. The expansion of the decision rule engine to incorporate IMCI clinical guideline support required minimal code changes and enabled the support of 27 distinct IMCI classifications and 69 potential treatment recommendations that were driven by an analysis of 50 symptoms.

```xml
<?xml version="1.0" encoding="UTF-8"?>
<!-- ============================================================ -->
<!-- CCM CLASSIFICATION RULES --> 
<!-- ============================================================ -->
<!-- Note: The <Symptom> element string must match a string -->
<!-- element in the 'symptom_ids.xml' file (otherwise an -->
<!-- will be thrown at runtime when parsing the -->
<!-- 'ccm_classification_rules.xml' document) -->
<ClassificationRules>
    <!-- ==================================================== -->
    <!-- Classification: Cough for 21 Days or more -->
    <!-- Type: DANGER_SIGN -->
    <!-- Symptoms: Cough -->
    <!-- AND -->
    <!-- Cough Started 21 Days ago or more -->
    <!-- ==================================================== -->
    <Classification>
        <Category>cough_for_21_days_or_more</Category>
        <Name>Cough for 21 Days or more</Name>
        <Identifier>CCM_COUGH_FOR_21_DAYS_OR_MORE_CLASSIFICATION</Identifier>
        <CcmTreatmentDisplayName>Cough</CcmTreatmentDisplayName>
        <Type>DANGER_SIGN</Type>
        <Priority>1</Priority>
        <SymptomRule rule="ANY_SYMPTOM">
            <Symptom value="yes">ccm_ask_initial_assessment_cough_symptom_id</Symptom>
            <!-- Cough : YES -->
        </SymptomRule>
        <SymptomRule rule="ANY_SYMPTOM">
            <Symptom value="yes">ccm_ask_initial_assessment_cough_duration_twenty_one_days_symptom_id</Symptom>
            <!-- Cough for 21 Days or more : YES -->
        </SymptomRule>
    </Classification>
```

Fig. 3. XML for Clinical Guidelines

3.3   Validation

Once the application was developed a workshop was performed with fourteen healthcare providers in Malawi, Africa on 12th November 2013. The researchers aimed for inclusion of the representative population (i.e. healthcare providers), which was achieved. During this workshop, patient scenarios were presented to the healthcare providers. Initially, the participants were required to deliver the necessary healthcare services using the CCM paper-based approach. Subsequently, the healthcare providers were required to complete a different patient scenario using the electronic version, eCCM. This enabled the developers and the research team to gather richer insights into the current paper-based process and identify any shortcomings of the electronic solution. All requirements were captured via a team of four observers during the process. To overcome this issue, the participants were allowed to play around with the application without following any particular patient scenario.

4   Conclusion and Future Work

This paper outlines the development of an XML-based clinical algorithm representing the IMCI and subsequently the CCM guidelines in Malawi. This algorithm forms the decision-base for the Supporting LIFE eCCM cloud-based mobile application which is currently in its fourth development cycle. The extensible, portable, customisable nature of XML provided the optimum technology to build the CCM rule engine. Figure 3 illustrates the accessibility of the XML schema developed for this study.

Testing the robustness of the clinical algorithm is a high priority for the Supporting LIFE project. The eCCM (and subsequently the underlying algorithm) is tested on a continual basis through the test-driven development methodology employed by the senior software developer, the testing undertaken by the clinical lead on the project and ongoing testing with HSAs on the ground in Malawi. The technical feasibility of the mobile application will be conducted in Spring 2015. While rigorous test procedures are in place, it is imperative that adequate training is provided to all HSAs in the utilisation of the CCM guidelines and use of eCCM. While it is highly unlikely, in the event that the eCCM recommends an incorrect treatment the HSAs need to be both knowledgeable and confident enough to use their own judgment and overrule the technology. In terms of integrating the technology into HSA practice, it is important that the eCCM is treated as a sophisticated support tool which should be used to enhance the existing knowledge base of the healthcare provider.

The importance of mobile technical solutions is high priority on the ICT4D agenda [6]. While some achievements have been made in the development of robust technical solutions in the health domain in low resources settings [20], ICT4D 2.0 aspires to move beyond standalone, hardcoded, inflexible mHealth solutions [6]. According to Tomlinson [22, p.4] *"We need to establish an open mHealth architecture based on a robust platform with standards for app development which would facilitate scalable and sustainable health information systems"*. Supporting LIFE eCCM endeavors to achieve this by providing a clinical algorithm that may be leveraged and extended to adapt to the changing IMCI/CCM guidelines in Malawi. Furthermore, there is the potential to use Supporting LIFE eCCM in other developing countries where similar WHO guidelines are in operation. In fact eCCM could be extended to incorporate an accessible user interface (UI) that would facilitate the input of updated or new CCM guidelines. These input forms could be utilised by clinicians to update the CCM algorithm without requiring access to or knowledge of the underlying XML code.

This paper presents a clear rational for adopting XML as part of an mHealth intervention. XML provides a robust and scalable approach when creating an agnostic clinical guideline decision-support rule engine for mHealth applications as it has the structure and flexibility for a wide variety of configurations. More importantly this paper clearly demonstrates that approaches such as XML should be considered when developing mHealth interventions within low resource setting, as it can provide the adaptability of rules from one system architecture to another. This will help ensure that mHealth initiatives become far more sustainable in the future

Acknowledgement. "The Supporting LIFE project (305292) is funded by the Seventh Framework Programme for Research and Technological Development of the European Commission www.supportinglife.eu".

References

[1] Venkatesh, V., Sykes, T.A.: Digital divide initiative success in developing countries: A longitudinal field study in a village in India. Information Systems Research 24, 239–260 (2013)

[2] Cline, G.B., Luiz, J.M.: Information technology systems in public sector health facilities in developing countries: the case of South Africa. BMC Medical Informatics and Decision Making 13, 13 (2013)

[3] Ssekakubo, G., Suleman, H., Marsden, G.: Issues of Adoption: Have E-Learning Management Systems Fulfilled their Potential in Developing Countries? In: Proceedings of the South African Institute of Computer Scientists and Information Technologists Conference on Knowledge, Innovation and Leadership in a Diverse, Multidisciplinary Environment, pp. 231–238. ACM (2011)

[4] Gupta, B., Dasgupta, S., Gupta, A.: Adoption of ICT in a government organization in a developing country: An empirical study. The Journal of Strategic Information Systems 17, 140–154 (2008)

[5] Gomez, R., Pather, S.L.: ICT Evaluation: Are we asking the right questions? The Electronic Journal of Information Systems in Developing Countries 50 (2011)

[6] Heeks, R.: ICT4D 2.0: The next phase of applying ICT for international development. Computer 41, 26–33 (2008)

[7] Sutinen, E., Tedre, M.: ICT4D: A computer science perspective. In: Elomaa, T., Mannila, H., Orponen, P. (eds.) Ukkonen Festschrift 2010. LNCS, vol. 6060, pp. 221–231. Springer, Heidelberg (2010)

[8] Unwin, T.: ICT4D: Information and communication technologies for development. Cambridge University Press, Cambridge (2008)

[9] Dhir, A., Moukadem, I., Jere, N., Kaur, P., Kujala, S., Ylä-Jääski, A.: Ethnographic Examination for Studying Information Sharing Practices in Rural South Africa. In: ACHI 2012, The Fifth International Conference on Advances in Computer-Human Interactions (2012)

[10] Blaya, J.A., Fraser, H.S., Holt, B.: E-health technologies show promise in developing countries. Health Affairs 29, 244–251 (2010)

[11] Gurman, T.A., Rubin, S.E., Roess, A.A.: Effectiveness of mHealth behavior change communication interventions in developing countries: a systematic review of the literature. Journal of Health Communication 17, 82–104 (2012)

[12] Motamarri, S., Akter, S., Ray, P., Tseng, C.: Distinguishing 'mHealth'from other healthcare services in a developing country: a study from the service quality perspective. Communications of AIS 34 (2013)

[13] Kleine, D., Unwin, T.: Technological Revolution, Evolution and New Dependencies: what's new about ict4d? Third World Quarterly 30, 1045–1067 (2009)

[14] Chib, A., van Velthoven, M.H., Car, J.: mHealth adoption in low-resource environments: A review of the use of mobile healthcare in developing countries. Journal of Health Communication, 1–53 (2014)

[15] Chib, A.: The promise and peril of mHealth in developing countries. Mobile Media and Communication 1, 69–75 (2013)

[16] Källander, K., Tibenderana, J.K., Akpogheneta, O., Strachan, D.L., Hill, Z., ten Asbroek, A., Conteh, L., Kirkwood, B.R., Meek, S.R.: Mobile health (mHealth) approaches and lessons for increased performance and retention of community health workers in low-and middle-income countries: a review. Journal of Medical Internet Research 15 (2013)

[17] Tomlinson, M., Rotheram-Borus, M.J., Swartz, L., Tsai, A.C.: Scaling up mHealth: where is the evidence? PLoS Medicine 10, e1001382 (2013)

[18] Lemaire, J.: Developing mHealth Partnerships for Scale. Geneva: Advanced Development for Africa (2013), http://www.healthunbound.org/node/22214

[19] Albrecht, U.V.: Transparency of health-apps for trust and decision making. Journal of Medical Internet Research 15, e277 (2012)

[20] Ruiz-Zafra, Á., Benghazi, K., Noguera, M., Garrido, J.L.: Zappa: An open mobile platform to build cloud-based m-health systems. In: van Berlo, A., Hallenborg, K., Rodríguez, J.M.C., Tapia, D.I., Novais, P. (eds.) Ambient Intelligence & Software & Applications. AISC, vol. 219, pp. 87–94. Springer, Heidelberg (2013)

[21] Marsh, D.R., Gilroy, K.E., Van de Weerdt, R., Wansi, E., Qazi, S.: Community case management of pneumonia: at a tipping point? Bulletin of the World Health Organization 86, 381–389 (2008)

[22] George, A., Menotti, E.P., Rivera, D., Marsh, D.R.: Community case management in Nicaragua: lessons in fostering adoption and expanding implementation. Health Policy and Planning 26, 327–337 (2011)

[23] Iwaya, L., Gomes, M., Simplício, M.A., Carvalho, T., Dominicini, C., Sakuragui, R., Rebelo, M., Gutierrez, M., Näslund, M., Håkansson, P.: Mobile health in emerging countries: A survey of research initiatives in Brazil. International Journal of Medical Informatics 82, 283–298 (2013)

[24] Chigona, W., Nyemba, M., Metfula, A.: A review on mHealth research in developing countries. The Journal of Community Informatics 9 (2012)

[25] Nhavoto, J.A., Grönlund, A.: Mobile Technologies and Geographic Information Systems to Improve Health Care Systems: A Literature Review. JMIR mhealth and uhealth 2, e21 (2014)

[26] DeRenzi, B., Lesh, N., Parikh, T., Sims, C., Maokla, W., Chemba, M., Hamisi, Y., Mitchell, M., Borriello, G.: E-IMCI: Improving pediatric health care in low-income countries. In: Proceedings of the SIGCHI Conference on Human Factors in Computing Systems. ACM (2008)

[27] Mitchell, M., Getchell, M., Nkaka, M., Msellemu, D., Van Esch, J., Hedt-Gauthier, B.: Perceived improvement in integrated management of childhood illness implementation through use of mobile technology: qualitative evidence from a pilot study in Tanzania. Journal of Health Communication 17, 118–127 (2012)

[28] Boonstra, E., Lindbæk, M., Ngome, E.: Adherence to management guidelines in acute respiratory infections and diarrhoea in children under 5 years old in primary health care in Botswana. International Journal for Quality in Health Care 17, 221–227 (2005)

[29] Horwood, C., Voce, A., Vermaak, K., Rollins, N., Qazi, S.: Experiences of training and implementation of integrated management of childhood illness (IMCI) in South Africa: a qualitative evaluation of the IMCI case management training course. BMC Pediatrics 9, 62 (2009)

[30] Wells, J.: Productively Testing the Scalability of MT and TM to the Limit! And Beyond? Translating and the computer (2001)

[31] WHO Report.: The Work of WHO in the African Region. Report of the Regional Director. Congo, September 2-6 (2013)

[32] Boley, H., Tabet, S., Wagner, G.: Design Rationale for RuleML: A Markup Language for Semantic Web Rules. SWWS (2001)

[33] Holzner, S.: Inside XML, New Riders Publishing, Indiana (2001)

[34] Bray, T., Paoli, J., Sperberg-McQueen, C., Maler, E., Yergeau, F.: Extensible markup language (XML). World Wide Web Consortium Recommendation REC-xml-19980210 (1998), http://www.w3.org/TR/1998/REC-xml-19980210

[35] Katehakis, D.G., Sfakianakis, S.: An infrastructure for integrated electronic health record services: the role of XML (Extensible Markup Language). Journal of Medical Internet Research 3 (2001)

Observations of Non-linear Information Consumption in Crowdfunding

Rob Gleasure and Joseph Feller

Department of Accounting, Finance, and Information Systems,
University College Cork, Ireland

Abstract. The number and scale of crowdfunding platforms has grown rapidly within a short period of time. While some of these platforms offer donors specific financial or material rewards, others ask donors to contribute to campaigns for charitable or philanthropic reasons. This fundamental difference means it is difficult to model interpersonal communication and information consumption in simple linear economic terms. Yet to date the dominant research paradigm for charitable crowdfunding has done exactly that. This study seeks to investigate and model non-linear information consumption based upon a field study of Pledgie.com, an established charitable crowdfunding platform. Quantitative analyses of data from over 5,000 individual crowdfunding campaigns reveal several curvilinear relationships between the information provided and the level of funding received. This suggests that information consumption by the donor community is more accurately modelled across two distinct stages, the first of which is discussion-based, the second of which is based on donation.

Keywords: Crowdfunding, Charity, Information consumption.

1   Introduction

Recent years have seen the emergence and rapid growth of crowdfunding, a phenomenon in which a community of funders provides money directly to individuals and organizations without relying upon traditional financial intermediaries (Rossi 2014). This phenomenon has grown from the broader crowdsourcing paradigm in the last decade (c.f. Howe 2008). However, what makes crowdfunding unique is that rather than providing crowdsourcers with access to knowledge and/or specific skills, crowdfunding communities provide the actual capital investment necessary to enable specific projects and outcomes. This has the potential to change much of the financial landscape, including how individuals manage their savings and investments (Gelfond and Foti 2012), how businesses create new products (Ordanini et al. 2011) and how entrepreneurs launch new businesses (Lasrado and Lugmayr 2013). Moreover, the emergence of crowdfunding has changed the economic landscape for philanthropic and charitable markets, which now possess a community of donors proactively seeking opportunities to donate (Heller and Badding 2012, Liu et al. 2012).

   Yet much of existing crowdfunding research assumes a view of the crowd in which relationships between information, interactions, and behaviours are linear in nature.

R. Prasath et al. (Eds.): MIKE 2014, LNAI 8891, pp. 372–381, 2014.

This view is not consistent with observations in other forms of crowdsourcing such as Wikipedia (e.g. Ciffolilli 2003), or open source software development (e.g. Ducheneaut 2005), where relationships and communication dynamics are recognized as evolving over time within the course of specific projects. This is because crowd-supported activities are often motivated by a range of less tangible outputs, a factor particularly salient for crowdfunding projects with strong philanthropic elements (c.f. Oomen and Aroyo 2011, Gambardella 2012).

Thus, this study investigates non-linear information consumption within a charitable crowdfunding market, namely Pledgie.com, and models this information consumption across two distinct but related stages. The next section briefly synthesizes existing crowdfunding literature relevant to non-linear information consumption. This identifies a set of variables that characterize key informational dimensions of a crowdfunding campaign. These variables are then coded in a field study of Pledgie.com and data are analysed to explore non-linear relationships. These findings inform the development of a two-stage model of charitable crowdfunding.

2   Existing Research on Crowdfunding

The term 'crowdfunding' is used to describe the array of web-enabled behaviours focused on harnessing the wealth of crowds. Yet these behaviours vary significantly in terms of the goals for both campaign administrators and donors. Fund-seekers in some crowdfunding platforms offer donors material rewards for contributing towards products and services (e.g. Kickstarter, Indiegogo, RocketHub). Most research on these rewards-based platforms has looked at linear factors such as ideological simi-larities between donors and campaign administrators (e.g. Oomen and Aroyo 2011), similarities to ongoing personal projects (Gambardella 2012), and geographical prox-imity (e.g. Agrawal et al. 2011). Yet there are several notable exceptions, including studies of how meaning is created within the ongoing dialogue between donors and fund-seekers (Beaulieu and Sarker 2013) and of reciprocity-effects benefitting fund-seekers who have previously donated funds to other campaigns (Zvilichovsky et al. 2013).

Other crowdfunding platforms facilitate donations in the form of commercial inter-est-based loans (e.g. Smava, Lending Club, PPDai) or as purchases of business equity (e.g. CrowdCube, EarlyShares). The vast majority of empirical research on these fi-nancially-oriented platforms has conceptualized crowdfunding behaviour in simple economic terms, whereby investment decisions are informed in linear terms according to specific quantifiable criteria (Gleasure and Feller 2014).

The final type of platform differs from the others, as donations are made to those in need with no promise of material reward (e.g. Razoo, Crowdrise, Kiva). Research on these platforms has focused on the motivations and biases of donors, including simi-larities in culture and gender (Riggins and Weber 2012), the presence of health or educational goals (Heller and Badding 2012), and whether beneficiaries are individu-als or groups (Galak et al. 2011). Due to the lack of material rewards, these platforms are arguably likely to manifest most complexity *vis-a-vis* how information is con-sumed and communicated. Yet studies of these charitable systems have largely adopted the same linear view of information consumption as research investigating loan and equity-based crowdfunding (Gleasure and Feller 2014). This has meant heavy reliance on linear and logistic regression analyses of large data sets, which may in isolation overlook important non-linear relationships (King 1986, Osborne and Waters 2002).

This study aims to expand upon this existing crowdfunding research by investigating and theoretically modelling dynamic within-campaigns information consumption across a range of informational factors already known to represent powerful predictors of funding. Moreover, we perform this investigation quantitatively in a manner that lends itself to contrast with those techniques already commonly employed in existing crowdfunding literature.

3   Method: A Field Study of Pledgie.com

Pledgie is selected as an appropriate environment for this study due to its ability to represent the multitude of charitable crowdfunding platforms reasonably faithfully. Firstly, unlike a site such as Razoo.com which has achieved breakaway market-leader status, the scale of Pledgie is representative of the majority of established charitable crowdfunding platforms. Secondly, the informational content and mechanisms for interaction are also reasonably standard. This is in contrast with sites such as Crowd-Tilt, which focus on enabling friends and family to pool money for some cause, or GiveCollege, which focuses on fundraising for student tuition.

Pledgie was established in 2007 to enable 'highly personal' charitable donations as part of online volunteerism. Since then, the website has hosted campaigns across a variety of causes. The basic mechanism for Pledgie is similar to many other crowdfunding platforms, i.e. a campaign administrator launches a campaign with some fundraising target in mind, then other users can donate towards that campaign either openly or anonymously. Each campaign has its own webpage where details are provided, and donors are invited to leave comments or ask questions.

Existing crowdfunding research suggests seven key sources of information that are likely to impact upon behaviour. Firstly, studies of peer-to-peer lending suggest that the amount of funding sought can have a negative relationship with funding behaviour, as funders may be discouraged when large amounts of money are required to achieve project or personal goals (Herzenstein et al. 2011). Conversely, studies of patronage suggest that funders may often increase or decrease funding to meet the demands of specific projects (Kuppuswamy and Bayus 2013). Thus, the fundraising target set by campaigns on Pledgie.com was recorded. Studies of peer-to-peer lending also showed a significant impact from pictures presented with fundraising campaigns (Pope and Syndor 2011). Hence the number of images was also recorded, as well as whether or not a video was included.

Several variables were also included in light of the findings already described from Beaulieu and Sarker (2013) concerning the importance of dialogue. To obtain a high-level measure of dialogue intensity, the number of donors' comments on each campaign was recorded, as well as a separate record of the number of comments made by campaign administrators (hereafter referred to as 'responses'). The proportion of donations made anonymously was also recorded, as levels of anonymity in donation behaviour have been shown to vary with social considerations (Burtch et al. 2013).

The number of pledges made by campaign administrators to other campaigns was also recorded, as this information is visible to potential donors who may be influenced by the potential for reciprocity (Zvilichovsky et al. 2013). Lastly, the number of actual donations was recorded to allow for the possibility that fewer donors contributed larger amounts, or vice-versa. An exhaustive record of individual Pledgie.com campaigns were obtained (N=18,615) and analysed along the variables described above.

4   Findings from Pledgie.com

Of the campaigns analysed only 5,736 (38.2%) attracted funding. Interestingly, while 35,433 different donors contributed during this time period, only 2,607 different members left comments. A five figure summary of the number of comments made by individual commenters [0, 1, 1, 1, 138] suggests that a small number of users are responsible for the majority of discussion. Similarly, of the 78,487 donations made, almost half were made anonymously (44.75%, N=35,121). A five figure summary of the remaining 43,336 is [1,1,1,1,71], suggesting most donors made a single donation while a small dedicated group made up the rest of non-anonymous donations.

All fundraising was converted to US Dollars at standard foreign exchange market rates as of 17th June 2014. Among the subset of 5,736 campaigns that received funding, the average target set was $31,708.58 and the average amount raised was $860.71. This came from an average of 13.41 pledges, of which 0.46 were made anonymously. Each campaign received an average of 0.61 comments, and an average of 0.14 responses to comments by campaign administrators. Only 3% of campaigns included a video in their campaign page, though an average of 0.65 images were included. A five figure summary shows a strong negative skew in funding levels [$00.01, $42, $165, $500, $72,367]. Because of this negative skew and because this funding represents the dependent variable for analysis, a logarithmic transformation was applied to these values to maintain the integrity of the statistical assumptions required for testing. Such transformation makes it difficult to draw conclusions about the size of specific relationships but it does not compromise the ability to detect the presence or absence of positive or negative relationships, nor the extent to which these relationships are linear in nature (Bartlett 1947, Osborne 2002).

A multiple linear regression analysis was performed to test the relationships between these 8 variables and the transformed values for the amount of funding raised (see Table 1). This test is significant (p <.001, adj. R^2 = .154***), though only 4 of the 8 sources of information possess significant relationships. 3 of these 4 sources of information possess positive relationships with fundraising, namely the number of donations, the number of images, and the number of responses to comments by administrators. Conversely the target set for fundraising had a negative relationship.

Several findings from this analysis are not intuitive. Firstly, the overall explanatory power of the combined variables is low. Secondly, the absence of relationships is surprising between the amount of fundraising and the number of comments, the proportion of donations made anonymously, the number of donations to other campaigns by the campaign administrator, and whether or not a video was included. To investigate the possibility that the low explanatory power of these variables is due to non-linear relationships, a series of hierarchical multiple regressions were performed with squared terms as covariates (c.f. Cortina 1993). The variables investigated using this method were the number of comments, the target set for fundraising, the number of pledges made by campaign administrators to other campaigns, and the number of responses to comments by campaign administrators. The proportion of donations made anonymously was also investigated, however because this variable presents a ratio, the square root (rather than the squared term) was used as a covariate. The impact of these covariates was then tested in a series of five moderated hierarchical regressions. Over the course of these tests, the impact of the number of administrators' responses to comments went from highly significant (p < .001, Beta = .079) to non-significant (p = .144, Beta = .026). Thus, the squared term for this value was also investigated within a sixth and final iteration of moderated hierarchical regression analysis.

Table 1. Results of analysis

	Beta
Number of comments	-.003 NS
Proportion of donations made anonymously	-.019 NS
Admin pledges to other campaigns	-.030 NS
Target set for fundraising	-.025*
Admin responses to comments	.079***
Number of images	.119***
Video included or not	-.020 NS
Number of donations	.359***
R^2	.154***
* = p < .10, ** = p < .05 *** = p < .001, NS = Not Significant	

The results of these tests are presented in Table 2. These results show significant relationships for the squared terms for the number of comments, the number of pledges made by campaign administrators to other campaigns, as well as the square root of the proportion of donations made anonymously. Furthermore, the addition of these terms increased the overall adjusted R^2 of the model to .292 (p<.001). This suggests that the ability of the recorded information to predict fundraising is almost doubled when these variables are modelled in non-linear terms. No significant relationships or increases in overall adjusted R^2 were observed from the addition of squared terms for the target set nor the number of responses to comments. Hence models 5 and 6 are abandoned.

Table 2. Results of hierarchical regression investigating curvilinear relationships

Model number	1	2	3	4	5	6
	Beta	Beta	Beta	Beta	Beta	Beta
Number of comments	-.001 NS	.130**	.090**	.102**	.102**	.094**
Number of comments squared		-.126**	-.091**	-.101**	-.101**	-.094**
Proportion of pledges anonymous	-.002 NS	-.001 NS	-1.205***	-1.174***	-1.174***	-1.172***
√Proportion of pledges anonymous			1.253***	1.205***	1.205***	1.203***
Admin pledges to other campaigns	-.024 NS	-.023 NS	-.033**	1.664***	1.663***	1.658***
Admin pledges squared				-1.696***	-1.695***	-1.690***
Target set for fundraising	-.047**	-.046**	-.042**	-.041**	-.080 NS	-.079 NS
Target squared					.040 NS	.038 NS
Responses to comments	.079***	.045**	.030*	.026 NS	.026 NS	.050*
Responses squared						-.028 NS
Number of images	.145***	.147***	.133***	.123***	.123***	.123***
Number of pledges	.345***	.333***	.290***	.246***	.247***	.247***
Video included or not	-.019 NS	-.020 NS	-.024*	-.017 NS	-.017*	-.018 NS
R^2	.154***	.157***	.265***	.292***	.292***	.292***

$* = p < .10, ** = p < .05 *** = p < .001$, NS = Not Significant

5   Discussion

This paper has reported on a field study of Pledgie.com, a charity-based crowdfunding platform, to investigate non-linear information consumption in crowdfunding campaigns. The data suggest that the relationship between the number of comments and the amount of funding received follows an inverted U-shape. This means that while more comments on a campaign are initially associated with more funding, once a critical mass of comments have been received additional discussion becomes predictive of less fundraising. The same is true of the number of donations made by campaign administrators to other campaigns. A higher number of donations to other campaigns by a campaign administrator predicts better fundraising only up to a point, beyond which it becomes negatively associated with funding. Conversely, the proportion of anonymous donations follows a traditional U-shaped relationship with fundraising. This means that a low proportion of anonymous donations is initially encouraging. However once sufficient named donations have been received, a higher proportion of anonymous donations becomes a positive predictor of fundraising.

One explanation for these findings is that, rather than being linear fundraising entities, crowdfunding campaigns on Pledgie.com follow a 2-stage process (see Figure 1). When a campaign is first presented to donors, the community evaluates the project to determine its worthiness. If it is passes some philanthropic threshold, investment then proceeds at a collective level. Such a 2-stage model makes sense given that an equal and uncompetitive distribution of funds would allow few projects to meet their goals. This explains why during the evaluation stage comments are beneficial to fundraising, as without such commentary the necessary discussion of the merits of a project cannot take place. During this stage, discussion is important if consensus over the worthiness of a project is to be reached among the donor community. Such an evaluation of 'deservingness' often impacts significantly on the nature of charitable donations (Eckel and Grossman 1996).

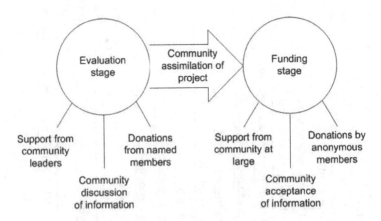

Fig. 1. A Two-stage model of crowdfunding on Pledgie.com

Further, the social effects of charitable donations are greater when they can be associated with specific individuals as expressions of exclusive 'kinship' (Komter 1996). This explains why named donations are initially more effective than anonymous donations on Pledgie.com in terms of building consensus. This also means that a campaign administrator with more existing social ties is likely to fair better than one with fewer ties. Yet as such social ties rely upon exclusionary boundaries to be truly effective (Kurzban and Leary 2001), an unusually high number of ties may actually predict a negative response, if they are seen to lack discretionary filtering. This further provides insights as to why, while some donations by a campaign administrator to other campaigns benefits a campaign, too many donations becomes a negative predictor.

As campaigns progress from evaluation to the funding stage, discussion of the project is likely to decrease when consensus has been reached. Furthermore, as this consensus spreads to other donors less central to the campaigns evaluation, the value of anonymous donations increases proportionally when compared with named donations. This is because exclusionary interactions between the campaign administrator and other donors become decreasingly appealing to the broader potential donor population.

6   Conclusions and Implications

These findings suggest that crowdfunding campaigns on Pledgie.com should not be viewed as linear fundraising entities. Rather, campaigns must first generate adoption from important sections of the community, only after which can large-scale fundraising occur. This is a significant finding for crowdfunding research for both theoretical and methodological reasons. Theoretically, this model serves to elucidate the difference between crowdfunding and traditional financing mechanisms. In particular, it demonstrates that the crowd is not simply a collection of individuals making donation decisions in isolation from one another. It is instead a complex decision-making network based which appears to use social and informational thresholds as cues for focused investment. The authors of this study call for this finding to be investigated further and replicated across other charity crowdfunding websites, as well as other types of crowdfunding platform, i.e. peer-to-peer lending, crowd patronage, and crowd equity investment. Methodologically, this study has implications in light of the dominance of statistical theory building/testing methods that have been applied within the crowdfunding research stream, many of which assume linear information/behaviour relationship. These studies have identified a number of significant relationships, the nature of which may differ considerably from what was hypothesized.

These findings also have important implications for crowdfunding practice, within which fundraising is often viewed as the result of unstructured democratic decision-making. The findings from this study contradict this view, and suggest that campaign administrators should consider individual investments from crowd members, not only in terms of quantity, but also in terms in social endorsement.

References

Agrawal, A., Catalini, C., Goldfarb, A.: Friends, Family, and the Flat World: The Geography of Crowdfunding. National Bureau of Economic Research Working Paper:16820s

Beaulieu, T., Sarker, S.: Discursive Meaning Creation in Crowdfunding: A Socio-Material Perspective. In: International Conference for Information Systems, Milan, Italy (2013)

Burtch, G., Ghose, A., Wattal, S.: An empirical examination of the antecedents and consequences of contribution patterns in crowd-funded markets. Information Systems Research 24(3), 499–519 (2013)

Chen, D., Han, C.: A Comparative Study of Online P2p Lending in the USA and China. Journal of Internet Banking & Commerce 17(2), 1–15 (2012)

Ciffolilli, A.: Phantom authority, self-selective recruitment and retention of members in virtual communities: The case of Wikipedia. First Monday 8(12) (2003)

Ducheneaut, N.: Socialization in an open source software community: A socio-technical analysis. Computer Supported Cooperative Work 14(4), 323–368 (2005)

Eckel, C.C., Grossman, P.J.: Altruism in anonymous dictator games. Games and Economic Behavior 16(2), 181–191 (1996)

Gambardella, M.: How to (Crowd-) Fund and Manage the (User-) Innovation: The Case of Big Buck Bunny. In: Proceedings of the Workshop on Open Source and Design of Communication, pp. 51–56. ACM, Lisbon (2012)

Gelfond, S.H., Foti, A.D.: Us$500 and a Click: Investing the "Crowdfunding" Way. Journal of Investment Compliance 13(4), 9–13 (2012)

Gleasure, R., Feller, J.: From the Wisdom to the Wealth of Crowds: A Metatriangulation of Crowdfunding Research. TOTO Research Project working paper V5 (2014)

Heller, L.R., Badding, K.D.: For Compassion or Money? The Factors Influencing the Funding of Micro Loans. The Journal of Socio-Economics 41(6), 831–835 (2012)

Herzenstein, M., Sonenshein, S., Dholakia, U.M.: Tell Me a Good Story and I Lend You Money: The Role of Narratives in Peer-to-Peer Lending Decisions. Journal of Marketing Research (48), S138–S149 (2011)

Howe, J.: Crowdsourcing: How the power of the crowd is driving the future of business. Random House (2008)

Komter, A.E.: Reciprocity as a principle of exclusion: Gift giving in the Netherlands. Sociology 30(2), 299–316 (1996)

King, G.: How not to lie with statistics: Avoiding common mistakes in quantitative political science. American Journal of Political Science 30(3), 666–687 (1986)

Kuppuswamy, V., Bayus, B.L.: Crowdfunding Creative Ideas: The Dynamics of Projects Backers in Kickstarter. SSRN Working Paper Series (2013), http://business.illinois.edu/ba/seminars/2013/Spring/bayus_paper.pdf

Kurzban, R., Leary, M.R.: Evolutionary origins of stigmatization: the functions of social exclusion. Psychological Bulletin 127(2), 187–208 (2001)

Lasrado, L.A., Lugmayr, A.: Crowdfunding in Finland–a New Alternative Disruptive Funding Instrument for Businesses. In: Proceedings of International Conference on Making Sense of Converging Media. ACM, New York (2013)

Liu, Y., Chen, R., Chen, Y., Mei, Q., Salib, S.: "I Loan Because..": Understanding Motivations for Pro-Social Lending. In: Proceedings of the Fifth ACM International Conference on Web Search and Data Mining, pp. 503–512. ACM, Seattle (2012)

Oomen, J., Aroyo, L.: Crowdsourcing in the Cultural Heritage Domain: Opportunities and Challenges. In: Proceedings of the 5th International Conference on Communities and Technologies, pp. 138–149. ACM, Brisbane (2011)

Ordanini, A., Miceli, L., Pizzetti, M., Parasuraman, A.: Crowd-Funding: Transforming Customers into Investors through Innovative Service Platforms. Journal of Service Management 22(4), 443–470 (2011)

Osborne, J., Waters, E.: Four assumptions of multiple regression that researchers should always test. Practical Assessment, Research & Evaluation 8(2), 1–9 (2002)

Pope, D.G., Sydnor, J.R.: What's in a Picture? Evidence of Discrimination from Prosper.Com. Journal of Human Resources 46(1), 53–92 (2011)

Riggins, F.J., Weber, D.M.: A Model of Peer-to-Peer (P2p) Social Lending in the Presence of Identification Bias. In: Proceedings of the 13th International Conference on Electronic Commerce, pp. 1–8. ACM, Liverpool (2012)

Rossi, M.: The New Ways to Raise Capital: An Exploratory Study of Crowdfunding. International Journal of Financial Research 5(2), 8–18 (2014)

Zvilichovsky, D., Inbar, Y., Barzilay, O.: Playing Both Sides of the Market: Success and Reciprocity on Crowdfunding Platforms. In: International Conference for Information Systems, Milan, Italy (2013)

The Linked Data AppStore

A Software-as-a-Service Platform Prototype
for Data Integration on the Web

Dumitru Roman, Claudia Daniela Pop, Roxana I. Roman, Bjørn Magnus Mathisen,
Leendert Wienhofen, Brian Elvesæter, and Arne J. Berre

SINTEF
Forskningsveien 1, Oslo, Norway
`dumitru.roman@sintef.no`

Abstract. This paper introduces The Linked Data AppStore (LD-AppStore) – a
Software-as-a-Service platform prototype for data integration on the Web.
Building upon emerging Linked Data technologies, the LD-AppStore targets
data scientists/engineers (interested in simplifying tasks such as data cleaning,
transformation, entity extraction, data visualization, crawling, etc.) as well as
data integration tool developers (interested in exploiting the use of their tools by
data engineers). This paper provides an overview of the architecture of the LD-
AppStore, the APIs of the basic data operations supported by the platform,
presents a set of data integration workflows, and discusses the current status of
the implementation.

Keywords: Open Data, Linked Data, Data Integration, APIs, SaaS, Data
Workflows, Prototype.

1 Introduction and Motivation

Linked Data [1] is a technological approach for representing and linking data on the
Web within and across domains with the purpose of making it more useful. It is primarily based on standards developed by W3C: URIs for identifying things, HTTP for
retrieving resources, RDF [2] for representing data, and SPARQL [3] for querying
and updating data. A primary application domain of Linked Data so far has been data
integration from disparate Web sources, and it is seen as a potential solution addressing the heterogeneity aspect of big data [4].

While in recent years a significant amount of datasets have been made available as
Open (and often as Linked) Data,[1] applications making use of such data have been
rather few. For example, as of end of 2014, the official EU public open data portal[2]
lists more than 48,000 datasets but has information on less than 80 applications

---

[1] See for example the open data portals listed at `http://www.datacatalogs.org/`
(currently 290).
[2] `http://publicdata.eu/`

R. Prasath et al. (Eds.): MIKE 2014, LNAI 8891, pp. 382–396, 2014.

making use of that data; the UK data portal[3] lists less than 20,000 datasets and 350 applications. The situation is not much different for other data portals. The reasons for relatively low usage of such open/linked data include: the technical complexity and economical cost of integration, publishing, interlinking and providing reliable access to the data both for data publishers and consumers; lack of simplified and unified solutions for data consumption; and lack of tools and infrastructures where datasets and 3rd party components can be made easily available to application developers to reuse, combine and develop novel data-driven and data-intensive applications. At present, Linked Data publishers and application developers need to rely on generic platforms (like the Amazon Web Services or Google App Engine cloud providers), and build, deploy and maintain complex Linked Data software and data stacks from scratch. Tools addressing various aspects of data integration process, though available in a Linked Data context, can barely be used together in a simplified manner for more advanced data integration tasks. This situation results in a high cost of data integration and a rather complicated and time consuming process, calling for new mechanisms for supporting open/linked data publishers and application developers.

To support open/linked data publishers and application developers, and to simplify the open/linked data integration process, this paper introduces a concept and proto-type–*The Linked Data AppStore (LD-AppStore)*–for data scientists/engineers aiming to enable them, in an effective way, the use of tools for tasks such as data cleaning, transformation, entity extraction, data visualization, crawling, etc. At the same time, the LD-AppStore aims to enable data integration tool developers to exploit the use of their tools by plugging them into the LD-AppStore.

LD-AppStore is primarily motivated by supporting and simplifying workflows of typical data integration tasks. Examples of such workflows include:

- o Extract and select data from a relational database, extract entities from a document, discover and create links between the extracted entities and se-lected data, create a new dataset with the links and the associated data/entities, store the new dataset, query the dataset and visualize the results.
- o Extract data from a CSV file, clean it (e.g. remove duplicates), link the cleaned data to an external dataset, create a new dataset with the links and the associated data/entities, query the dataset and visualize the results.
- o Identify event patterns in a dataset, create query templates from the identified patterns, apply the query templates on real time data, and visualize the results.

Whereas it is possible to achieve such integration scenarios by using and combin-ing existing approaches and tools on a small scale, the complexity associated with creating and maintaining an environment where such data analysis can be performed at a large scale goes beyond the capabilities of existing systems. Many tools require substantial prior expertise which is not easily available to data scientists/engineers interested in a simplified process of data integration workflows. Moreover, data inte-gration tool developers do not have an easy way to integrate the functionalities of their tools with functionalities provided by other tools, that when chained together in a data integration workflow can create added-value services for data integration. The

---

[3] http://data.gov.uk/

LD-AppStore introduced here aims to address simplification of the data integration process for data scientists/engineers and offer data integration developers means to expose the functionalities of their tools to a wider data engineers base. This paper reports on the status of the LD-AppStore, with a focus on its architecture, APIs, data integration workflows, and tools that have been integrated into the platform.

The rest of this paper is organized as follows. Section 2 briefly discusses relevant related work. Section 3 provides an overview of the LD-AppStore introducing its architecture, basic data operations supported by the platform together with their APIs, and a preliminary design of examples of data integration workflows aimed to be supported by the platform. Section 4 describes the current implementation of the LD-AppStore and the functionalities of the tools that have been integrated into the platform. Section 5 summarizes the paper and suggests further steps for improving the current implementation.

2   Related Work

The LD-AppStore approach follows the research line of bundling well-established technologies and tools for publishing, provisioning, and consuming Linked Data in order to simplify integration of disparate data sources on the Web. Notable approaches developed in this area include toolchains such as the Linked Data Stack[4] and the LarKC platform[5].

The Linked Data Stack [5] provides a number of loosely coupled tools for managing various operations on Linked Data such as data extraction, storage, querying, linking, classification, search, etc. The focus of the Linked Data Stack is on the development of the actual tools supporting various operations. Although the tools are bundled in a Debian package and a Web application can be deployed to offer a central access to all the tools via a Web interface[6], the Linked Data Stack does not provide an as-a-service hosted solution where 3rd party tool developers can plug-in their implementations for different data operations and where data publishers can configure and execute workflows of data operations implementations on their data --- which is what LD-AppStore targets. Furthermore, LD-AppStore takes a top-down approach to data operations in the sense that it prescribes a set of Web APIs for various data operations (that existing tools can implement), enabling thus easier ways for creating workflows of data operations independent of their actual implementations.

The LarKC platform [6] aimed to provide an experimental platform for massive, distributed and incomplete reasoning on the Web by developing an environment to execute 'plug-ins'. There are different categories of plug-ins such as transforming data from one representation to another, applying different kinds of reasoning (deductive, inductive), etc., which can be assembled in to a 'pipeline', where the output of each plug-in feeds in to the input of the next plug-in in sequence. LarKC is a generic platform for reasoning on the Web (not specifically designed to support publication of Linked Data) and not meant to be provided as-a-Service.

---

[4] http://stack.linkeddata.org/
[5] http://www.larkc.eu/
[6] http://demo.lod2.eu/lod2demo

DaPaaS[7], COMSODE[8], and LinDA[9] are a number or recent EU funded research projects addressing the problem of simplifying access, integration, and usage of open data based on Linked Data technologies, primarily focusing on data publication and consumption aspects. The projects are in early stages of development with their approaches not entirely defined at this stage, however ideas from the LD-AppStore are finding traction in the DaPaaS project.

Relevant commercial products and services include Azure Data Marketplace[10], Socrata[11], GoodData[12] and Datameer[13]. They are SaaS solutions for data integration, publication and consumption, however none of them focus on the use of Linked Data.

3   The LD-AppStore Platform Overview

3.1   Architecture

The LD-AppStore targets primarily two types of users:

- o *Data Scientists/Engineers*, interested in analysing data through a set of data operations, such as data cleaning, transformation, storage, linking, visualization, etc., and various combinations of such operations.
- o *Linked Data tool developers*, interested in developing tools for implementing various data operations that data scientists/engineers may be interested in.

The LD-AppStore was designed as a service where data engineers can get access to various types of data operations that they can apply on their data, and have access to different implementations of those data operations – implementations provided by developers. It is meant to serve as a registry of data operations and their implementations, offering data engineers enhanced features for data integration/analysis in terms of chaining various data operations.

Figure 1 provides a high level overview of the LD-AppStore architecture. On the upper side of the figure, components for basic data operations are depicted: RDF-ization of relational data (mapping relational tables to RDF graphs), data visualization (visualization of RDF graphs), entity extraction (extracting entities from various sources), data storage (storage of RDF data manipulated in the platform), link discovery (finding links between data in RDF graphs), crawling (searching through RDF graphs), and data streaming (querying streams of RDF data). A set of Web APIs have been designed for these data operations and are introduced in Section 3.2. The tools that implement these basic data operations are made available through the registry functionality of the platform (lower right part of Figure 1). When using a specific data operation, the data engineer may select which implementation of that operation she

---

[7] http://project.dapaas.eu/
[8] http://www.comsode.eu/
[9] http://linda-project.eu/
[10] http://datamarket.azure.com/
[11] http://www.socrata.com/
[12] http://www.gooddata.com/
[13] http://www.datameer.com

wants to use. The Linked Data tool developers have access to the platform for registering their implementations, i.e. the implementations of the Web APIs corresponding to the data operations APIs.

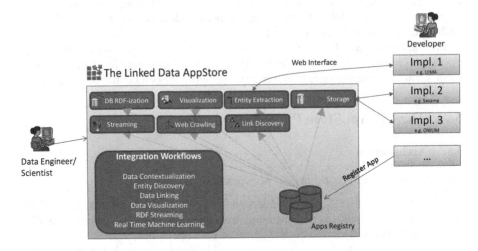

Fig. 1. LD-AppStore Architecture Overview

The lower left side of the picture depicts a set of data integration workflows meant to seamlessly combine the basic data operations in data workflows/pipelines (configurable by the data engineers) that can eventually provide further useful insights into the data on which they are applied. A set of workflows have been designed and are briefly introduced in Section 3.3.

Section 4 provides an overview of the current implementation of the LD-AppStore together with the tools that have been modified to implement the APIs of the operations considered in the LD-AppStore.

3.2   Basic Data Operations Web APIs

The platform offers six different types of basic data operations. In the following we briefly introduce the Web APIs (JSON request/response parameters) designed for them.

DB-RDFization. The purpose of this operation is to map data from relational databases to RDF, for exposing relational data as a Linked Data.

A request to the DB-RDFization operation consists of the following attributes: *mode* (the type of conversion, either R2RML [7] or Direct Mapping [8]); *database_name; database_pass; database_user; base_uri; database_url; sparql_output; r2rml_file; sparql; format.*

The response contains the following attributes: *uri* (the URI of the result file); *status* (the execution status of the operation; if any errors occur the message errors will be stored in the status attribute).

The interface can be accessed at a URI having the following form: http://[host:port]/db2triples/rdfization.json. The format is:

```
{"mode":null,"database_name":null,"database_pass":null,"database_us
er":null,"base_uri":null,"database_url":null,"sparql_output":null,"
r2rml_file":null,"sparql":null,"format":null}
```

For the R2RML operation a R2RML configuration file is required. For this, a POST request has to be made, containing as attachment the R2RMLfile.

Entity Extraction. This operation is used to extract relevant entities from various sources. In order to have access to the source file from which the entities are to be extracted, a POST request containing the source file has to be made in order to store the file.

The interface for posting the file can be accessed at a URI having the following form: http://[host:port]/uima/savefile.json.

After the POST operation with the attached file has been made, the service can be accessed by with another POST request consisting of the following attributes: *url* (the URL of the file to process); *filename* (name of the file to process), *annotator* (describes the annotator (Java class) used for entity searching).

The response consists of the following attributes: *uri* (the URI of the resulted file); *status* (this represents the execution status of the operation; if any errors occur the message errors will be stored in the status attribute).

The interface can be accessed at a URI having the following form: http://[host:port]/uima/entityExtraction.json. The format is:

```
{"url": null,"filename": null,"annotator": null}
```

Data Visualization. Understanding the RDF data is easier if visualized in an interactive and simple way. Provided with a resource address it offers the possibility for viewing a representational graph of the data. Since typically visualization of data runs in the client's browser, currently this operation does not have a Web API.

Storage. This operation offers the possibility to store the data manipulated through the LD-AppStore platform, and query it. There are three sub-operations as follows.

Create Repository. This provides the possibility to create a local repository where data can be stored.

A typical request for this operation consists of the following attributes: *name* (the name of the repository); *type* (the type of repository); *query* (not used for this operation, can be null); *filename* (not used for this operation, can be null); *url* (not used for this operation; can be null).

The interface can be accessed at a URI having the following form: http://[host:port]/sesame/createRepo.json.

Add Data to Repository. This is used to add RDF data to a repository. In order to store a RDF file, a POST operation with the file attachment must be made.

The interface for posting the file can be accessed at a URI having the following form: http://[host:port]/sesame/savefile.json.

After the POST operation with the attached file has been made, the service can be accessed by performing another POST operation with a request consisting of the attributes: *name* (the name of the repository); *type* (the type of repository); *query* (not used for this operation, can remain null); *filename* (the name of RDF file which is going to be stored); *url*: the URL of the RDF file which is going to be stored.

The interface can be accessed at a URI having the following form: http://[host:port]/sesame/addRdf.json.

Query Repository. This is used for querying the repository. The request consists of the following attributes: *name* (the name of the repository); *type* (the type of repository); *query* (SPARQL query); *filename* (not used for this operation, can remain null); *url* (not used for this operation, can remain null).

The interface can be accessed at a URI having the following form: http://[host:port]/sesame/query.json.

The response for all the 3 sub-operations described above consists of the following attributes: *filename* (file containing the results); *results* (string list of results); *status*.

Streaming. This operation allows querying over streams of data. The request contains the following attributes: *url* (URL of the streaming location); *query* (C-SPARQL query); *queryName* (a unique name of the query).

The interface can be accessed at a URI having the following form: http://[host:port]/csparql/streaming.json.

The response contains: *urlViewResults* (URL of the location where results can be viewed); *urlPostedResults* (URL of the location where results were posted); *status*; *serverstatus* (status of the C-SPARQL server: started, shut down, errors).

Link Discovery. This is a data operation for discovering relations between different data sources.

A configuration file needs to be provided for this operation. In order to save the configuration file, a POST operation with the file attachment must be made. The interface for posting the file can be accessed at a URI having the following form: http://[host:port]/silk/savefile.json.

After the POST operation with the attached file has been made, the service can be accessed by performing another POST operation with the request containing the attributes: *url* (URL of the configuration file to be imported); *linkageid* (id found in the configuration file); *filename* (name of the imported file).

The interface can be accessed at a URI having the following form: http://[host:port]/silk/datalinkage.json.

The response contains: *filename* (of the file containing the results); *status*.

Web Crawling. Crawling the Linked Data web is related to search operation. The request contains the following attributes: *uri* (URI containing Linked Data to be crawled).

The response consists of the attributes: *uri* (URI of the file containing the results); *status*.

The interface can be accessed at a URI having the following form: http://[host:port]/ldspider/webcrawling.json.

3.3 Data Integration Workflows

Whereas the data operations introduced above are useful elementary operations one may need to deal with when doing data integration in a large scale context, seamless combination of these (and other) operations in more complex workflows / pipelines, easily configurable by the users, can provide further useful insights into the data on which they are applied. This section provides examples of such workflows of basic data operations, aimed to be supported by the LD-AppStore platform. Most of the workflows take data from various sources and apply some processing, resulting in data that is supposedly more valuable and more easily accessible.

The following figures each represent a workflow. Boxes represent tasks in the workflows and correspond to processing operations on the data. Each task is introduced with a name, followed in parentheses by examples of tools that can be used for the operations. The arrows represent the flow of operations, i.e. data resulting from one operation serve as input for another operation. Figures 2 to 4 consist of workflows related to publication of RDF data through a set of improvements applied to the original (primarily non-RDF) data. Figure 5 is related to visualization of data resulting from the previous workflows. Figure 6 and 7 are related to management of real time data.

Figure 2 introduces the Data Contextualization workflow where the basic idea is to provide basic context to data extracted from a multitude of formats (pdf, Excel, relational tables, etc.). Data is first brought into a common syntactical format (CSV) after which it can be integrated with external data, cleaned, exported to RDF and finally stored, being made accessible for querying at a later stage.

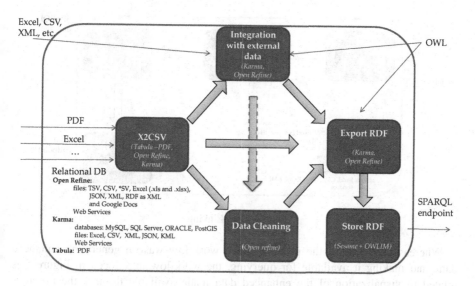

Fig. 2. Data Contextualization

Figure 3 introduces the Entity Discovery workflow where the basic idea is extracting entities from a variety of data source, providing context to the data and make it available for querying. Note that this workflow makes use of the previous workflow.

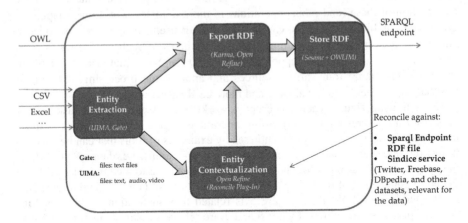

Fig. 3. Entity Discovery

Figure 4 introduces the Data Linking workflow where the idea is to apply link discovery techniques to data from various sources and formats. Data is first RDFized and stored, after which the link discovery operation is applied for discovering and creating links in the RDFized data sources. The links are then stored together with the data and made available for querying.

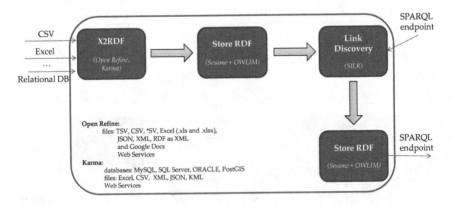

Fig. 4. Data Linking

Whereas the focus of the previous three workflow was on generating enhanced data, and making it available for querying, the workflow introduced in Figure 5 is related to visualization of the enhanced data made available through the previous workflows.

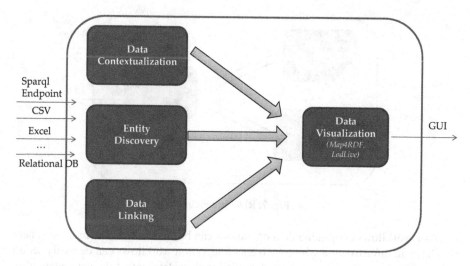

Fig. 5. Data Visualization

The last two workflows introduced here deal with management of real time data.

Figure 6 proposes the Real-time Machine Learning workflow where the idea is to use outputs from the workflows introduces above, in combination with real-time data (eventually RDF-ized) to identify patterns that in turn can produce query templates.

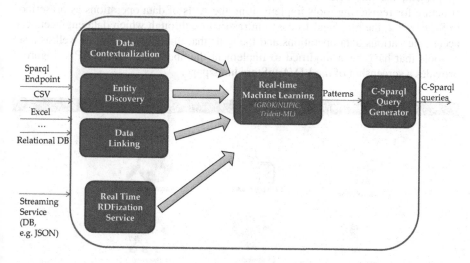

Fig. 6. Real-time Machine Learning

Figure 7 introduces the RDF Streaming workflow which is tightly related to the previous workflow. The query templates produced by the previous workflow are applied to real time data to produce real-time answers.

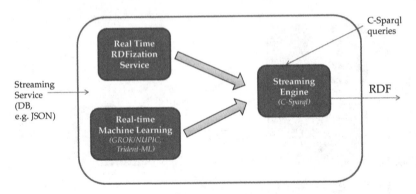

Fig. 7. RDF Streaming

Other workflows combining data operations can be envisioned. The basic idea here is to have an infrastructure where such data integration workflows can be easily set up and configured by data engineers, and applied to their data so that the data integration process is significantly simplified.

4   The LD-AppStore Prototype

At present, the implementation of the LD-AppStore consists of the backend infrastructure for registering tools implementing the APIs of data operations as prescribed in Section 3.2, the graphical frontend infrastructure through which data engineers can access the various data operations and the tools that implement them, as well as a set of tools that have been modified to implement the above mentioned APIs. Figure 8 provides a screenshot of the LD-AppStore homepage.

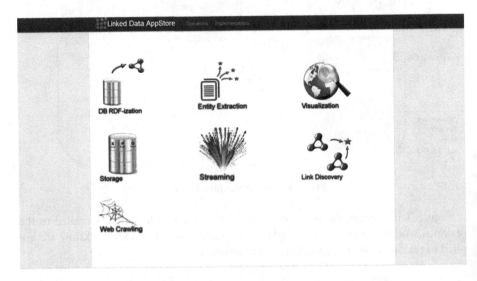

Fig. 8. Screenshot of the LD-AppStore homepage

Technologies. The programming language used for implementing the platform was Java and the framework used is Spring MVC.[14] Maven has also been used as a core technology of the platform and also for the integrated tools wrapper implementations. For the database server, MySQL was chosen. The deployment of the platform as well as for the tools integrated was done with Tomcat 7.

For a better structuring and understanding of the code the layers pattern[15] has been used. The repository layer has access to the entity layer. The service layer has access to the repository and entity layers. The controller layer has access only to the service layer and entity layer. The API layer contains the classes (representing operations) which are mapped to the JSON interface for each operation.

Integrated Tools. The platform offers the possibility to register new tools as implementations for various operations. For each of the already registered tools a programmatic Web interface has been made which follows the one for its corresponding operation. In this way, implementation independence has been obtained, as long as each of the newly added tools implement the operation's interface. For now, the interface of each operation coincides with the interface of the tool corresponding to the interface. New integrated tools would not need all the attributes, so they will expose to the end user only the needed attributes and let the others empty. The following tools have been integrated in the current prototype.

DB2Triples. DB2Triples[16] is an implementation of the R2RML and Direct Mapping specifications. It has been registered with the DB-RDFization operation. Therefore a programmatic Web interface which complies with the DB-RDFization interface has been implemented in DB2Triples. The request attributes which must be provided are: *mode* (the type of conversion, either R2RML or Direct Mapping); *data-base_name; database_pass; database_user; base_uri; database_url; sparql_output; r2rml_file; sparql; format.*

UIMA. The Unstructured Information Management Architecture (UIMA) is an architecture and software framework for creating, discovering, composing and deploying a broad range of multi-modal analysis capabilities and integrating them with search technologies.[17] UIMA has been registered with the Entity Extraction operation. The attributes which must be provided from the Entity Extraction operation interface are: *url* (URL of the file to process); *filename* (name of the file to process); *annotator* (the annotator (java class) used for entity searching).

LodLive. LodLive[18] is a Web-based application for graphical browsing of RDF data. In the LD-AppStore prototype it is used for the visualization operation (this is not wrapped with Web services, since it is runs in the client's browser).

---

[14] Spring MVC - an open source web Model-View-Controller application framework and inversion of control container for the Java platform.

[15] Layers Pattern - a design pattern, which aims to organize the code into a set of logical layers.

[16] https://github.com/antidot/db2triples

[17] https://uima.apache.org/

[18] http://en.lodlive.it/

Sesame. OpenRDF Sesame[19] is a framework for processing RDF data and includes parsers, triplestores, and SPARQL querying capabilities. Sesame has been used in the LD-AppStore prototype for performing the following storage operations: Create Repository, Add Data to Repository, Query Repository. The request attributes needed for Create Repository operation with Sesame are: *name* (the name of the repository); *type* (the type of repository); *query* (can remain null); *filename* (can remain null); *url* (can remain null). The attributes needed for the add operation with Sesame are: *name* (the name of the repository); *type* (the type of repository); *query* (can remain null for this operation); *filename* (the RDF file which is going to be stored); *url* (the URL of the RDF file which is going to be stored). Finally, the attributes needed for Query Repository operation with Sesame are: *name* (the name of the repository); *type* (the type of repository); query (SPARQL query); *filename* (can remain null); *url* (can remain null).

C-SPARQL. Continuous SPARQL (C-SPARQL)[20] is a language and engine for continuous queries over streams of RDF data. C-SPARQL is the streaming tool integrated in the prototype for the streaming operation. The request attributes which need to be provided are: *url* (URL of the streaming location); *query* (C-SPARQL query); *query-Name* (a unique name of the query).

Silk. The Silk framework[21] is a tool for discovering relationships between data items within different Linked Data sources. The Silk framework is the tool registered for Link Discovery operation in the LD-AppStore prototype. The request attributes needed are: *url* (URL of the configuration file to be imported); *linkageid* (id found in the configuration file); *filename* (name of the imported file).

LDSpider. LDSpider[22] is a Web crawling framework for RDF data. For crawling the web of data (Web Crawling operation) LdSpider has been integrated in the prototype. The request parameter which need to be provided is *uri* (URI containing linked data to be crawled).

Further details about the tools that have been selected for inclusion can be found in the report of the BigFut project [9].

5   Summary and Outlook

This paper introduced the concept of an applications store for data integration on the Web and reported on the prototype that has been developed so far. The target users for such an infrastructure are data scientists/engineers that are interested to use, in a rather simplified manner, tools for tasks such as data cleaning, transformation, entity

---

[19] http://www.openrdf.org/
[20] http://streamreasoning.org/download/
[21] http://wifo5-03.informatik.uni-mannheim.de/bizer/silk/
[22] https://code.google.com/p/ldspider/

extraction, data visualization, crawling, etc. At the same time, data integration tool developers have the possibility of exploiting the use of their tools by plugging them into the LD-AppStore.

On a longer term, to make the LD-AppStore a successful concept and be deployable at a large scale, with a large user base, several things need to be enhanced and further evolved:

- o *More Tools to Be Included in the AppStore.* Currently a number of relevant tools have been included in the prototype, however more would need to be included to offer data engineers choices for implementations of data operations.
- o *Implementation of the Data Integration Workflows.* The work so far has been focused on the APIs of the basic data operations and their implementation by some of the existing tools, however significant benefits come when the users are able to create, use and configure their own workflows of basic operations.
- o *Improvement of the Prototype.* The current prototype is meant to demonstrate the concept of an AppStore for Linked Data, however to be able to deal, for example, with large data sets and large number of concurrent users, a more reliable infrastructure is needed.

For more capabilities of generating RDF data from legacy data, other tools would have to be integrated in the platform. On one hand, this will offer a wider range of options for the user (e.g. could convert data from an Excel file to RDF), on the other hand, it will encourage other developers to add their implementations to the platform. One tool which can be relevant is OpenRefine[23] for working with messy data, cleaning it up, transforming it from one format into another, extending it with Web services, and linking it to databases. Another example is Karma [10] which is a tool[24] that enables users to quickly and easily integrate data from a variety of data sources including databases, spreadsheets, and delimited text files, XML, JSON, KML and Web APIs. A third option is Grafter[25] (a Clojure library/framework) that allows users to define pipelines for importing and transforming tabular data to RDF.

A graphical domain-specific language (DSL) to support data transformation and cleaning, which builds upon Grafter and incorporates ideas from both OpenRefine and Karma, is now being developed as a tool that will be offered in the LD-AppStore. Currently the LD-AppStore implementation is closed source, but an open source version is planned for the future.

Acknowledgements. This work was partly funded by the following projects: BigFut (SINTEF internally funded project 102003299), DaPaaS (FP7 610988)[26], SmartOpenData (FP7 603824)[27], and InfraRisk (FP7 603960)[28].

---

[23] http://openrefine.org/
[24] http://www.isi.edu/integration/karma/
[25] http://grafter.org/
[26] http://project.dapaas.eu/
[27] http://www.smartopendata.eu/
[28] https://www.infrarisk-fp7.eu/

References

[1] Linked Data. Introduction at http://www.w3.org/standards/semanticweb/data.

[2] RDF. Introduction at http://www.w3.org/TR/rdf11-concepts/.

[3] SPARQL. Introduction at http://www.w3.org/TR/sparql11-overview/.

[4] I. Mitchell and M. Wilson. Linked data – Connecting and exploiting big data, 2012, Fujitsu White paper. Available at http://www.fujitsu.com/uk/Images/Linked-data-connecting-and-exploiting-big-data-%28v1.0%29.pdf.

[5] S. Auer et al. Managing the Life-Cycle of Linked Data with the LOD2 Stack. ISWC, Springer, 2012.

[6] D. Fensel et al. "Towards LarKC: a platform for web-scale reasoning.", IEEE International Conference on Semantic Computing, 2008.

[7] R2RML: RDB to RDF Mapping Language, W3C Proposed Recommendation 14 August 2012, Available via http://www.w3.org/TR/2012/PR-r2rml-20120814/.

[8] A Direct Mapping of Relational Data to RDF, W3C Proposed Recommendation 14 August 2012, Available via http://www.w3.org/TR/rdb-direct-mapping/.

[9] D. Roman, C. D. Pop, R. I. Roman, B. M. Mathisen, L. Wienhofen, and B. Elvesæter. The Linked Data AppStore v1. SINTEF Internal Report BigFut SEP 102003299, November 2013.

[10] Gupta, Shubham, et al. "Karma: A system for mapping structured sources into the Semantic Web." 9th Extended Semantic Web Conference (ESWC2012).(May 2012). 2012.

Geospatial Decision Support Systems:
Use of Criteria Based Spatial Layers
for Decision Support in Monitoring of Operations

Shanmugavelan Velan

Indian Institute of Technology Kharagpur, WB 721302, India
velan_vs@iitkgp.ac.in

Abstract. This paper brings out a conceptual approach for geospatial analysis for decision making process while monitoring operations. It capitulates on the ability of GDSS to support GIS layers. The decisions, decision making processes and the time involved are all broken down to a function of criteria over a set of information. This information is recommended to be stored in layers in special criteria based spatial form which reduces the processing requirements to simple manipulation of specific independent layers. This enables incremental decisions based on adequate set of data as and when required, which is akin to anytime algorithm. The human and machine analyses are integrated in a geo-visual analytical model. A case study of evacuation of an injured person from a mine using helicopter is presented. In this example all advantages of the concept are proven.

Keywords: geospatial decision support, spatial decision support, anytime algorithm, monitoring of operations, geo intelligence generation.

1 Introduction

Geospatial Decision Support Systems (GDSS) provides us tools to handle complex operational problems that managers are faced with in their day to day management of operations. Spatial Decision Support System (SDSS) is defined as " software that is intuitively obvious to use, solves their specific problems efficiently, and delivers immediate results" [9]. It is different from Geospatial Information System (GIS) though it uses GIS technology [4]. Desham P. J. explains it as 'it can be viewed as spatial analogues for decision support systems (DSS) to address business problems' [6] and can be extended to operations research [6]. Anytime algorithm concept, which originated from the study on time dependent planning [5], provides us a means of arriving at a solution within the given time and information constraint. It is specifically useful for management of operations where time uncertainty exists since, the "quality of results improves gradually as computation time increases" [13].

1.1 Literature Review

From the available literature is emerges that selection of data model for GDSS is very important [2]. While explaining the Tolomeo System of decision support, [1] the

R. Prasath et al. (Eds.): MIKE 2014, LNAI 8891, pp. 397–406, 2014.
© Springer International Publishing Switzerland 2014

authors have brought out the importance of Visual Interaction Model (VIM) and integration of Artificial Intelligence (AI) in DSS. This has been further emphasized by G. Andrienko et al while summarizing the workshop on Geovisual analytics in 2006 [3]. Thus a good GDSS has to have a combination of AI, VIM and Geo-data Model. Harms et al have brought out a four tiered approach for building intelligence into GDSS [7]. They have also analysed the importance of spatio temporal analysis in DSS. However in their 'Knowledge Layer' they have not outlined the process of generation of knowledge. Any analysis using geo data, typically use large data, also complicated and varied processing have to be done repeatedly for arriving at the desired result. Decisions are based on the results of the analyses which should verify or confirm certain required criteria [7]. This involves efficient management of intermediary data sets in a time and computationally constrained environment where computational resource and the time to compute are restricted.

1.2   Approach

This paper presents a novel approach to overcome the requirement of management of intermediary datasets and the requirement of large computational resources. We present a model of linking the intelligence criteria with data layers. The data layers have the required information data sets either in spatial form, like lines, points and areas or in attribute forms linked to relational databases. The spatial form is generally static or constant and the attribute is generally dynamic or filled during run time. By this process, the verification or confirmation of the required criteria becomes a function of the layer itself and not individual data sets. Since layers are distinct and can be handled separately in geospatial environment, the linked decision making also can be handled in smaller incremental model. The required processing can be carried out in the back ground and provided as pre-processed data layer to the user along with the spatio-temporal state obtained at the time of calculation. Added advantages are: there is no requirement of large computational machines at field, the information is handled just in time, and information can also be made to be handled by the concerned persons only, providing a level of inherent security of information.

2   Conceptual Model

Consider a time constrained operation O, which is to be completed in time T. Let, the time required in the present manual and other processes, T_m, be unacceptably greater than time T. Hence, there is a requirement of finding a means to keep $T_m \leq T$. Time T_m consists of T_D, T_E Where T_D is the time required for taking decision and T_E is the time required for executing the Operation O. Generally in the manual environment; $T_m = T_D + T_E$, this requires all analyses be carried out before the execution of the operation and does not permit changes as new situations develop during execution.

Geo data needs long time to process and needs specialized geo environment to process, both of which are generally not available in the field. To overcome these, we propose a model that will enable processing using smaller data sets up to a predetermined level with known criteria, to take certain basic decisions. Other decisions are taken as higher details are processed and more criteria come into consideration. Such a process is feasible using anytime algorithms and if the datasets are arranged in specific a model suiting such processing. This approach also attempts to reduce the uncertainty in the decision making and execution processes which gives a possibility of taking better decisions and execution in shortest time.

2.1   Decision Making

Decisions for and during the execution of operations are taken by many actors. Each actor passes the decision to the people below him in the hierarchy and gives information about the decision to the people above him in the hierarchy. Each actor takes a set of decisions for any specific operation. Each of these decisions has a set of criteria to be satisfied. Each criterion is the result of certain analysis based upon information available at that time. It can be seen that, T_D, the total time taken for the decisions is the summation of all of these decisions. The overall execution depends on the correctness of each of these decisions. Each of these decisions contributes incrementally to the successful execution. So instead of taking one final decision at the beginning of the execution it would be beneficial if the decisions are taken as the execution progresses in specific incremental bits. But the increments have to be programmed at the commencement itself, hence they are of the specificity [3] type as formulated by Zilberstein.

Let, the set of actors involved in any one operation be $A^1, A^2, A^3 \ldots A^n$, and $D_1^1, D_2^1, D_3^1 \ldots D_n^1$, be the decisions to be taken by actor A^1, and $D_1^2, D_2^2, D_3^2 \ldots D_n^2$, be the decisions to be taken by actor A^2, so on. Let, each of this decision D_r^n has criteria $C_1, C_2 \ldots C_q$. Here, each of the criteria C_q is a result of an analysis function of the group, f_r which uses the data set consisting of collection $i_1^r, i_2^r \ldots i_n^r$. The final execution of the operation will depend on all decisions taken by each actor.

2.2   Strategy to Handle the Problem

Geospatial environment permits storing and handling data in layers. So, If the closely related data in the data sets $i_1^r, i_2^r \ldots i_n^r$ are grouped in to layers $l_1, l_2 \ldots l_n$, then the functions set $f_1, f_2 \ldots f_r$ becomes function between layers and criterion $C_1 = f_1(l_1, l_2)$, $C_2 = f_2(l_3, l_4) \ldots$ and $C_n = f_n(l_{n-1}, l_n)$.

Based on the above, now each of the decision in the decisions set, $D_1, D_2 \ldots D_n$, becomes a AND operation like $D_1 = (C_1 \text{ AND } C_2 \text{ AND } C_3 \ldots \text{AND } C_n)$. Here the time taken to arrive at the decision D_1 is T_{D1} , D_1 is T_{D2}, etc.

Here, $T_D = T_{D-init} + \sum_{i=1}^n (T_{Di})$, where T_{D-init} is the minimum in escapable time that is required initially to commence the operations. Now, the overall time required becomes, $T = T_{D-init} + T_E$. Here T_E is not fixed and keeps changing as per decisions taken which are now based on larger information which may not have been available at the commencement of the operation and hence reduces the uncertainty in yet another way than given in [2].

Each actor A^i from the set $A^1, A^2, A^3 \ldots A^n$ takes his own set of decisions sequentially. The output quality of the anytime algorithms based solution and the related performance profile [10], are ensured by correct sequencing of the decisions based on user experience. The key in this approach is to group all geo spatial data in correct spatial model and lay down layers in correct structure through a dedicated Spatial Data Model Structure (SDMS) and to handle them just in time [8].

3   Case Study

The above concept can be used in operations monitoring processes in geo resources industrial activity such as mining. Here we discuss the model in light of casualty evacuation process in a mine that has a Geo Enterprise Resource Management solution in place for monitoring of the mining operations.

3.1   The Setting

The area taken up for the case study is a hypothetical mine based on actual terrain conditions available in iron ore mines located within Bonai Synclonorium region of India. The terrain is generally semi-mountainous and under developed with restricted roads and tracks. Figure 1 below gives the sketch of the area. There is a city nearby, the mining HQ is located in nearby town which has a Helipad and the miners' colony is next door. There is an airport in the city which has helicopters stationed for rescue operations. There is a Helipad in the town.   The city has a good hospital, the town a section hospital and the miners' colony has an emergency medical room. The sketch is not to scale.   During the daily operation a dumper has tripped and the driver is seriously injured.

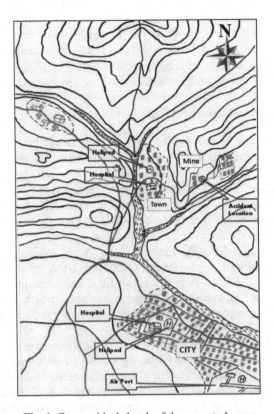

Fig. 1. Geographical sketch of the case study area

3.2   Parameters

The typical approximate time parameters for decision making and evacuation in the case are given at Table 1. The survivability time factor is given at Table 2. The times given are the minimum required time.

Table 1. Time required for decision and evacuation

Ser No	Actions	Time
1	Immediate rescue, assess causality and reporting	10 mins
2	Move of medical staff from miners colony to the mine	10 mins
3	Move of doctors from town hospital to accident site	20 mins
4	Casualty move in ambulance from mine to town hospital	30 mins
5	Time to administer resuscitation and medical care	10 mins
6	Decision for evacuation by helicopter	10 mins
7	Intimation to airport and making helicopter ready	05 mins
8	Pilot briefing and take off	05 mins
9	Known flight time from city airport to town helipad	20 mins
10	Emplaning	03 mins
11	Known flight time from town helipad to hospital helipad	15 mins
12	Deplaning and move from helipad at hospital to operation theatre	03 mins
13	Time for analysis of helicopter landing pad within 1 KM from the mine	10 mins
14	Time for analysis of helicopter landing pad within 2 KM from the mine	20 mins

Table 2. Survivability time factor for grievously wounded casualty

Ser No	Time to reach operation table (within)	Survivability
1	10 mins	100%
2	30 mins	75%
3	60 mins	50%
4	90 mins	25%
5	After 120 mins	0%
Survivability increases by 25% if resuscitation if given before 30 mins.		

From Tables 1 and 2 it emerges that the minimum time, T_m, by which the injured person reaches the operation theatre is 71 mins, by which time the survivability is below 50%. Even if resuscitation is administered within first 30 mins, the survivability is almost the same in the present manual environment. It is also evident that the casualty cannot be taken by road to the town and evacuated from there, as it will add another 20 mins. Hence it will be advantageous if the casualty is moved directly from the mine to the city hospital. For this we need to identify a location nearby where the helicopter can land to pick-up the casualty. This location can be either inside or outside the mine. A sample set of criteria to be considered while selecting location for landing a small helicopter like Alouette III which is used world over for casualty evacuation is given at Table 3. The priority of consideration is also given at Table 3.

Table 3. Criteria for helipad selection

Criteria No	Parameters	Value / size	Priority
C_1	Open field with grass or very low shrubs.	100m x 100 m	1
C_2	Main landing pad be devoid of any vegetation.	25 m x 25 m	2
C_3	landing pad should have hard standing.	25 m x 25 m	2
C_4	Distance from power lines along path	100 m	5
C_5	Distance from telephone lines along path	100 m	5
C_6	Slope of ground	1 in 15	1
C_7	Accessibility from road	500 m	4
C_8	Distance from loose earth / blast piles	100 m	5
C_9	Visibility	5000 m	3

3.3   Actors

Actors in this situation are; A^1, the mine operations manager, A^2, the hierarchy that gives approval for evacuation by helicopter, A^3, the pilot of the helicopter.

3.4   Decisions

The decisions in this case study can be mapped in a directed acylic graph (DAG) tree structure [11]. All actions will commence when the mine operations manager takes a decision to evacuate the casualty by air.

The decisions that need to be taken by A^1, is; D_1^1, should the person be evacuated by air?

The decisions that need to be taken by A^2, is; D_1^2, can the evacuation be permitted?

The decisions that need to be taken by A^3, are; D_1^3, should the town helipad be used?, D_2^3, where to land the helicopter for evacuation?, D_3^3, is the information adequate for landing and takeoff?

3.5   Criteria

The criteria for decision D_1^1 are; the type of casualty, the state of casualty, the personal details of the worker etc. The criteria for decision D_1^2 are; the resources availability, location of hospitals, availability of helicopters etc. The criteria for decision D_1^3 are; time required for evacuation from Table 1, survivability from Table 2, location of helipads, location of hospitals. The criteria for decision D_2^3 are given at Table 3.

The decision D_3^3 is to be taken while approaching the landing pad and all criteria listed at Table 3 are verified.

3.6   Geo Spatial Data

The spatio-temporal data that are required for taking decision and monitoring this casualty evacuation are; land use pattern, standing crop details, land cover, aviation details, height data, electricity lines alignment, telephone lines alignment, weather details, road network, water ways, road network state, waterways state, casualty details, mine personal details, mine vehicle state, location of hospitals, contact details at hospitals, contact details at airport, contact details at etc. The related information can be grouped under the following categories and stored as geo spatial layers:-

- l_1, Height data layer, consisting of height of all surveyed points, place holder for height of each square meter of the terrain is created and kept ab-initio.
- l_2, Land Use Land Cover (LULC) layer, consisting of land cover data, land use data, details of crops, ownership of land, land extends, etc.
- l_3, Road network layer, consisting of types, bridges, classification of roads etc.
- l_4, Water ways layer, consisting of, water level, bridges, crossing points etc.
- l_5, Hospital layer consisting of locations, facilities available, contact details etc.
- l_6, Aviation layer, consisting of airport locations, helipad locations, contact details, procedure for alerting and requisitioning aircrafts, weather details, visibility details, air routes, etc.
- l_7, Electricity network layer, consisting of alignment of power lines, location of pylons, location of sub-stations, contact details etc.
- l_8, Telephone network layer, consisting of alignment of overhead telephone lines, location of telephone towers, contact details of liaison personal etc.
- l_9, the ore dump layer, consisting of areas with loose earth dumps.
- l_{10}, the operations layer, consisting of current run time data linked to the locations. This also is a dynamic layer. The location of casualty is marked in this layer.

3.7   Geo Spatial Analysis

The geo spatial analyses for selection of helipad location are listed in this section as per each criterion. The analysis should progress as per the priority laid at table 3.

The verification of criteria C_1 can be done by query of layerl_2. Criteria C_2 can be verified by querying 25 m x 25 m area in the results of earlier query by restricting the detailed search to 25m x 25m samples at a time. Standard geospatial algorithms support this type of analyses. Criteria C_3can be verified from layer l_4, by removing the area that is within 100 m from rivers and streams, assuming that it will be marshy during rainy season. A nearness search to the rivers and streams will give the desired result. After these query, the system will retain all 100 m x 100 m area with has 25 m x 25 m clear are in center and is away from the rivers.

Verification of criteria C_6 is a lengthy procedure. The height data matrix of the area of interest can be generated from the Shuttle Radio Topographical Mission (SRTM) data maintained by the United States Geological Survey (USGS). The steps to be followed for verifying of the 100 m x 100 m area is within 1:15 slope are:-

- Load the 90 m x 90m height data of the required area from USGS website or from the local database.
- Interpolated to 1m x 1m data grid array for the whole area.
- Now sample 100 m x 100 m areas from the North-West corner moving 25 m at a time both towards East and South.
- Find height difference between the highest point and lowest point in the current sample being considered. Let this be h. Find the distance between these two points. Let this be d.
- If verify if $\tan h/d$ is less than 0. 0012. If yes, retain the area for further calculations else drop it and move to the next area.
- Consider the next area, if it has zones covered by any of the retained area in the previous step then move to the next area else carry out the last two steps again.
- Highlight all retained areas in different colour.

Criteria C_9 can be verified from the weather data which can be either maintained within the system or can be queried from the weather department. Criteria C_7 can be ensured by buffer zone search of layer l_3. Those areas that falls within the buffer are to be retained. Similarly criteria C_4 C_5 and C_8 can be verified by buffer zone analysis of the layers l_7, l_8, and l_9 respectively. However in this analysis, the areas that fall within the buffer zone should be eliminated.

3.8   Functioning of the System

The system functions as per the sequence listed below.

- Step 1. The on duty manager of the mine fills in the casualty details in the system. The system shows the location of the hospital and other medical facilities. The information about the casualty is sent to the mine medical room and all others concerned.
- Step 2. The medical personal of the mine query details about the injury and fill in the type of evacuation to be done. This is sent to all concerned. The mine manager takes decision on the type of evacuation to be done.
- Step 3. The mine manager applies for evacuation by helicopter to the higher management, informs medical to arrange for resuscitation from the town hospital.
- Step 4. The helicopter based evacuation is approved. Details sent to airport, city hospital and all others linked. The contact details of the personal in these are shown in the screens of the medical personal and the mines personal for information.
- Step 5. The pilots are briefed about the location of the casualty and on the aviation details on the system. Based on the initial data the pilots take decision to land nearby accident location casualty to pick up the casualty.

- Step 6. The helicopter is airborne; on request from the pilots the details of open grounds near the accident location are shown.
- Step 7. The system carries out the slope analysis and shortlists the open patches that meet the slope criteria.
- Step 8. The system carries out the further filtration based on criteria C_2, C_3 and shows those areas that match.
- Step 9. Should the pilots want, the system carries out search for criteria C_7 and lists that meet it.
- Step 10. Should the pilot want, the system carries out further filtering based on criteria C_4 C_5 and C_8. If at this stage, there are no available areas then, the alignment of the electricity line, telephone lines and the location of loose earth dumps is shown to the pilots and the mines manager along with other relevant data. The mines manager can take actions like cutoff the power in the lines, cutoff the lines, spray water, clear the area with dozer, etc.
- Steps 6 and 7 are mandatory. Steps 8, 9 and 10 are carried out only if the pilots want it. Since the results are shown at every step and inputs are taken for further analysis, the processing can be stopped at any point in time.
- All through the system, the survivability factor of the injured person is shown to all concerned.
- The details about the injury, the injured person, medicines administrated and the current condition of the patient are updated in layer l_9 by the medical team. This data can be used by the medical team at the city hospital to prepare the operation theatre from the reception of the patient. Support measures like provisioning of blood of the correct type, specialists required can be carried out independent of the progress of the evacuation.

3.9   Analysis of the Case Study

The case study brings out utility of the data layers for decision making and monitoring of operations especially when geo-spatial data is involved. The main advantage of the system is, it takes into account both the time and processing constraints. For example, in the above case study, the computers on board the helicopter may not be able to take on geo processing and the communication setup may not support the required data connectivity, but the calculations can be done in a ground based computer in a geo environment and only results can be sent to the display in the helicopter. It also presents a set of decision options to the pilots as they approach the accident location. The pilots are free to use their experience of flying over that area to select optimal location from the set presented by the system even before the system completes all processing. At the same time the system gives full support for a pilot flying for the first time over the area.

The system enables simultaneous processing by all actors independent of others. The required information goes to the correct person just-in-time to enable decisions. The key to this is the correct identification of the information that are required and correct modeling them into layers and interlinking them in a structure that considers the processing and decision requirements also as against mundane machine number crunching suggested in other multi criteria geospatial analysis[2].

4   Conclusion

In this paper we have presented a novel approach to using anytime algorithm in decision making and monitoring processes. The concept brings out a new SDMS which is decision oriented as against the common information storage data models. In any operations monitoring process, the actors, their decisions, the linked criteria for making decision and the information required are to be identified. All related information and data sets are to be grouped into layers that are analytically independent. Once this is achieved, then the processing in the geospatial domain becomes a simple mathematical function whose result is used to evaluate and satisfy the criteria based on which decisions will be taken.

The case study brought out how the time and processing constraints in the field can be overcome using specificity model of anytime algorithm. It also brought out how this model can present the actors with a set of decision making choices that are based on accurate calculations for that level of details.

References

1. Angehrn, A.A., LüthiSource, H.-J.: Intelligent Decision Support Systems: A Visual Interactive Approach. Interfaces 20(6), 17–28 (1990)
2. Armstrong, M.P., Densham, P.J.: Database organization strategies for spatial decision support systems. International Journal of Geographical Information Systems 4(1) (1990)
3. Andrienko, G., Anddrienko, N., Janowski, P., Keims, D., Hraak, M.J., Maceachren, A., Wrobel, S.: Geovisual analytics for spatial decision support: Setting the research agenda. International Journal of Geographical Information Science 21(8), 839–857 (2007)
4. Crossland, M.D., Wynne, B.E., Perkins, W.C.: Spatial decision support systems: An overview of technology and a test of efficacy. Decision Support Systems 14, 219–235 (1995)
5. Dean, T.L., Boddy, M.: An analysis of time–dependent planning. In: Proceedings of Seventh National Conference on Artificial Intelligence, Minneapolis, Minnesota, pp. 49–54 (1998)
6. Densham, P.J.: Spatial decision support systems. Geographical information Systems: Principles and Applications 1, 403–412 (1991)
7. Harms, S.K., Deogun, J., Goddard, S.: Building Knowledge Discovery into Geospatial Decision Support Systems. In: SAC 2003, ACM, Melbourne (2003) 1-58113-624-2/03/03
8. Horsch, M.C., Poole, D.: An anytime algorithm for decision making under uncertainty. In: Proceedings of the Fourteenth Conference on Uncertainty in Artificial Intelligence, Morgan Kaufmann Publishers Inc (1998)
9. Cooke, D.F.: Spatial Decision Support System: not just another GIS. Geo. Info. Systems 2(5), 46–49 (1992)
10. Sholomo, Z.: Using Anytime Algorithm in Intelligent Systems. AI Magazine, 73–83 (Fall 1996)
11. Sholomo, Z.: Approximate Reasoning Using Anytime Algorithm, Computer Science Department Faculty Publication Series, University of Massachusetts – Amherst, Paper 226 (1995)

Malware Detection in Big Data
Using Fast Pattern Matching:
A Hadoop Based Comparison on GPU

Chhabi Rani Panigrahi[1], Mayank Tiwari[1], Bibudhendu Pati[2],
and Rajendra Prasath[3]

[1] Dept. of Information Technology, C.V. Raman College of Engineering
Bhubaneswar, Odisha, 752 054, India
{panigrahichhabi,mayanktiwari09}@gmail.com
[2] Dept. of Computer Science and Engineering
C.V. Raman College of Engineering, Bhubaneswar, Odisha, 752 054, India
bpatimilu@gmail.com
[3] Business Information Systems, University College Cork, Cork, Ireland
drrprasath@gmail.com

Abstract. In big data environment, hadoop stores the data in distributed file systems called hadoop distributed file system and process the data using parallel approach. When the cloud users store unstructured data in cloud storage, it becomes very important for cloud providers to secure those data. To provide malware security, cloud service providers should scan the whole contents of the database, which is a very time intensive job. It may even take days to complete the tasks. The main aim of the proposed work is to reduce the processing time by introducing Graphics Processing Unit (GPU) in hadoop cluster. The proposed work integrates two text pattern matching algorithms with the map-reduce programming model for faster detection of malware in big data. The results of our study indicate that use of GPU decreases the processing time of text pattern matching algorithms in big data hadoop.

1 Introduction

Storing data in cloud storages brings new challenges to the security of data. These security challenges includes steganoanalysis, detecting malwares, appropriate authentications techniques etc. Hadoop stores big volume of data in Hadoop Distributed File System (HDFS).

Malwares include viruses, ransomwares, spywares, adwares, scarewares and other type of malicious programs. Since unstructured data is not stored in traditional row-column format, they are most prone to malwares. Examples of unstructured data include e-mail messages, videos, audio files, photos, web-pages, and many other kinds of business documents. Most of the time unstructured data also includes executables files, zip files, compressed files etc. These types of files require to be searched with malwares signatures. Most of the time malwares detection also requires their behavior to be checked.

R. Prasath et al. (Eds.): MIKE 2014, LNAI 8891, pp. 407–416, 2014.

Pattern matching algorithms check wheather a given text contains a set of patterns or not. One of the typical pattern matching approach is the automation approach which is introduced by Aho *et* al. [1]. They proposed a linear time algorithm for multi pattern search using finite-state machine. The performance of the algorithm is not affected by the large patterns, but it requires a significant amount of memory due to state explosion. In contrast, Boyer Moore's algorithm as proposed in [2] is based on heuristics approach.

The proposed approach mainly focuses on faster detection of malwares in a huge collection of unstructured data stored in HDFS. The detection process is a content scanning algorithm, where the whole contents of the data is matched with malware signature database. The proposed technique uses Graphics Processing Unit (GPU) to solve the malware detection problems and uses existing pattern matching algorithms to search the malwares in malware data base and implements it on map-reduce programming model. When the computational part is large, our approach offloads the computational intensive parts to GPU.

The next section explains the proposed contributions and Section 3 gives an overview of the pre-requisites required to understand our approach. Section 4 describes the related work and our proposed approach is explained in detail in Section 5. Section 6 gives the experimental setup required to implement our approach and the results and its analysis is given in Section 7. Section 8 concludes the paper.

2   Contributions

- We propose a map reduce model in hadoop for faster detection of malware in malware data base.
- We implement two existing pattern matching algorithms: Wu-Manber's algorithm [3] and Bloom Filter's algorithm [4] in map reduce model for searching malware signature.
- We compare the performance of the implemented pattern matching algorithms in terms of both CPU and GPU.

3   Pre-requisites

In this section, we describe in detail a few basic concepts that we have used in our approach.

3.1   Malware Signatures

Signatures are different for different types of files such as the zip or compressed files have a different sets of signatures than a mp3 or video file. Generally, we have three major categories of signatures such as **basic, MD5 and regular expression signatures** [5]. In these three categories, the basic signatures are hexadecimal strings and are scanned with the full contents of files. The MD5 signatures are the MD5 checksum of a target file. Lastly, regular expression signature are considered as higher version of basic signatures. This in addition to basic signatures, also includes different forms of wildcards. There are a few other signatures developed and are used for a variety of extended functions [6]. An approximate of 66 signatures exist to examine archive metadata along with it, and 167 signatures exist to examine anti-phishing [6]. Taking the average, the percentage of basic signatures is 52.9, MD5 signatures is 43.7, regular expression is 3.3 and others is 0.16 [7]. The CPU overhead in scanning is very high in basic signatures and regular expression signatures.

3.2 Pattern Matching Algorithms

Malware detection in big data environment is mainly required for stored unstructured data in distributed file systems. Detecting malwares in big data Hadoop requires map-reduce algorithms. The two basic pattern matching algorithms includes Wu-manber's algorithm and Bloom Filter algorithm and are explained as follows.

Bloom Filter and Wu-Manber Algorithm

Software bloom filters were first introduced in 1970 [4]. This algorithm matches whether a given text or element is a part of a given pattern set or not. It has a constant time and it uses several hash functions along with bit vector. This algorithm is space efficient and can filter a set of nearly 30,000 patterns by consuming a memory of 34.76 KB. When bloom filter algorithm rejects an element, it cannot guarantee that the element does not belong to the given pattern set, or in many cases it gives false alarms, and is a major drawback of this algorithm. This algorithm uses K different hash functions to hash the signatures and set the long vector of v bits to 1 or 0. In the matching phase, it hashes the input elements with the same hash function and checks the bit vector, if a match is found, it enters exact matching phase, else rejects the element.

Wu-Manber algorithm is based on the principles of both shifting and hashing [3]. This algorithm uses three tables that is shift table, hash table and prefix table for pattern matching. It uses Boyers Moore's [2] shift tables but is not identical. It has a search window which examines the given pattern for a text in right to left direction, the search window shifts its window based on the greatest value of prefix and shift table. Once it finds a candidate position, it enters exact match phase and in the matching phase, if a match is found, it alarms.

4   Related Work

Literature review indicates that there exist a number of work in big data environment to detect malwares, but they are not based on content matching, instead they are based on system call or monitor[10,11-12]. These approaches focused on having a database of behaviors of commonly known malwares and then monitoring the data for such behaviors and predicting malwares. These approaches basically focussed on monitoring or inspecting the behavior. Since the behavior database size is not big, the detection process is quite fast but at the same time, it cannot guarantee accuracy of detecting malwares or in many cases it gives false alarms. There are generally four common behavior types that can be used to detect malware and a number of work has been proposed to detect them [8,7,9]. These four behaviors include *replication and propagation* behavior, *privacy invasion* behavior, *malicious code injection* behavior and *persistent* behavior. In replication and propagation behavior, a malware or running program tries to replicate programs code, whereas in privacy invasion behavior, it is important to track the flow of sensitive information processed by web browsers. Code injections are important attack methods used by malwares, to filter it, methods that can inspect whether a Dynamic Link Library (DLL) copied to the memory space of a target process is malicious are used. In case of persistent behavior, Auto-Start Extensibility Point (ASEP) [7] techniques are used to detect a malware, and any process with ASEP modification behavior is considered as malware.

Other work which are based on content matching mainly focussed on speeding up the existing pattern matching algorithms. The word *pattern* in this context refers to the hexadecimal string in the virus signature database. Many pattern matching algorithms have been proposed to solve the problem of slow virus-scanning system problem [10,11]. Many of them uses shift based algorithms, which originates from a classic single pattern matching algorithm BM [2]. The BM utilizes information from the pattern to quickly shift the text during searching or scanning to reduce the number of compares as many as possible.

5   Proposed Approach

In this section, we present the detailed Map-Reduce implementation of signature based malware detection algorithm and we named it as Map-Reduce Antivirus (*Map-Reduce AV*); especially, we present how our proposed scheme is applied to the map-reduce programming model. Here we apply a single map-reduce approach but we tested it with two existing pattern matching algorithms. Our proposed technique as shown in Fig.1 has one mapping scheme and one reducing scheme and can be applied to all pattern matching algorithms. In our technique, the input is given to the mapping phase and then the *Mapper* calls the GPU kernel, collects the results from GPU and forwards the collected results in terms of key-value pairs to the *Reducer*. The *Reducer* receives the key-value pairs, examines which key is determined to have the matched signatures through its value, and writes the output to a file in distributed file system.

The detailed steps involved in *Map-Reduce AV* are described as follows:

Step1: In this step, the custom input format reads the data from HDFS in form of blocks of size 50 MB. Here the Key is the file name and value is the block containing several lines of size 50 MB.

NOTE: Traditional or existing input formats reads the data line by line format. Here the block is actually a line of size 50 MB.

Step2: The map function computes the key value pairs for different files and sends the values to GPU, where GPU does the actual content matching for malware detection.

Step3: The malware signature file is cached to the hadoop's distributed cache. While calling the GPU, the map method checks the extension of the file via the received key and calls the GPU by passing its corresponding value along with its signature set. Here for different types of files, different signatures are used.

Step4: During this step, the map function collects the results from the GPU, the results contains information about the block where a pattern match occurs.

Step5: This result is emitted to the reducer method through key value pairs. Here the key is same as the file name and value is binary 1 or 0. 1 indicates the block contains the signatures and 0 indicates no signature is found.

Step6: In this step, *Reducer* receives the key value pairs, and examines which values are 1. Then it emits the keys and values having 1. These values are written as result to the distributed file system.

NOTE: The virus signatures sent to the GPU contain the signature in first column and its corresponding hash value in the second column. The hash value can be calculated using any one of the hashing function. Pre-computation of hash values reduces the overall computation of GPU. The proposed technique uses Java's multithreading concept to calculate the hash table and bit vector from the cached signature file.

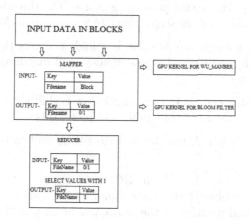

Fig. 1. The Proposed Map Reduce Approach

Next, we explain the steps involved for two existing pattern matching algorithm with map-reduce model.

GPU Kernel Implementation of Wu-Manber's Algorithm

The different steps involved in GPU Kernel Implementation of Wu-Manber's algorithm are described as follows:

Step1: In this step, CUDA device properties such as global memory size, constant cache size, constant memory size are collected from the method calling the GPU kernel. This is done so that GPU kernel can be called with appropriate threads and blocks.

Step2: The virus signatures are sent to GPU's constant cache or read-only cache, where as the actual data to examine is sent to the Global memory.

Step3: Now each thread has to take part of the data from global memory to shared memory, then take the signatures from constant cache and start processing.
    NOTE: Here if the data from global memory gets divided, then it may miss some of the patterns at the border. To avoid this, we assume that the threads of a block while taking the data from the global memory to shared memory, should also copy the part of data, which is to be copied to other blocks shared memory or common data for two blocks shared memory. Appropriate indexing must be used to implement it. We propose the length of the common data to be copied is equal to maximum length of the signature-1.

Step4: During this step, each thread computes the prefix table and the shift table for each signature, stores these tables in the shared memory. Here the prefix table and shift table are two dimensional matrices which accommodate every signature.

Step5: In this step, the actual processing starts. The threads start the search through search window and the steps from 6 to 8 are repeated for every signature.

Step6: Whenever a candidate position is found, it enters the exact matching phase by calculating the hash of search window and matching it with the hash value of the signature.

Step7: If a mismatch occurs, the search window is shifted to n positions based on the greater value from the shift table and prefix table.

Step8: If a match is found, the thread writes the result to the global memory result array.

Step9: This result is sent back from GPU to CPU.
    Here, the total number of threads is equal to the number of rows of data to be copied per shared memory of the particular block and the total number of blocks is the total number of rows of the whole data divided by total number of threads.

GPU Kernel Implementation of Bloom Filter Algorithm

The Bloom Filter algorithm uses a k bit long vector to examine the search window. The k bit long vector is already processed and then sent to the GPU.

To fill the bit vector, the technique uses three hash functions, hashes the prefix of the signatures, and sets position of the bit vector to 1 corresponding to that hash value of the signature. The different steps involved in GPU Kernel Implementation of Bloom Filter algorithm are described as follows:

Step1: Collect the CUDA device properties, such as global memory size, constant cache size, constant memory size, from the method calling the GPU kernel. This is done so that this properties helps us to decide number of threads and blocks.

Step2: The bit vector is sent to the global memory along with the data. The signatures are sent to the constant cache or read only cache.

Step3: Now each thread has to take part of the data from the global memory to the shared memory, then take the signatures from constant cache and start processing.

NOTE: Here if the data from global memory gets divided, then it may miss some of the patterns at the border. To avoid this, we assume that the threads of a block while taking the data from the global memory to shared memory, should also copy the part of data, which is to be copied to other blocks shared memory or common data for two blocks shared memory. Appropriate indexing must be used to implement it. We propose the length of the common data to be copied is equal to maximum length of the signature-1.

Step4: The threads start shifting the search window and the Steps from 5 to 8 are repeated for every signature.

Step5: The thread hashes the prefix of the text pattern or the malware signature, using three hash functions such as *fnv*, *murmur* and *fowler-noll-vo*, which are non cryptographic.

Step6: Based on the hash value, the thread checks clearance of the hash values in the bit vector.

Step7: If the bit vector position corresponding to that hash value is found to be 1, then it enters exact matching phase.

Step8: Again the hash of the element is calculated and searched in the hashed table and if a match is found the result is written to the global memory.

Step9: This result obtained is sent back to CPU from global memory.

Here, also the number of threads is same as the number of rows of data to be copied to the shared memory. And the number of blocks is the total number of rows in the data divided by total number of threads. This algorithms gives the worst case scenario for time consumption, as this algorithm scans the input data from the start to end. But there are many other specific file formats where the signatures need to be examined only with the header or tailor part of the file.

6   Experimental Setup

Hadoop provides a framework to process the stored data in a parallel fashion using map-reduce programming model [12]. Storing data in HDFS is easy as it internally handles data efficiently at a very large scale. HDFS indexes the input data and breaks the big data to blocks of default size of 64MB and stores it in different cluster nodes. To provide data backups, hadoop has a replication factor, which replicates the blocks to different nodes.

We deployed our proposed map-reduce model on a hadoop cluster of 2 nodes, each with a GPU. In cluster, we used NVIDIA's Ge-Force cards which were connected to every node in the cluster. In the cluster, each node has a Core i3 @ 2.20 Ghz CPU and the NVIDIA's Ge-Force GT540M GPU, having two streaming-multiprocessors and 96 CUDA cores. The used Operating system is Ubuntu 12.04 LTS. We downloaded the virus signatures from Clam AV site [5]. The total signatures from Clam AV is about 150000, and the surprising factor is that at certain interval of time, this data increases it size in folds. Java's multithreading is used during the hash table and bit vector calculations in the mapping phase which helps in utilizing every core of CPU.

During our experimentation, real world data set including executable and compressed files, having extensions bzip2, gzip, tar etc is used for testing. We used 15000 signatures and data of different volumes as depicted in the graphs. Generally, the malwares are mainly embedded on executable files or stored in compressed files. So, it is required to scan the whole contents of the executable, compressed files with the virus signatures. For each different type of file, a different set of virus signature is needed. According to our algorithm, if any match is found with the malware signatures and contents of file, then the result is written to a separate file. This file contains results in form of file name followed by an indicator, indicating a pattern match or not. For other types of files, it is not required to scan the complete contents of the file but the perfect scanning can be done by just matching the headers and footers of the contents of the file. So to handle huge numbers of files, we require to execute the map-reduce program using an executable or scripts. Hadoop's streaming allows to do the same.

7   Result Analysis

In our experimentation, we compare the time taken by both CPU and GPU in our two implemented pattern matching algorithms (Bloom Filter and Wu-Manber) and the results of time comparisons is shown as in Fig.2. The time comparison graphs for Bloom Filter and Wu-Manber algorithms are shown in Fig.2(a) and Fig.2(b) respectively. The results indicate that Wu-Manber algorithm is faster as compared to Bloom Filter algorithm and is due to the shifting property of Wu-Manber algorithm.

We have also compared the time taken by GPU in both the implemented pattern matching algorithms and is shown as in Fig.3. The graph shows that GPUs give more speedups in both the algorithms. The hadoop uses java as its base

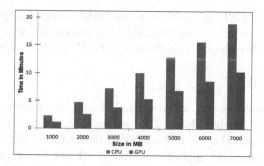

Fig. 2. Time comparison graph - Bloom-Filter Algorithm

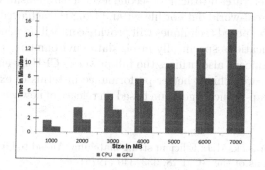

Fig. 3. Time comparison graph - Wu-Manber Algorithm

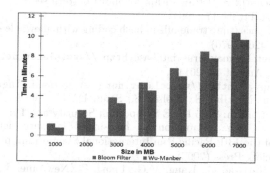

Fig. 4. Time Comparison in GPU of Bloom Filter vs. Wu-Manber Algorithm

programming language, to integrate it with NVIDIA's CUDA. We used JCUDA which involves use of Java Native Interface (JNI). The use of JNI adds extra overhead every time whenever the GPU kernel is called. Another reason is that we have different mappers and reducers for the input and output sets. So we cannot make the GPU kernel call asynchronous, this adds the major overhead in our approach, as every same signature set has to be copied every time to the GPU. Due to no asynchronous operation, it is very difficult to hide the memory copy latency.

To compare it with CPU, we used Java's multithreading concept so that the full power of CPU can be compared. From the obtained results, we can conclude that even using GPU, Wu-Manber algorithm is faster than Bloom-Filter algorithm.

8 Conclusion

Through the proposed work, we have tried to solve the content based signature matching scheme for big data. The proposed technique reduces the time consumption through the use of high performance GPU. We implemented two famous pattern matching algorithms that is Bloom Filter and Wu-manber's algorithm to search for signatures in the content of files. As we move to big data for testing, we may find a high gap between CPU and GPU times. The use of GPU's in HDFSs elevates it to achieve second level of parallelism. We perfomed our experiment on a real-world data or file set and from the achieved exploration, we believe that our proposed techniques will provide compelling benefits in the fields of antivirus applications specifically in big data environment. The use of Java's multithreading concept also utilizes the full power of CPU by engaging its every core. In this way, we achieve better performance in terms of execution time by optimizing our map-reduce algorithm based on release of new versions of Hadoop.

References

1. Aho, A.V., Corasick, M.J.: Efficient string matching: An aid to bibliographic search. Communications of the ACM 18, 333–340 (1975)
2. Boyer, R.S., Moore, J.S.: A fast string searching algorithm. Communications of the ACM 20 (1977)
3. Wu, S., Manber, U.: A fast algorithm for multi-pattern searching, Univ. Arizona, Tucson, Report TR 94–17 (1994)
4. Bloom, B.H.: Space/time trade-offs in hash coding with allowable errors. Commun. ACM 13, 422–426 (1970)
5. ClamAV project: Clamav virus database, http://www.clamav.net/download.html (last accessed: August 15, 2014)
6. Kojm, T.: Clam-av, http://www.clamav.net (last accessed: August 15, 2014)
7. Christodorescu, M., Jha, S., Seshia, S., Song, D., Bryant, R.: Semantics-aware malware detection. In: 2005 IEEE Symposium Security and Privacy (2005)
8. Dai, S.Y., Kuo, S.Y.: Mapmoon: A host-based malware detection tool. In: Proceedings of the 13th Pacific Rim International Symposium, pp. 349–356. IEEE Computer Society Press (2007)
9. Brumley, D., Hartwig, C., Kang, M.G., Liang, Z., Newsome, J., Poosankam, P., Song, D., Yin, H.: Automatically identifying trigger- based behavior in malware. Botnet Detection 36, 65–88 (2008)
10. Xu, B., Zhou, X., Li, J.: Recursive shift indexing: a fast multi-pattern string matching algorithm. In: Proc. of the 4th International Conference on Applied Cryptography and Network Security (ACNS), pp. 64–73. IEEE Computer Society Press (2006)
11. Fisk, M., Varghese, G.: An analysis of fast string matching applied to content-based forwarding and intrusion detection, Technical Report CS2001-0670, University of California San Diegoy (2002)
12. Wikipedia: Map-reduce programming, wikispace, http://map-reduce.wikispaces.asu.edu (last Accessed: August 15, 2014)

Design and Implementation of Key Distribution Algorithms of Public Key Cryptography for Group Communication in Grid Computing

M. Ragunathan and P. Vijayavel

Department of Information Technology
V.H.N.S.N College (Autonomous), Virudhunagar-626 001,Tamil Nadu, India
{mtragunathan,pvijayavel}@gmail.com

Abstract. The group communication involves the association of various nodes to perform various tasks that depend on communication and resources sharing. In a group communication secured resource sharing should be ensured. In order to have a secured communication a simple and competent security methods are vital. The centralized authentication is required for preventing the hackers from intruding the group. To prevent the different attack in the network and to determine the effectiveness of the key management algorithms like access control polynomial are already in use. This paper proposes design and implementation of key distribution algorithm for secure communication in Grid Computing. This paper gives details of implementation of algorithms using Euclidian and prime number based key generation. The time and security analysis for group communication using the proposed algorithms have been carried out and reported.

Keywords: Group communication, Grid computing, Projection profile, Overlapping lines.

1 Introduction

Grid computing refers to pool of heterogeneous system with optimal workload management utilizing an enterprise's entire computational resources. Grid is a type of geographically distributed autonomous resources dynamically at runtime depending on their availability, capability, performance, cost and users quality of service requirements. Grid computing is providing an environment for collaboration among a wider audience. The grid can offer a resource balancing effect by scheduling grid jobs on machines with low utilization. One of the important issue in grid computing is security, it consist of authentication, authorization, integrity, confidentiality numerous hackers continuously explore the security holes, hackers they are mainly focus on active attacks, like masquerading, replaying and denial of services. Authentication process it verifying the particular user in a grid or not and authorization process verify that particular are allowed to perform requested operation on data objects. Integrity process ensure the original data received, confidentiality means it prevent the reading message from other persons. Grid technologies promise to make it possible for scientific collaborations

R. Prasath et al. (Eds.): MIKE 2014, LNAI 8891, pp. 417–424, 2014.

to share resources on an unprecedented scale, and for geographically distributed groups to work together in ways that were previously impossible. The main aim of this paper is to develop least encryption and decryption time, efficient algorithm for group communication and prevent the public message to outsider of the grid.

1.1   Group Key Management

In a group communication several member are participated and multicast communication is efficient way of message sharing. Communication among the group, data integrity should be ensured. Practically the group was heterogeneously so group member can easily join and leave. For adding the user inform to the KDM after authenticate the user KDM will assign a Personal Identification Number to the member, for leaving again inform to the KDM it will revoke the particular member and generate new secret key for the group.

2   Related Works

Grid Infrastructure provides reliability, security privacy and authentication for group message transferred over the network in addition to the facilities to securely traverse the distinct organization that a part of collaboration[1]. Xukai Zoua, Yuan-Shun Dai and Xiang Rana have proposed an elegant Dual level key management mechanism using Access Control Polynomial and one way functions. The first level provided flexible and secure group communication whereas the second level offered hierarchical access control [2]. Jabeenbegum and Purusothaman have proposed cluster based hierarchical key distribution protocol for secure group communication. This approach uses prime number addition for member joining and leaving [4].vinothkumar and Seenivasan have proposed and simple and secure key distribution in grid computing[10]. This approach uses prime number multiplication for member joining and leaving. Venkatesulu and Kartheeban have proposed Euclidian algorithm based key computation protocol for secure group communication in dynamic grid environment[3]. This approach uses prime number division and modulo operation for key computation. Li et al have proposed an authenticated encryption mechanism for group communication in terms of the basic theories of threshold signature and basis characteristics of group communication in grid[11]. In this mechanism, each member in the signing group can verify the identity of the signer, and the verifying group keeps only private key.

3   Euclidian Based Prime Number Computation Algorithm

When a member is willing to join the grid, the member send a request to Key Distribution Manager (KDM). The KDM received the request and generate the Personal Identification Number for the member. Before member joining the host certificate, service certificate and user certificate are authenticated.

3.1   Level I

Step 1. Consider there are 'n' members in the grid $(m_1, m_2, m_3...m_n)$

Step 2. Assign a personal Identification number PIN to each member in grid the PIN is large prime random number.

Step 3. The KDM select a random prime number for group key $k > PIN$

Step 4. For all i for group G and computes the message (a_i, G_i, D_i, b_i) combination in the following manner.

$$G_i = (m_1 * m_2 * m_3... * m_n) + K \tag{1}$$

$$D_i = (for all m without PIN_i) \tag{2}$$

$$A_i = K/PIN_i \tag{3}$$

$$B_i = K mod PIN_i \tag{4}$$

It is assumed that the member mi would recognize the combination (a_i, G_i, D_i, b_i) meant for it. Finally KDM publishes (a_i, G_i, D_i, b_i) from this public information, any group member m_i can get the key by computing

Step 5. By receiving the group message the member has to calculate the following

$$K = a_i * PIN_i + b_i \tag{5}$$

$$P_i = (G_i - K)/D_i \tag{6}$$

If PIN_i and P_i is equal means the receiving group message is correctly receive without any disruption.

If PIN_i and P_i is not equal the Masquerade attacks happened (that means some attacker attacks the Group message and modified it).

3.2   Level II

Suppose if any members wish to join the grid system

The new member request to KDM that it wish to join the grid system. Then the KDM assign new Personal Identification number m_{n+1}, before the above three type of certificate authenticated.

a)   Then KDM generate the new group message $G' = (m_1 * m_2 * m_3...m_n) * m_{n+1} + K$ and $D' = G'$ without $m_{n+1} - K$

b)   Then the new member has to follow the step 3, 4 and 5 in order to receive their key.

3.3   Level III

Suppose if any member willing to withdraw from the grid system they should inform to the KDM. Then the KDM calculate the new group message as follows

a)   select a new Group Key (large prime number where $k > m_i$ for all i)

b) $G'' = (m_1 * m_2 * m_3 * \ldots * m_{j-1}) + K'$ Memberm_k other than m_i cannot get the hidden group key K using (a_i, G_i, D_i, b_i)

4   Illustrative Example

Let four users are send request to join the grid,KDM generate the Personal Identification number $PIN_1 = 5, PIN_2 = 7, PIN_3 = 11, PIN_4 = 14$ and group key $K = 17$

$$G_1 = (m_1 * m_2 * m_3 * m_4) + K \qquad (5 * 7 * 11 * 13) + 17 = 5022$$

$$D_1 = G_1 without PIN_1 \qquad (7 * 11 * 13) = 1001$$

$$A_1 = K/PIN_1 \qquad (17/5) = 3$$

$$B_i = K mod PIN_1 \qquad (17 mod 5) = 2$$

$$K = a_1(G_1/D_1) + b_1 \qquad 3(5005/1001) + 2$$

Similarly, for other members, the combination $(a2_i, G_2, D_2, b_2) = (2, 5022, 715, 3)$ $(a_3, G_3, D_3, b_3) = (1, 5022, 455, 6)(a_4, G_4, D_4, b_4) = (1, 5022, 385, 4)$

Member1 computes the key by using $K = a_1 * PIN_1 + b_1 = 3 * 5 + 2 and P_1 = (G_1 - K)/D_1 = (5022 - 17)/1001$ then check if the $PIN_1 == P_1$ correct

Member2 computes the key by using $K = a_2 * PIN_2 + b_2 = 2 * 7 + 3 and P_2 = (G_2 - K)/D_2 = (5022 - 17)/715$ then check if the $PIN_2 == P_2$ correct

Member3 computes the key by using $K = a_3 * PIN_3 + b_3 = 1 * 11 + 6 and P_3 = (G_3 - K)/D_3 = (5022 - 17)/455$ then check if the $PIN_3 == P_3$ correct

Member4 computes the key by using $K = a_4 * PIN_4 + b_4 = 1 * 13 + 4 and P_4 = (G_4 - K)/D_4 = (5022 - 17)/385$ then check if the $PIN_4 == P_4$ correct Our proposed algorithm is compared with Access Control Polynomial for key computation time and results are given below

5   Security Analysis

Given K and PIN, it is easy to compute $a = K/PIN, \quad b = K mod PIN, \quad G = (m_1 * m_2 * \ldots m_n) + K and D = G$ without PIN But given a and b it is very difficult to compute K, without knowing P such that K=aP+b, even the sender side group message is also (a_i, G_i, D_i, b_i) server combination and all the G and D is very large prime number (1024 to 4096 depends upon our need to increase

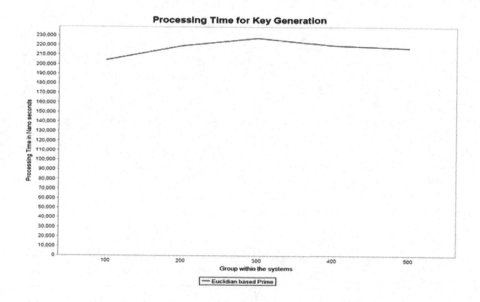

Fig. 1. Processing Time for Key Generation by Euclidian based Prime

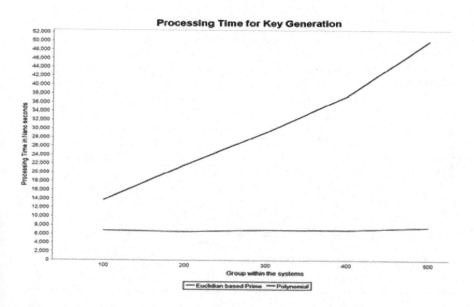

Fig. 2. Processing Time for Key Generation by Euclidian based Prime and Polynomial

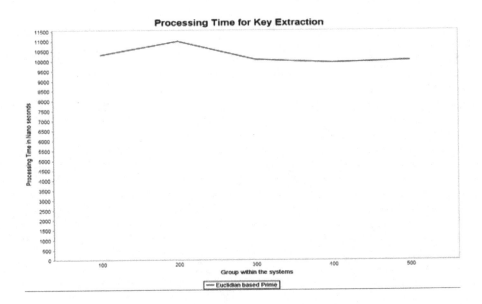

Fig. 3. Processing Time for Key Extraction by Euclidian based Prime

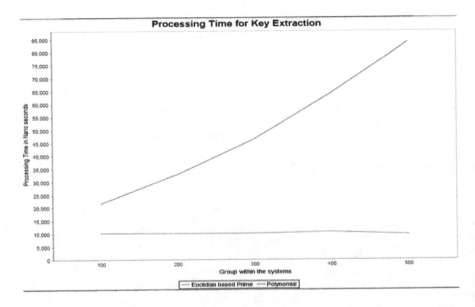

Fig. 4. Processing Time for Key Extraction by Euclidian based Prime and Polynomial

the complexity), it is very difficult to compute K, unless the corresponding PIN are known

1. $PIN_1, PIN_2, PIN_3, PIN_4$ are not know

2. Masquerade attacker try to modify the group message, receiver side Check $PIN_1 == P_1$ so this algorithm prevent from this type of attack

3. Relationship between the a,G,D,b is little so through the four argument we cannot find K

4. If any user withdraw from the grid, KDM change group message and group Key that means absent user can have different key.

5. In a grid each member have own PIN and different public message so internal attack also not possible.

6   Conclusion

In this paper, we proposed a Euclidian based prime number computation algorithms analyzed its security and performance. Here we compared the time analysis with DLKM [2], this algorithms requires minimum time for key computation and key extraction and we compare the security analysis with EAB [3] this algorithm is more secure and prevent the public message. We wish to extend this algorithm to dynamic grid environment.

References

1. Dhanya, D., Ramakrishnan, A.G., Pati., P.B.: Script Identification in printed bilingual documents. Sadhana 27, 73–82 (2002)
2. Foster, I., Kesselman, C.: The Grid: Blueprint for a new computing Infrastructure. Morgan Kaufmann, San Francisco (1999)
3. Zoua, X., Dai, Y.-S., Rana, X.: Dual-Level Key management for secure grid communication in dynamic and hierarchical groups. Future Generation Computer System 23(6), 776–786 (2007)
4. Venkatesulu, M., Kartheeban, K.: Euclidian algorithm based key computation protocol for secure group communication in dynamic grid environment. International Journal of Grid and Distributed Computing 3(4) (December 2010)
5. Jabeenbegum, S., Purusothaman, T., Karthi, M., Balachandran, N., Arunkumar, N.: An Effective key computation protocol for secure group communication in Heterogeneous Network. International Journal of Computer Science and Network Security, 313–319 (February 2010)
6. Hwang, K., Fox, G.C., Dongarra, J.J.: Distributed and Cloud Computing. Morgan Kaufmann (2012)
7. Bellare, M., Yee, B.: Forward security in private-key cryptography. Topics in Cryptology (2003)

8. Irwin, D.E., Grit, L.E., Chase, J.S.: Balancing Risk and Reward in a Market-Based Task Service. In: Proc. 13th IEEE Int'l Symp. High Performance Distributed Computing, pp. 160–169 (2004)
9. Buyya, R., Abramson, D., Giddy, J., Stockinger, H.: Economic Models for Resource Management and Scheduling in Grid Computing. Concurrency and Computation: Practice and Experience 14, 1507–1542 (2007)

Author Index